"Blood and Homeland"

"Blood and Homeland"

Eugenics and Racial Nationalism in Central and Southeast Europe, 1900–1940

Edited by
Marius Turda and Paul J. Weindling

C E U PRESS

Central European University Press
Budapest   New York

Published in 2007 by

Central European University Press
An imprint of the
Central European University Share Company
Nádor utca 11, H-1051 Budapest, Hungary
Tel: +36-1-327-3138 or 327-3000
Fax: +36-1-327-3183
E-mail: ceupress@ceu.hu
Website: www.ceupress.com

400 West 59th Street, New York NY 10019, USA
Tel: +1-212-547-6932
Fax: +1-646-557-2416
E-mail: mgreenwald@sorosny.org

ISBN 963 7326 77 4 cloth
978-963-7326-77-6

Library of Congress Cataloging-in-Publication Data
Blood and homeland : Eugenics and racial nationalism in Central and Southeast Europe,
1900-1940 / edited by Marius Turda and Paul J. Weindling.
p. cm.
Includes index.
ISBN-13: 978-9637326776
1. Eugenics—Europe, Central—History—20th century. 2. Eugenics—Balkan
Peninsula—History—20th century. 3. Racism—Europe, Central—History—20th century. 4. Racism—Balkan Peninsula—History—20th century.
5. Nationalism—Europe, Central—History—20th century. 6. Nationalism—Balkan
Peninsula—History—20th century. I. Turda, Marius. II. Weindling, Paul. III. Title.

HQ755.5E8B56 2006
363.9'2—dc22

2006031799

Printed in Hungary by
Akadémiai Nyomda, Martonvásár

Contents

Introduction

Part I. Ethnography and Racial Anthropology

Part II. Eugenics and Racial Hygiene in National Contexts

Acknowledgments

The majority of the papers included in this volume were presented at the international workshop *Technologies of Race: Eugenics, Biopolitics and Nation-Building in Interwar Europe*, held at the Central European University in Budapest between 28 and 29 June 2004. The workshop was generously sponsored by Pasts, Inc. Centre for Historical Studies in Budapest, in collaboration with the History Department of the CEU, the Wellcome Trust, and the Department of History, Oxford Brookes University.

First and foremost, I should like to express my gratitude to Sorin Antohi, the director of Pasts, Inc., for his support, generosity and guidance before, during and after the workshop. Equally important, he is responsible for initiating the idea of the publication of this volume. His advice and encouragement of this project has proved invaluable.

Thanks are also due to László Kontler, the then head of the History Department at CEU, for his support. Without Zsuzsanna Macht, the administrative assistant in the History Department, and Dávid Marno, former assistant to the director of Pasts, Inc., the workshop would not have been a success.

I owe especial thanks to the copy-editor of this volume, Katya A. M. Kocourek from the School of Slavonic and East European Studies in London (SSEES), who graciously accepted the difficult, and at times seemingly insurmountable, task of copy-editing the first version of the manuscript. A generous grant from the Institute of Historical and Cultural Studies at Oxford Brookes University made Katya's editorial work possible. I should also like to express my gratitude to Elisabeth Jay, the Institute's director, for her assistance.

Matt Feldman deserves special gratitude for his careful reading of the final version of the manuscript. His suggestions and comments have helped to improve both the clarity and the quality of most papers included in this volume. I am also grateful to Diane B. Paul for her constructive criticisms and comments.

Linda Kunos, as Assistant Editor at CEU Press, deserves further gratitude for her patience and kindness during the publication.

Finally, I should like to acknowledge the generous financial support I have received as Marie Curie Fellow, which has enabled me to conduct research for this volume as well as to enjoy the intellectually stimulating environment at Oxford Brookes University.

Marius Turda
London, September 2006

List of Contributors

Margit Berner is Curator at the Department of Anthropology, Natural History Museum, Vienna.

Maria Bucur is Associate Professor and John V. Hill Chair in East European History, Indiana University.

Herwig Czech is a PhD candidate at the University of Vienna and a Researcher at the Documentation Center of Austrian Resistance.

Magdalena Gawin is Assistant Professor at the Institute of History, Polish Academy of Sciences.

Roger Griffin is Professor of Modern History, Department of History, Oxford Brookes University.

Aristotle Kallis is Senior Lecturer in European Studies, Lancaster University.

Ken Kalling is Lecturer in the History of Medicine, University of Tartu.

Egbert Klautke is Lecturer in Modern History at the School of Slavonic and East European Studies, University College London.

Monika Löscher is Researcher at the Institute of History, Vienna.

Răzvan Pârâianu is Junior Research Fellow at Pasts Inc. Centre for Historical Research, Central European University, Budapest.

Attila Pók is Deputy Director of the Institute of History, Hungarian Academy of Sciences.

Christian Promitzer is an Academic Assistant in the Department for Southeast European History, University of Graz.

Michal Šimůnek is a Research Assistant in the History of Biology, Charles University Prague.

Maria Teschler-Nicola is Associate Professor of Physical anthropology, University of Vienna, and Head of the Department of Anthropology, Natural History Museum, Vienna.

Sevasti Trubeta is a Researcher at the Institute of East European Studies, Free University of Berlin.

Marius Turda is Marie Curie Fellow and Academic Fellow in Central and Eastern European Bio-Medicine, Department of History, Oxford Brookes University.

Kamila Uzarczyk is Assistant Lecturer and Research Fellow, Medical University of Wroclaw.

Paul J. Weindling is Wellcome Trust Research Professor in the History of Medicine, Department of History, Oxford Brookes University.

Rory Yeomans is an independent scholar affiliated to the Centre for South East European Studies at the School of Slavonic and East European Studies, University College London.

Central and Eastern Europe, 1925
(*The Times Atlas of European History,* 1994, London, p. 173)

Eugenics, Race and Nation in Central and Southeast Europe, 1900–1940: A Historiographic Overview

Marius Turda and Paul J. Weindling

In the concluding chapter to *The Wellborn Science: Eugenics in Germany, France, Brazil and Russia* (1990), Mark B. Adams complained about the lack of diversity in the comparative history of eugenics: "We are beginning to know something of Russian eugenics, but what of the Austro-Hungarian Empire and Slavic eastern Europe—Czechoslovakia, Hungary, Bulgaria, and the Ukraine? As a Catholic Slavic country, Poland should be an especially intriguing test case. Lemaine, Schneider, Clark, and others are clarifying the character of eugenics in France; what of other Latin cultures of Europe, what of eugenics in Italy, Spain, Portugal, Romania?"[1] After the collapse of Communism in 1989, topics such as eugenics, anti-Semitism and racism were resurrected as scholarly areas of interest, and researchers were given access to materials previously controlled by Communist regimes. As a result, a number of recently published monographs have quickly become essential readings of eugenic movements in Romania, Austria and Poland.[2] Yet studies and monographs are still lacking on the history of eugenic movements in other Central and Southeast European countries.[3] However, as this volume demonstrates, substantial analytical effort has been recently devoted to compensate for the lack of historiographic interest in these topics.

It should not be assumed that comparative histories of eugenic movements in Central and Southeast Europe have never preoccupied eugenicists and scholars of eugenics. In 1921, the Hungarian eugenicist Géza von Hoffmann (1885–1921) wrote an article under the title "Eugenics in the Central Empires since 1914," which constitutes the first analysis of various eugenic movements in Central Europe. Hoffmann compared the activities of various eugenics societies, including the Berlin Society for Racial Hygiene; the German Society for Racial Hygiene in Munich; the International Society for Racial Hygiene; the Austrian Society for the Study of the Science of Population; the Czech Society for Eugenics; and the Hungarian Society for Racial Hygiene and Population Policy.[4] In 1924, the American eugenicist Samuel J. Holmes (1868–1964), professor of zoology at the University of California, published *A Bibliography of Eugenics*, which is arguably the first attempt to produce a comprehensive review of the main themes related to eugenics since the late nineteenth century. In contrast to Hoffmann, Holmes offered a more technical perspective on the achievements of Central European eugenics. In addition to American, British, French and German eugenicists, he cited several Central European supporters of eugenics under the following subheadings: "Eugenics and Works of a General Character" (János Bársony, Ladislav Haškovec and Géza von Hoffmann); "Genealogy" (Géza von Hoffmann); "The Problem of Degeneracy" (Emil Mattauschek); "The Birthrate" (Géza Vitéz); "Selective Influence of War" (János Bársony); "Immigration and Emigration as Related to Racial Changes" (Géza von Hoffmann); and, finally, "Negative Eugenics, Sterilization, Segregation, etc." (Géza von Hoffmann).

Holmes' intention was to enumerate rather than to comment upon works on eugenics included in his anthology. With the exception of Hoffmann's *Die Rassenhygiene in den Vereinigten Staaten von Nordamerika* (Racial Hygiene in the United States of America), which was regarded as "the most comprehensive work on the subject," Holmes did not insist on any eugenic study from Central or Southeast Europe. Nevertheless, the thematic arrangement of his book meant that the general interests of Central European eugenicists were clearly identifiable. They were preoccupied not only with the historiography of eugenics, but also with critical social and medical issues, including degeneracy, decline in birthrates and sterilization.[5] One question, therefore, is appropriate: Did Holmes' comparative survey of eugenic literature reflect the practical objectives of the eugenics societies and organizations in Central and Southeast Europe?

Eugenics Societies and Programmes of Social Hygiene

Following the precedent set by the Society for Racial Hygiene (1905) in Germany and the Eugenics Education Society (1907) in Britain, eugenics societies flourished in Central and Southeast Europe, starting with Prague and Vienna in 1913 and followed by Budapest in 1914.[6]

Towards the end of the First World War, such organizations increased in number and scope. The Hungarian Society for Racial Hygiene and Population Policy, and the Polish Society for the Struggle against Race Degeneration were both established in 1917 (the latter was renamed the Polish Eugenics Society in 1922).[7] After the war, eugenics and social hygiene received increased financial support. In 1919, an Institute of Hygiene and Social Hygiene was created in Romania. In 1923, the active Czech Society of Eugenics founded an Institute of Eugenics Research[8], while Austrian promoters of eugenics established societies for racial hygiene in Linz (1923) and Vienna (1925). The Bulgarian Society for Racial Hygiene, and the Austrian League for Racial Improvement and Heredity were both founded in 1928. As the case studies included in this volume demonstrate, the institutionalization of eugenics in Central and Southeast Europe may be understood as part of an international movement to establish eugenics societies and national research institutions.[9]

The eugenic programmes advocated by these societies were largely influenced by national contexts; however, this national distinctiveness does not mean that individual countries pursued radically different social and medical policies. In the interwar period, the countries of Central and Southeast Europe faced similar problems in the fields of racial and social hygiene. In 1929, the Swiss eugenicist Marie-Thérèse Nisot discussed this convergence in eugenic methods, attitudes and policies in Central and Southeast Europe in the two-volume survey of eugenic movements *La Question Eugénique dans les divers pays* (The Eugenics Question in Various Countries).[10]

With respect to the eugenic movement in Austria, for instance, Nisot assessed different eugenic methods pursued by Austrian eugenicists and social hygienists, such as birth control, the legalization of abortion, the regulation of marriage, and various measures of social hygiene (including the protection of the infant and maternity, the struggle against tuberculosis, mental maladies, and venereal diseases), as well as methods to combat alcoholism.[11] On the other hand, Estonia, which had had a eugenics society since 1924,[12] received a less detailed analysis than Greece, which did not have a eugenics society but compensated with a strong social programme of medical protection.[13] Moreover, eugenicists in Hungary were preoccupied with concerns similar to their Austrian counterparts, most notably the supervision of birth control and the regulation of marriage.[14]

Among Central European countries, eugenic movements in Poland, Romania and Czechoslovakia were cited as exemplary. The assessment of Polish eugenics was, for example, divided into two chapters. The first chapter described the activities of the Polish Eugenic Society (its members, main goals and publications); the second focused on various eugenic measures introduced in Poland, including the protection of the infant and maternity; and prophylactic measures against tuberculosis, venereal diseases, mental maladies and alcoholism.[15]

The analysis of the eugenic movement in Romania was also divided into two chapters: one dealt with general issues like eugenics, and biometrical and statistical research promoted by the Institute of Hygiene and Social Hygiene in Cluj; the other highlighted social hygiene measures introduced in Romania, especially the struggle against tuberculosis and venereal diseases, and the rehabilitation of individuals with disabilities.[16]

In particular, Nisot praised the achievements of the eugenic movement in Czechoslovakia. First, she emphasized the role of Bohemia as the bastion of the Czechoslovak eugenic movement. Nisot then discussed eugenic organizations in Czechoslovakia, such as the Czechoslovak Eugenics Society, the Czechoslovak Institute of Eugenics, and the Eugenics Committee, the latter affiliated to the *Masaryk Work Academy*. Finally, the eugenic measures recommended by Czechoslovak eugenicists were described. In addition to the regulation of marriage, physical culture, social hygiene, and the rehabilitation of people with disabilities, special attention was devoted to the necessity of prenuptial medical certificates, which, according to Nisot, were considered as particularly important by Czechoslovak eugenicists.[17]

This comparative history of eugenics in Central and Southeast Europe can be usefully read in conjunction with studies on national eugenic movements written by the Central European eugenicists. In 1913, for example, in a short note published in *Archiv für Rassen- und Gesellschaftsbiologie* (Journal of Racial and Social Biology), Géza von Hoffmann announced the creation of a Eugenics Committee in Hungary; he later also notified *The Journal of Heredity*.[18] Hoffmann's analysis of the Eugenics Committee in Hungary was expanded, and, in 1918, it became an article, "Rassenhygiene in Ungarn" (Racial Hygiene in Hungary), the first detailed study of the eugenic movement in Hungary.[19]

Ladislav Haškovec (1866–1944), the founder of Czech neurology and one of the most prominent Czech eugenicists, similarly outlined the contribution of regional eugenic movements in Central Europe. In "The Eugenic Movement in the Czechoslovak Republic," a paper presented to the Second International Congress of Eugenics hosted by the American Museum of Natural History in New York in 1921, Haškovec described the state of eugenic research in Bohemia, which he dated "independently of American and English efforts, from the year 1900."[20] The Czech biologist Vladislav Růžička (1872–1934) reinforced Haškovec's analysis of eugenics in Czechoslovakia through his report to the Ninth Conference of the International Federation of Eugenics Organizations (1930).[21] Eventually, in 1935, the geneticist and secretary of the Czechoslovak Eugenics Society Bohumil Seckla (1901–1987) summarized these developments in a short review published in *The Eugenics Review*.[22] These reviews and articles shared one common denominator: eugenics was perceived as a symbol of the constructive process in building a modern nation-state.

Yet alongside these texts, less favourable analyses of eugenic movements in Central Europe were also published. One anonymous contributor to *The Eugenics Review* argued, for example, that in 1935 in Austria "the general Press shows no interest in eugenical problems; nor does there exist in the whole of Austria a single scientific or popular journal devoted totally or partially to eugenics."[23] The author of the review adopted a defensive strategy, one that would vindicate eugenics in Austria by marginalizing its importance rather than by exposing its fascination with the racial hygiene policies of the Nazi regime. The author was concerned that the relationship between Austrian and German racial hygiene movements, considered natural at the time, would, however, later stigmatize the scientific achievements of Austrian eugenics between the wars, creating the erroneous image of an inextricable and homogeneous Austro-German racial hygiene and eugenic movement. Key figures, like the racial hygienist Heinrich Reichel (1876–1943)—who made a significant contribution to the development of Austrian eugenics and then eschewed membership in the NSDAP—is one notable example of Austrian resistance to the model of Nazi racial hygiene.[24] As with the controversial, and at times complicated, relationship between Nazi racial policies and eugenics, a similar question should be posed when discussing the history of eugenics in Central and Southeast Europe between 1900 and 1940: Did eugenics follow a separate scientific agenda, or was it instead an instrument in various projects of national rejuvenation announced by racial nationalism?

New Locations, Old Ideas

It is customary for historians to trace the origins of the closely related ideological currents of Social Darwinism and eugenics to Britain, and especially to the visionary statistician Francis Galton (1822–1911). Since the American civil rights movement from the mid-1960s, and particularly from the 1980s—a period dominated by discussions about disability rights and the nature of totalitarian continuities after the Second World War—historians have focused on eugenics and sterilization in Britain, North America and Germany. New concerns with professionalism and social welfare meant that eugenics became more than a variation of anti-Semitism and racism. It is thus generally agreed in the scholarship that Britain, the USA and Germany provide the theoretical and practical models which other eugenic movements, directly or indirectly, emulated. Such stereotyping is common not only in those countries—notably Sweden, Norway and Switzerland—located in the academic vicinity of such traditional centres of eugenic thinking as Britain and Germany,[25] but also in histories of eugenics dealing with regions outside the orbit of European eugenics, such as Latin America and China.[26]

How does the diffusion of eugenic concepts impact the topics addressed by this volume? If the study of British, American and German eugenics serves as the barometer of "excellence" against which other eugenic movements continue to be evaluated, it is reasonable to assume that the hegemonic narratives developed for British, German and American eugenics provide powerful interpretative models for scholars dealing with this topic in Central and Southeast Europe. On the one hand, as some of the contributions to this volume clearly show, this hermeneutic strategy is used to help explain the dissemination of eugenics in these regions. On the other hand, however, if one focuses solely on the reception of foreign influences, it does not reveal much about the Central and Southeast European contribution to eugenics and racial thought between 1900 and 1940. Accordingly, the historiographic approach proposed here emphasizes four aspects: the creativity of local eugenic movements; the relationship between eugenicists and the nation-state; the role of professionals and expert knowledge on race; and, finally, the influence exercised by other eugenic movements, such as British eugenics and German racial hygiene.

First, racial anthropologists and eugenicists in Central and Southeast Europe utilized many channels in order to present their programs of national rejuvenation and scientific success. Take, for instance, international congresses organized in Central Europe in the interwar period. Notable here are scholarly events like the International Health Conference organised in 1922 in Warsaw; the Anthropological Congress held in Prague in 1924 under the auspices of L'Institut International d'Anthropologie; the Second International Congress of Catholic Physicians, convened in 1936 in Vienna; or the XVIIe Congrès International d'Anthropologie et d'Archéologie Préhistorique organized in 1937 in Bucharest—all scientific gatherings used by Central and Southeast Europeans to inform outside observers about their achievements in eugenics and racial anthropology, as well as policies of public health and preventive medicine.

Second, the successes and failures of eugenic movements in Central and Southeast Europe were not characterized by the social experimentation of a liberal intelligentsia pursuing solutions to the crises brought about by modernity (Britain); the social change engendered by immigration and racial segregation (United States); and, generally, the traumatic human experiences generated by the First World War (France and Italy). Rather, eugenic movements in Central and Southeast Europe reflected the aspirations of a segment of trained professionals dependent upon the state for funding and legitimacy, and whose main goal was the strengthening of their newly created national states.

In the interwar period, the idea of a homogeneous national community figured prominently in Central and Southeast Europe. Many diverse solutions were proposed, including the creation of a liberal democracy based on the Western model; a peasant state, according to the indigenous nature of much of Central and Southeast European society; or a corporatist state modelled on fascist ideology.[27] Diversity of opinion notwithstanding, all theories shared a common axiom: the state was a nation-state, and the ethnic majority therein represented the nation. Defining the nation in interwar Central and Southeast Europe became synonymous with justifying the domination of a given ethnic majority.[28]

In this context, eugenics and racial nationalism offered one of the most compelling definitions of the nation, one based on the biological laws of heredity. The state, eugenicists argued, should become modern not only in terms of infrastructure, economic performance and political institutions, but also in terms of health policies. These policies, however, should be conducted to preserve the "biological capital" of the nation. In turn, this appropriation of racial nationalism by eugenics changed the way the nation-state was represented in eugenic discourses. Ultimately, eugenics and racial nationalism transformed the state from an indistinct entity governed by impersonal laws, to the guardian of the nation governed by biological laws. Eugenics, it was claimed, was the best strategy to achieve a "healthy body politic" and a strong nation-state. For example, the first treatise on eugenics published in Romania after 1918 was called *Igiena Naţiunii* (The Hygiene of the Nation) and subtitled *Eugenia* (Eugenics). The book encapsulated the essential relationship between eugenics and the protection of the nation; indeed, for its author Iuliu Moldovan (1882–1966), the most important Romanian eugenicist of the interwar period, the two were identical.

Third, in contrast to historical narratives postulating the analytical centrality of American, German and British eugenics, an alternative interpretation argues that eugenic movements in Central and Southeast Europe pursued research programmes and theoretical questions that arose locally, from indigenous intellectual and social conditions. To be sure, imitation of other eugenic paradigms did not lack supporters, but the momentum of eugenic movements was largely dictated by local realities rather than external developments, however similar their theoretical outlook. To give one example: in Poland, the first proposals to introduce compulsory sterilization of the mentally and physically handicapped were discussed as early as 1918, independent of other eugenic movements in Europe.

However, with the rise of Nazism in the late 1920s, many supporters of eugenics in Central and Southeast Europe favoured an orientation towards German racial hygiene. Eugenic regulations introduced by the Nazi regime, such as the sterilization laws of 1933, were often received sympathetically in Central and Southeast Europe. Bulgarian supporters of eugenic sterilization, for instance, were strongly encouraged by the introduction of sterilization laws into Germany in their own efforts to legalize sterilization in Bulgaria. Nevertheless, for many Central and Southeast Europeans, Nazi sterilization regulations confirmed the efficiency of the German eugenic movement instead of simply providing a stimulus for the introduction of similar legislation in their own countries. In Hungary, for example, the debate on eugenic sterilization occurred in the early 1930s and, as in the case of Romania, those who opposed it were victorious.[29]

The acceptance and introduction of eugenic sterilization in Central and Southeast Europe existed nevertheless. For instance, on 1 April 1937, a sterilization law was introduced in Estonia. In Greece and Yugoslavia, on the other hand, most participants in the debate on sterilization embraced the idea of voluntary sterilization, although supporters of compulsory sterilization, such as Konstantinos Moutousis, holder of the chair of Hygiene at the University of Athens, and Stevan Ivanić, the director of the Central Institute for Hygiene in Belgrade, received significant academic and political support during the 1930s.

One should not assume, however, that the influence of German racial hygiene in Central and Southeast Europe was exclusively related to the fascination with Nazism.[30] In many cases, because of its association with Nazism, eugenicists in Central and Southeast Europe rejected German racial hygiene, as illustrated by the Czechoslovak and Polish cases discussed in this volume. At the same time, there was an established tradition of academic reciprocity between Germany and countries in Central and Southeast Europe that made the German model of racial hygiene particularly attractive. A majority of Central and Southeast eugenicists were educated in Germany and Austria, and some of them—like the Hungarian racial hygienists Géza von Hoffmann and Lajos Méhely; the Romanian eugenicists Iordache Făcăoaru and Petru Râmneanţu; the Greek anthropologist Ioannis Koumaris; and the Bulgarian eugenicist Stefan Konsulov—had strong connections with German racial hygiene. In 1926, for example, the founding programme of the *Deutsche Gesellschaft für Blutgruppenforschung* (German Society for the Research of Blood Groups) mentioned the following Central and Southeast Europeans as external members: Ioannis Koumaris, professor of Anthropology at the University of Athens; Frigyes Verzár, director of the Physiological Institute in Debrecen; Lajos Méhely, professor of Zoology at the University of Budapest; two Czechs, Oscar Bail and Jindrich Matiegka; and one Bulgarian, Vasil Mollov.[31]

As a new field of research in the emerging Central and Southeast European states, eugenics required an additional pillar of support in order to establish itself as a respected as well as an innovative, scientific discipline; and it found this pillar within the medical profession. It is not surprising, therefore, that racial scientists and eugenicists in Central and Southeast Europe were in most cases medical doctors or benefited from some degree of medical training in their education. In the name of traditional autochthonous racial values, a new medical interpretation of the national body was proposed. Eugenicists, for instance, argued that new medical services should be introduced as part of the programme of national rejuvenation, a programme which should discourage the survival of the "unfit"—including not only the mentally disabled and other "defective" lineages of human breeding, but also those of different ethnic origin. To prevent the "degeneration" of the nation, eugenicists claimed additional rights over the proliferation of "genetically inferior individuals." It was suggested that only a eugenic state could save the nation from internal and external dangers.

How, then, did these internal developments in Central and Southeast Europe compare to the broader eugenic strategies envisioned by Western European states in interwar Europe? The victorious powers of the First World War, especially France, favoured the establishment of welfare states as a means of stabilizing the new post-war political entities created in Central and Southeast Europe. International agencies, such as the League of Nations Health Organization and the Rockefeller Foundation, offered financial support for the consolidation of health policies and medical systems. The Rockefeller Foundation, for example, was instrumental in developing public health administrations in Poland, Hungary and Czechoslovakia, which replaced German bacteriological methods with multifaceted forms of public health. In this context, eugenics became an integral part of public health in terms of strategies aimed at tackling chronic degenerative diseases, mental illness and sexually transmitted diseases. By and large, the endeavours of the Rockefeller Foundation in Central and Southeast Europe were motivated by American concepts of modernization and democratization, thus largely overlooking regional ethnic tensions. It was hoped that eugenicists in Central and Southeast Europe could create regional frameworks for collaboration in the sphere of public hygiene and preventive medicine.[32]

On the other hand, during the 1920s both Germany and the Soviet Union developed vigorous eugenic movements, and these states soon began to collaborate in areas of hygiene, social medicine and racial studies. The Kaiser Wilhelm Institute of Anthropology, Human Heredity and Eugenics, and the German-Soviet Racial Laboratory (*Forschungsstätte für Rassenforschung*) in Moscow, were both established in 1927.[33] For a while it seemed that ideological differences between the two centres of power in Central and Southeast Europe were disregarded as trivial in the name of science. The relationship between racial scientists and eugenicists in Central and Southeast Europe was nowhere near the level of German-Soviet collaboration, although regional scientific projects were not altogether missing. *Rassen im Donauraum* (Races in the Danube Region), the short-lived journal edited in 1935 by the Hungarian doctor and anthropologist János Gáspár (1899–198?) is an exceptional example of regional collaboration. The scientific ambitions and the regional dimensions of the journal were reflected in its subtitle, "Beiträge zur Rassenkunde, Erbbiologie und Eugenik der Donauvölker" (Contributions to the Racial Studies, Hereditary Biology and Eugenics of Danubian Nations). The first issue concentrated on such topics as "Racial Research and Racial Care in Bulgaria;" "Racial Research in Yugoslavia;" "On the Racial Issue as a Cultural Problem;" "Contributions to the Racial History and Racial Biology of the Romanians;" and "Racial Research in Hungary," authored by leading eugenicists and racial anthropologists from Bulgaria, Yugoslavia, Romania and Hungary.[34]

Although the German model of racial hygiene was one of the main sources of inspiration, eugenicists in Central and Southeast Europe did not neglect alternative scientific models. One was the French programme of pro-natalism and *puericulture*, terms denoting medical intervention in the case of maternal and infant health. Seen in binary "racial terms," the "Latin"—as opposed to the Nordic—model of "puericulture" and social assistance (practised in France and Fascist Italy) found supporters in Romania, as illustrated by Gheorghe Banu (1889–1957), the editor of *Revista de igienă socială* (Journal of Social Hygiene).[35] The "Latin" model of eugenics differed from racial selection and sterilization policies, which took their clearest forms in Scandinavia and Nazi Germany but were also encouraged in Switzerland.

Seen in terms of health and population policies, a conceptual and ideological divide separates the 1920s, when eugenic movements flourished in Europe, from the 1930s, when both Stalin, with his termination of programs for eugenic research and persecution of leading geneticists, and the papacy, with its encyclical *Casti Connubii*, "On Christian Marriage," took a clear ideological stance against eugenics. At the opposite end of the political spectrum was Nazi Germany, which viewed eugenics as the core of its racial and national policy.

The gradual transformation of eugenics from its initial, "positive," preoccupation with social and medical assistance, to the propagation of "negative" measures such as sterilization and clinical confinement, also marked the progress of eugenic thinking in Central and Southeast Europe. During the early 1920s, "racial hygiene" comprised the scientific model of hygiene and public health. Two illustrative examples of this state of affairs are the immunology and blood-group analysis of the Polish bacteriologist Ludwik Hirszfeld (1884–1954), first pioneered in multiethnic Salonika (Thessalonica) at the end of the First World War; and the attempts to extend bacteriology to the study of the inherited constitution by the Viennese social hygienist Julius Tandler (1869–1936).[36] These forms of hygiene greatly influenced the conceptualization of eugenics in Central and Southeast Europe, as illustrated by the Romanian and Bulgarian schools of social hygiene, for example.

The history of eugenics represents, however, just the first aspect of a broader research agenda endeavoured in this volume: racial nationalism constitutes the second. This volume aims not only to excavate hitherto unknown eugenic movements in Central and Southeast Europe, but also to explain their relationship with racism, nationalism and anti-Semitism. Both externally, in conjunction with similar developments in Western Europe, and internally, as a direct response to local conditions, eugenicists in Central and Southeast Europe campaigned for the implementation of a professionally controlled and biologically defined form of national belonging. Ultimately, the appropriation of eugenics by the state produced a new type of ideologue: the national scientist who wished not only to interfere in the life of the individual, but also to shape the physical *body* of the nation.

"Blood and Homeland": The Ethnic State

In 1943, a leading Romanian eugenicist of the interwar period, Petru Râmneanţu (1902–1981), delivered an inaugural speech to the faculty of medicine in Sibiu, under the title "Sânge şi glie" (Blood and Homeland).[37] Undoubtedly, the idea that biological concepts are necessary ingredients in shaping national identity and political phenomena received its strongest formulation in the interwar period. Accordingly, the nation was portrayed as a living organism, functioning in a biological fashion. Furthermore, the main characteristics of the national community were depicted in racial terms. In Central and Southeast Europe, the peasantry was considered the "racial repository" of the nation. The apparent contradiction between conservative (preservation of the peasantry) and technologically progressive (eugenics) ideas did not deter

eugenicists like Râmneanțu from advocating the "ethno-biological unity of the nation." This formula, advocated by nationalist eugenicists
and racial anthropologists in Central and Southeast Europe, was often
transformed into the slogan "national revolution" by the political discourse of the extreme right.[38]

The "blood and soil" mythology, in addition to a whole range of
modern techniques aimed at improving the health of the nation, helped
to create a new political biology, whose purpose was to prepare the
"chosen race" (Croats, Romanians, Hungarians, and so forth), at the
expense of others (Serbs, Vlachs, Jews), for the onset of racial utopia:
the ethnic state. The central theme here was not the attempt to define
race in terms of "blood," as serological research has advocated, but
rather in terms of the supposed racial value of blood groups.[39] As the
Hungarian eugenicist and racial nationalist, Lajos Méhely (1862–1953),
noted in his 1934 study "Blut und Rasse" (Blood and Race), the national
ideal should be "the strict protection of racial borders," according to
the laws of heredity.[40] "Blood," as a symbol of national belonging,
transcended science; it operated vertically, unifying the nation with its
mythical projection into the future.

The most radical component of this political biology was undoubtedly the racial principle, upon which the nation should be restructured.
Racial nationalism, combined with a scientific pretension to objectivity,
aimed at purifying the nation of any "unworthy" or "dangerous" elements.
As the Romanian eugenicist Iuliu Moldovan argued in *Statul etnic* (The
Ethnic State), the new nationalist discourse should fuse the science of
eugenics with nationalist assumptions about the existence of a "racial
core," whose protection was deemed vital for the future of the nation.[41]

One feature revealing the extent of the relationship between racial
nationalism and eugenics is the anthropometric debate about the "racial
origin" of various ethnic groups in the multiethnic regions of Central
and Southeast Europe. As ethnic minorities did not fit into the ideal
picture of homogeneous national states, they were either treated as
diasporas from the "mother nations," or simply as foreign groups in
the regions they inhabited. Following the First World War, eugenicists
and racial nacionalists debated the ethnic minority question and proposed various measures, including birth control and sterilization, as
well as transfer of populations, as possible solutions.

Romanian eugenicists, for instance, devoted much of their activities
to the issue of ethnicity. The case of Transylvania is typical of the con-

flict between these competing narratives on ethnicity. Before 1918, Romanians used arguments of historical, cultural and linguistic continuity in order to justify claims on Transylvania. After the creation of Greater Romania in 1918, Romanian eugenicists in Transylvania refocused their arguments on what seemed a more irrefutable basis: science. Mathematical formulae, statistics, and lab analyses of blood groups formed a corpus of evidence that eugenicists hoped would demonstrate that most Hungarians in Transylvania were of Romanian descent.[42]

Interest in race and eugenics was, therefore, more widespread in Central and Southeast Europe than historians have recognized. In the interwar period, the concept of race became part of the vocabulary of most political groups, from the left to the right of the political spectrum. After all, many intellectuals in Central and Southeast Europe shared a common concern about the future of their nation and their ethnic belonging, and this was coupled with the expectation that their states were obliged to protect the "racial qualities" of the nation. The infusion of racial nationalism with eugenics between 1900 and 1940 is identifiable within three clusters of ideas and ideological commitments: a) the professionalization of medicine; b) the emergence of "scientific" versions of nationalism; and c) the fusion of *völkisch* biomedical ideology with anti-Semitism.

The final category poses additional problems of interpretation. In many respects, eugenicists and racial anthropologists concerned with the "Jewish Question" in Central and Southeast Europe reproduced social and biological schemes already implemented by Nazi Germany. However, they were also innovative and creative in the theories they suggested. As anti-Semitism and racial nationalism became inextricably linked in bio-political and eugenic discourses, the projection of the alleged "Jewish degeneracy" was understood in political as well as medical terms. The concept of degeneration became a central term in interwar racial anti-Semitism, not to mention the political and medical categorization of the Jews. The new medical and racial order advocated by anti-Semitic eugenicists was based upon the "purification of the nation," namely, the elimination of all Jews categorized as being "alien" and "degenerate."

As new theories of heredity gained influence in explaining the diverse causes of mental disorders and psycho-pathological phenomena, eugenic and psychiatric practices shifted from a progressive understanding of society to a reactionary, and even totalitarian, conception. As a result, the notion of *kollektive Entartung* (collective degeneration) gradually became accepted in both nationalist and medical rhetoric during the interwar period. Degeneration supposedly threatened the "Volk," the "Race," and most importantly the "Nation." One example was the increasing attractiveness of concepts about "the degeneration of the Jewish race." In Romanian anti-Semitic discourse the notion of Jewish "racial decomposition" was graphically reiterated in numerous books, like *Degenerarea rasei evreiești* (The Degeneration of the Jewish Race), published in 1928 by the Romanian pathologist Nicolae Paulescu (1869–1931).[43]

The scientific language supplied by eugenics was fused with a populist vocabulary and, ultimately, informed anti-Semitism. Medical terminology derived from the disciplines of psychiatry and pathology allowed for the characterization of ethnic minorities as physically and mentally distinct from national ethnic majorities. The political atmosphere of emerging authoritarian regimes in the late 1930s encouraged eugenicists to seek to prohibit mixed marriages between members of minority groups and those of the dominant ethnic groups. They also attempted to use new methods of medical research in their assessment of racial affiliation. [44]

The biological definition advocated by eugenics and racial nationalism thus became the norm, rather than the exception, in many countries in Central and Southeast Europe. Facing a world war, the biological rhetoric of the 1940s intensified in nationalist tone. Of prime importance were the survival of "genetic capital" and the maintenance of the potential of the nation, alongside instruments for eliminating the dysgenic groups, be they defined socially or ethnically.

Conclusions

Recent scholarship on eugenics and fascist movements has complemented the results of research into the history of eugenics and racial nationalism.[45] Moreover, a number of scholars, most prominently Roger Griffin and Emilio Gentile, have emphasized the *palingenetic* nature of fascist ideology as well as its modern character, and have convincingly demonstrated the multifarious relationship between racial nationalism, eugenics, and radical politics during the interwar period.[46] This volume abandons the monolithic interpretative model of eugenics and racial nationalism, which, on the one hand, deals almost obsessively with the "uniqueness" of German racial hygiene and, on the other, views eugenics and racial movements in Central and Southeast Europe as insignificant imitations of German and, to some extent, British models.[47]

The time has come to reconsider the ways in which the histories of interwar Central and Southeast Europe are written. As this volume indicates, a new generation of scholars dealing with fascism, Nazism, racism and eugenics has emerged and acknowledged that the field of eugenics and racial nationalism in Central and Southeast Europe must be repositioned within a wider European perspective, which is itself, much like the academic profession, in the process of continuous trans-formation and thus subject to constant challenges from both compara-tive and national historiographies.

Endnotes:

[1] Mark B. Adams, ed., *The Wellborn Science: Eugenics in Germany, France, Brazil and Russia* (Oxford: Oxford University Press, 1990), 225.

[2] Maria Bucur, *Eugenics and Modernization in Interwar Romania* (Pittsburgh: Pittsburgh University Press, 2002); Brigitte Fuchs, *'Rasse', 'Volk', 'Geschlecht'. Anthropologische Diskurse in Österreich, 1850–1960* (Frankfurt am-Main: Campus Verlag, 2003); Kamila Uzarczyk, *Podstawy ideologice higieny ras i ich realizacja na przykładzie Śląska w latach 1924–1944* (Toruń: Wydawnictwo Adam Marszatek, 2003); and Magdalena Gawin, *Rasa i nowczesność. Historia polskiego ruchu eugenicznego, 1880–1952* (Warsaw: Wydawnicwo Neriton, 2003); Heinz Eberhard, Wolfgang Neugebauer, eds., *Vorreiter der Vernichtung? Eugenic, Rassenhygiene und Euthanasie in der österreichischen Diskussion vor 1938* (Vienna: Böhlau Verlag, 2005); Gerhard Baader, Veronika Hofer, Thomas Mayer, eds., *Eugenic in Österreich: Biopolitischer Methoden und Strukturen vor 1900–1945* (Vienna: Czernin Verlag, forthcoming).

[3] Jean Gayon, Daniel Jacobi, eds., *L'éternel retour de l'eugénisme* (Paris: Presses Universitaires de France, 2006) provides an interesting comparison between past and present eugenic practices and discourses; and Ruth Clifford Engs, *The Eugenics Movement: An Encyclopedia* (Westport, CT.: Greenwood Press, 2005), which focuses mainly on the United States, Britain and France. See also Peter Weingart, "Science and Political Culture: Eugenics in Comparative Perspective," *Scandinavian Journal of History* 24, 2 (1999), 163–177; and Deborah Barett and Charles Kurzman, "Globalizing Social Movement Theory: The Case of Eugenics," *Theory and Society* 33, 5 (2004), 487–527.

[4] Géza Hoffmann, "Eugenics in the Central Empires since 1914," *Social Hygiene* 7, 3 (1921), 285–296.

[5] Samuel J. Holmes, *A Bibliography of Eugenics* (Berkeley: University of California Press, 1924).

[6] (Anonymous), "Eugenics in Austria," *The Eugenics Review* 5 (1913–1914), 387; and (Anonymous), "Eugenic Research in Bohemia," *The Journal of Heredity* 7, 4 (1916), 157.

[7] (Anonymous), "Eugenics in Poland," *Eugenical News* 19 (1934), 9–10.

[8] Bohumil Seckla, "Eugenics in Czechoslovakia," *Eugenical News* 20 (1935), 101–103.

[9] See Paul J. Weindling, "International Eugenics: Swedish Sterilization in Context," *Scandinavian Journal of History* 24, 2 (1999), 179–197.

[10] Marie-Thérèse Nisot, *La Question Eugénique dans les divers pays*, 2 vols. (Brussels: Libraire Falk Fils, 1927 and 1929). The first volume discusses the eugenic movement in Great Britain, USA and France.

[11] Nisot, *La Question Eugénique dans les divers pays*, vol. 2, 90–108.

[12] Nisot, *La Question Eugénique dans les divers pays*, vol. 2, 281–282. See also Theophil Laanes, "Eugenics in Estonia," *Eugenical News* 20 (1935), 103–104.

[13] Nisot, *La Question Eugénique dans les divers pays*, vol. 2, 291–296.

[14] Nisot, *La Question Eugénique dans les divers pays*, vol. 2, 298–304.

[15] Nisot, *La Question Eugénique dans les divers pays*, vol. 2, 418–437.

[16] Nisot, *La Question Eugénique dans les divers pays*, vol. 2, 433–437.

[17] Nisot, *La Question Eugénique dans les divers pays*, vol. 2, 526–553.

[18] Géza Hoffmann, "Ausschüsse für Rassenhygiene in Ungarn," *Archiv für Rassen- und Gesellschaftsbiologie* 10, 6 (1913), 830–831; and Géza Hoffmann, "Eugenics in Hungary," *The Journal of Heredity* 7, 3 (1916), 105.

[19] Géza Hoffmann, "Rassenhygiene in Ungarn," *Archiv für Rassen- und Gesellschaftsbiologie* 13, 1 (1918), 55–67.

[20] Ladislav Haškovec, "The Eugenic Movement in the Czechoslovak Republic," in Charles Davenport, et. al., eds., *Scientific Papers of the Second International Congress of Eugenics* (Held at the American Museum of Natural History, New York, September 22–28, 1921) (Baltimore: Williams & Wilkins Company, 1923), 435–436.

[21] Vladislav Růžička, "Czechoslovakia: Report on Eugenic Work and Advance in Various Countries Adhering to the Federation," in *Report of the Ninth Conference of the International Federation of Eugenic Organisations (Farnham, Dorset, September 11th to 15th 1930)* (London: I.F.E.O., 1930), 70–71.

[22] Bohumil Seckla, "Eugenics in Czechoslovakia," *The Eugenics Review* 28, 2 (1936), 115–117.

[23] (Anonymous), "Eugenics in Austria," *The Eugenics Review* 26, 4 (1935), 259–261.

[24] See, for example, Heinrich Reichel, *Die Hauptaufgaben der Rassenhygiene in der Gegenwart* (Vienna: Veröffentlichungen des deutschösterreichischen Staatsamtes für Volksgesundheit, 1922).

[25] See Gunnar Broberg and Nils Roll-Hansen, eds., *Eugenics and the Welfare State: Sterilization Policy in Denmark, Sweden, Norway and Finland* (East Lansing, Mich.: Michigan State University Press, 2005).

[26] See Nancy Stepan, *'The Hour of Eugenics.' Race, Gender and Nation in Latin America* (Ithaca and London: Cornell University Press, 1991); and Frank Dikötter, *Imperfect Conceptions: Medical Knowledge, Birth Defects and Eugenics in China* (London: Hurst, 1998).

[27] Daniel Chirot, ed., *The Origins of Backwardness in Eastern Europe* (Berkeley: University of California Press, 1989); and Kenneth Jowitt, ed., *Social Change in Romania, 1860–1940. A Debate on Development in a European Nation* (Berkeley: Institute of International Studies, 1978).

[28] Katherine Verdery and Ivo Banac, eds., *National Character and National Ideology in Interwar Eastern Europe* (New Haven: Yale University Press, 1995).

[29] (Anonymous), "Sterilization in Hungary," *Eugenical News* 19 (1934), 142. See also (Anonymous), "Sterilization in Poland," *Eugenical News* 20 (1935), 13.

[30] As suggested by the contributors to the lavishly illustrated and extremely interesting catalogue edited by the United States Holocaust Memorial Museum, *Deadly Medicine. Creating the Master Race* (Washington: United States Holocaust Memorial Museum, 2004).

[31] (Anonymous), "Die Deutsche Gesellschaft für Blutgruppenforschung

(erläßt folgenden Aufruf)," *Archiv für Rassen- und Gesellschaftsbiologie* 18, 4 (1926), 446–450.

32 See Gábor Palló, "Make a Peak on the Plain: The Rockefeller Foundation's Szeged Project," in William H. Schneider, ed., *Rockefeller Philanthropy and Modern Biomedicine. International Initiatives from World War I to the Cold War* (Indiana: Indiana University Press, 2002), 87–105; Paul J. Weindling, "Public Health and Political Stabilization: Rockefeller Funding in Interwar Central and Eastern Europe," *Minerva* 31, 3 (1993), 253–267; Benjamin B. Page, "The Rockefeller Foundation and Central Europe: A Reconsideration," *Minerva* 40, 3 (2002), 265–287; and Željko Dugac, "New Public Health for a New State: Interwar Public Health in the Kingdom of Serbs, Croats, and Slovenes (Kingdom of Yugoslavia) and the Rockefeller Foundation," in Iris Borowy and Wolf D. Gruner, eds., *Facing Illness in Troubled Times. Health in Europe in the Interwar Years 1918–1939* (Frankfurt-am-Main: Peter Lang, 2005), 277–304.

33 Loren Graham, "Science and Values: The Eugenic Movement in Germany and Russia in the 1920s," *American Historical Review* 82, 5 (1977), 1133–1164. See also Susan Gross Solomon, John F. Hutchinson, eds., *Health and Society in Revolutionary Russia* (Bloomington: Indiana University Press, 1990); Paul J. Weindling, "German-Soviet Medical Co-operation and the Institute for Racial Research, 1927–ca.1935," *German History* 10, 2 (1992), 177–206; Paul J. Weindling, *Epidemics and Genocide in Eastern Europe* (Oxford: Oxford University Press, 2000); Mark B. Adams, Garlan E. Allen, and Sheila Faith Weiss, "Human Heredity and Politics: A Comparative Institutional Study of the Eugenics Record Office at Cold Spring Harbor (United State), the Kaiser Wilhelm Institute for Anthropology, Human Heredity, and Eugenics (Germany), and the Maxim Gorky Medical Genetics Institute (USSR)," *Osiris* 20, 1 (2005) 232–262; and Nikolai Krementsov, *International Science between the World Wars. The Case of Genetics* (London: Routledge, 2005).

34 See Chr. Seisow, "Rassenforschung und Rassenpflege in Bulgarien," *Rassen in Donauraum* 1, 1 (1935), 3–7; Bozo Skerlj, "Rassenforschung in Jugoslawien," *Rassen in Donauraum* 1, 1 (1935), 8–11; Svetan Stefanovic, "Über die Rassenfrage als Kulturproblem," *Rassen in Donauraum* 1, 1 (1935), 12–15; George Banu, "Beiträge zur Rassengeschichte und Rassenbiologie der Rumänen," *Rassen in Donauraum* 1, 1 (1935), 16–20; and Emil Wiktorin, "Rassenforschung in Ungarn," *Rassen in Donauraum* 1, 1 (1935), 21–28.

35 See also Virginie Alexandresco, "Enseignement officiel et particulier de la puericulture et vulgarisation de l'hygiène enfantile en Roumanie," *Annales de médecine et de chirurgie infantile* 2 (1907), 474–479.

36 See Ludwig Hirschfeld and Hanka Hirschfeld, "Serological Differences between the Blood of Different Races," *The Lancet* 197, 2 (18 October 1919), 675–679; and Julius Tandler, *Ehe- und Bevölkerungspolitik* (Vienna: Perles, 1924).

37 Petru Râmneanțu, "Sânge și glie," *Buletin eugenic și biopolitic* 14, 11–12 (1943), 370–392. Literally, *glie* means "soil," but in Romanian it is used to express figuratively the "homeland." Râmneanțu's idea of "Blood and Homeland" partly resembles the Nazi idea of "Blut und Boden" in that it fuses

the *völkisch* tradition of the nation with archaic mythologies of belonging. It also departs from the Nazi slogan in that it never promoted a policy of extermination of "less valuable" (*Minderwertigen*) races. See Richard Walther Darré, *Das Bauerntum als Lebensquell der nordischen Rasse* (Munich: J. F. Lehmanns Verlag, 1928); and Richard Walther Darré, *Neuadel aus Blut und Boden* (Munich: J. F. Lehmanns Verlag, 1930).

[38] See Emilio Gentile, "The Myth of National Regeneration in Italy: From Modernist Avant-Garde to Fascism," in Matthew Affron and Mark Antliff, eds., *Art and Ideology in France and Italy* (Princeton: Princeton University Press, 1997), 25–45.

[39] See Pauline M. H. Mazumdar, "Blood and Soil: The Serology of the Aryan Racial State," *Bulletin of the History of Medicine* 64 (1990), 187–219. Romanian and Hungarian eugenicists, Râmneanţu included, have produced numerous serological interpretations of the ethnic groups in Transylvania. See Petru Râmneanţu and Petru David, "Cercetări asupra originii entice a populaţiei din sud-estul Tranilvaniei pe baza compoziţiei serologice a sângelui," *Buletin eugenic şi biopolitic* 6, 1–3 (1935), 36–76; and Lajos Csík and Ernő Kállay, *Vércsoport vizsgálatok Kalotaszegi községekben* (Kolozsvár: Minerva, 1942).

[40] Lajos Méhely, "Blut und Rasse," *Zeitschrift für Morphologie und Anthropologie* 34 (1934), 244–257, especially 257.

[41] Iuliu Moldovan, *Statul etnic* (Sibiu: Tip. 'Cartea Românească din Cluj,' 1943).

[42] See Bucur, *Eugenics and Modernization in Interwar Romania*, 145–148.

[43] See Marius Turda, "Fantasies of Degeneration: Some Remarks on Racial Anti-Semitism in Interwar Romania," *Studia Iudaica* 3 (2003), 336–348.

[44] In 1940, with the adoption of a new constitution, the Minister of Justice, Ioan V. Gruia (1895–195?) declared: We consider Romanian blood as a fundamental element in the founding of the nation." See Ioan V. Gruia, "Expunere de motive la decretul lege nr. 2650/1940 privitor la reglementarea situaţiei juridice a evreilor din România," *Monitorul Oficial* 183 (9 August 1940). Reproduced in *Martiriul evreilor din Romania, 1940–1944. Documente şi mărturii* (Bucharest: Hasefer, 1991), 14–21.

[45] A good example of this new scholarship is Margit Szöllösi-Janze, ed., *Science in the Third Reich* (Oxford and New York: Berg, 2001).

[46] See Roger Griffin, *The Nature of Fascism* (London: Routledge, 1993); Emilio Gentile, *The Sacralization of Politics in Fascist Italy* (Cambridge, Mass.: Harvard University Press, 1994); and Paul J. Weindling, *Health, Race and German Politics between National Unification and Nazism, 1870–1945* (Cambridge: Cambridge University Press, 1989); and Edward Ross Dickinson, "Biopolitics, Fascism, Democracy: Some Reflections on Our Discourse about 'Modernity'," *Central European History* 37, 1 (2004), 1–48.

[47] For a similar endeavour, although conducted from a different analytical perspective, see Marek Kohn, *The Race Gallery. The Return of Racial Science* (London: Jonathan Cape, 1995). See also Amir Weiner, ed., *Landscaping the Human Garden. Twentieth-Century Population Management in a Comparative Framework* (Stanford: Stanford University Press, 2003).

Part I

ETHNOGRAPHY AND RACIAL ANTHROPOLOGY

German "Race Psychology" and Its Implementation in Central Europe: Egon von Eickstedt and Rudolf Hippius

Egbert Klautke

"Race psychology" claims to explain the characteristics, cultural abilities, and mental traits of nations and peoples by analysing their racial composition. It postulates that these characteristics or mental traits are linked to races in a hereditary and naturally determined fashion, thus existing independently of "external," social factors. From this perspective, the physical characteristics of people, in which traditional physical anthropology was predominantly interested, are perceived as indicators of mental and intellectual qualities. For proponents of "race psychology," the specific mental quality of a nation constitutes its identity; at the same time, mental differences constitute the essential differences between nations. Thus defined, "race psychology" formed the core of scientific racism which dominated disciplines such as anthropology and psychology in the first half of the twentieth century. Fritz Lenz (1887–1976), who in 1923 became the first associate professor of racial hygiene in Germany at the University of Munich and later a departmental director of the Kaiser Wilhelm Institute for Human Heredity, Anthropology, and Eugenics in Berlin, never described himself as a "race psychologist," and indeed rarely used the term at all. Yet in the most important German textbook on "Human Heredity" (*Menschliche Erblehre*), Lenz insisted that: "if it was only about physical racial differences (...) then the whole question of race would be meaningless."[1] In this text, Lenz dedicated a long chapter to the "inheritance of mental traits," thus demonstrating his belief that the main principles of "race psychology" were the core of all racial studies.

Lenz's position is indicative of the general attitude of academics towards the field of "race psychology" during the Third Reich. While its principles formed the basis of almost all academic and political theories of race—including those of the best-known Nazi ideologues—

most scholars and academics were reluctant to establish a new discipline under the banner of "race psychology" at university level. The institutionalization of "race psychology" made only slow progress during the 1930s. There were a number of individual attempts and pioneering studies which sought to establish "race psychology" as a discipline, but no "school of race psychology" was founded, and no chair established at a German university. At the Kaiser Wilhelm Institute for Anthropology in Berlin, one of the centres devoted to racial research in Nazi Germany, a Department for Hereditary Psychology under Kurt Gottschaldt (1902–1991) was created in 1935, but the research conducted there was concerned with individual heredity rather than the psychology of races.[2]

Instead, from the early 1920s, and increasingly so during the Third Reich, the formulation of theories of "race psychology" was left to two popular authors, Hans F. K. Günther (1891–1968) and Ludwig Ferdinand Clauss (1892–1974). Both were active in the "Nordic Movement" and, judging by the print-run of their books, became the most successful racial theorists in interwar Germany.[3] The justification of "race psychology" that Günther gave in his most comprehensive study of the racial makeup of the German people bears a strong similarity to Fritz Lenz's statement, quoted above: "If the human races differed only in their physical hereditary traits, then the study of racial appearances would be of much less interest. The mental hereditary differences of the human races cause the obvious differences in habit and appearance, in the deeds and works of individual peoples."[4]

With the help of the National Socialists, Günther and Clauss were able to pursue academic careers in the 1930s. Aided by the National Socialist state government of Thuringia, Günther was made professor at the University of Jena in 1930, and moved on to a chair at the University of Berlin in 1933. Clauss became a lecturer at the University of Berlin soon afterwards, but lost his job in 1943 because he had employed a Jewish research assistant (whom he saved from execution). Despite their academic careers under the Nazi regime, both Günther and Clauss remained outsiders in relation to the established scientific community. With backgrounds in the humanities—Günther had been a secondary-school teacher of German language and literature, and Clauss was a philosopher by training and onetime research assistant to Edmund Husserl (1859–1938) at the University of Freiburg—they were usually looked upon by anthropologists and psychologists with unease and suspicion.[5]

Both Günther and Clauss promoted the idea that the European nations were comprised of six distinct racial groups, each of which displayed typical physical and mental traits; they popularized typologies of these European racial groups based on photographs held to be typical representatives of these racial groups.[6] Although Günther claimed to work on a sound scientific basis and presented his writings as serious research, he relied almost entirely on secondary literature and the interpretation of rather arbitrarily chosen pictures, including paintings and drawings, alongside photographs. The Nordic race evidently constituted an ideal for him and served as the yardstick by which all other racial groups were to be judged: "If one studies the talents of different races by looking at the number of creative (*schöpferische*) individuals [they produced], then the Nordic race is exceptionally gifted."[7] In contrast to other anthropologists, Günther and Clauss made no qualms about calling their studies "race psychology"—or, in Clauss's case, *Rassenseelenkunde*, the term *Seelenkunde* being a means of avoiding the un-German term *Psychologie*. Their academic influence, however, was ambiguous and limited. Neither succeeded in establishing a school of race psychology, and despite the enormous success of their books, the scientific community adopted an ambivalent and awkward attitude towards their ideas.

Nevertheless, there were a number of "proper" academics who were convinced that "race psychology" was a desideratum to be developed further. These scholars attempted to strip race psychology of its political-populist character and introduce it into the scientific mainstream. One of these academics was Egon von Eickstedt (1892–1965), professor and director of the Institute of Anthropology and Ethnology at the University of Breslau from 1931 until 1945. Eickstedt was the head of the so-called Breslau school of anthropology that was in competition with the school of Eugen Fischer (1874–1967), based at the Kaiser Wilhelm Institute for Anthropology in Berlin. Like most German anthropologists, Eickstedt had studied medicine, thereafter specializing in social anthropology as a student of Felix von Luschan (1854–1924). In the 1920s, he became an expert on ethnic groups in South Asia, and a member of the German expedition in South Asia organized by the Research Institute for Social Anthropology in Leipzig.[8] In 1934, he published a comprehensive *Rassenkunde und Rassengeschichte der Menschheit* (Racial Study and Racial History of Humanity) as well as the study *Die rassischen Grundlagen des deutschen Volkstums* (Racial

Foundations of the German People). From 1935, he edited the *Zeitschrift für Rassenkunde* (Journal for Racial Studies). In 1936, he published a programmatic research essay on the *Grundlagen der Rassenpsychologie* (Foundations of Race Psychology) that was meant to establish his version of anthropology as the general approach to the field.[9]

Eickstedt's ambition as head of the Breslau school was to define and establish anthropology as a "holistic" science. This new approach would provide explanations of the physical as well as the psychological characteristics of races by combining and integrating the findings of the humanities, the social sciences, and the disciplines of medicine and biology. In this way, Eickstedt believed, it would be possible to overcome the scientific "positivism" of the nineteenth century, which had "atomized" the sciences, thereby restricting, rather than advancing, scholarship. Anthropology, Eickstedt claimed, needed to shake off this negative legacy in order to adopt the findings of all disciplines engaged in the "research of man." To achieve this aim, Eickstedt called for more systematic research on the psychological aspects of anthropology: "Within races, the same causality operates as within individuals. So quite logically, the physical racial form finds its equivalent in a mental racial form."[10]

Eickstedt defined races as "those zoological and biological living groups of body forms whose members show similar normal and hereditary traits." In accordance with popular and academic definitions, he made a clear distinction between "race" and "people" (*Volk*) as a cultural-traditional community. Peoples were "based on races, and races represented themselves in peoples," but the two categories were not to be confused.[11] Similarly, Günther's starting point for his racial studies was the differentiation between "race" and "people," or "nation"; his main reason for introducing a conception of six European racial groups was to abolish the idea of a "Germanic," or a "Slavonic," race. According to Günther, all European nations were mixtures of the six racial groups that he had defined; hence the idea of a "Germanic race" was misleading, because it lumped together the ideas of race and nation.[12]

Eickstedt's study on the *Grundlagen der Rassenpsychologie* was meant to set the research programme of a "holistic" anthropology and establish "race psychology" as an integral part of it. He put special emphasis on the introduction of the so-called "race formula" that would enable the researcher to define the degree of mixtures of racial groups in given populations. After 1939, the race experts of the Race and Settlement Main Office (RuSHA) of the SS used their own version of a "race formula" to determine which parts of the population in the territories occupied by the Germans were to be resettled. The SS' "race formula" resembled Eickstedt's proposal of 1936; whether the RuSHA was directly influenced or inspired by Eickstedt's proposals remains unclear.[13] Eickstedt believed that, by introducing the "race formula," he had developed sound scientific methods with which to prove common racial typologies. Hence, despite his criticism of the inadequacy of the methods employed in Hans F. K. Günther's studies, Eickstedt adhered to the racial typologies that Günther had popularized.[14]

Eickstedt's ambivalent attitude towards the work of Günther was representative of German academics in the Third Reich. Most anthropologists and psychologists applauded Günther for his intuitive insights into the racial composition of European nations and used varieties of his typology but criticized his intuitive and hermeneutic approach (*Wesensschau*), which, they contended, ought to be replaced by proper scientific methods. In his empirical work, Eickstedt followed this general attitude and applied Günther's typology, especially his nomenclature: Eickstedt's work was based on the assumption that a "Nordic race group" really existed alongside Eastern, Eastern-Baltic, Dinaric, and Western groups, albeit in mixed forms within a given population.[15]

Eickstedt's search for adequate scientific methods within race psychology drew him to the American version of race psychology. He showed particular interest in a comprehensive study published in 1931 by Thomas Russell Garth (1872–1939), under the title *Race psychology*.[16] Garth, a graduate of Yale University who had become professor of psychology at the University of Denver, had summarized the findings of more than one hundred empirical studies on the psychical differences between racial groups in the United States, conducted since the time of the First World War. Although the evidence of empirical material that Garth reported on had shaken his confidence in a close correlation between "race" and intelligence—a lack of confidence which Eickstedt did not share—the German professor showed a keen interest in the methods of American test psychologists. American "race psychology" had, he became convinced, found a means of proving beyond doubt the psychical differences between racial groups; it followed that German psychologists should make use of the American school of race psychology and adopt its quantitative methods. In his own work, however, Eickstedt did not adopt the research methods developed by American psychologists; instead, he stuck to the traditional study of physical characteristics found in the anthropological variety.[17]

The most important research project conducted by Eickstedt's Breslau school in the 1930s was the *Rassenuntersuchung Schlesiens* (The Race Study of Silesia).[18] This study was a large-scale research project for the racial screening of the Silesian population. Eickstedt's and his co-workers' aim was to document the racial characteristics of the entire population of Silesia in order to prove the predominantly "Nordic" character of the population in this contested region. Crucially, however, Eickstedt's research team restricted their sample to "healthy and normally built male persons aged between 20 and 50 years." Thus the study excluded women and the urban population, since these would include "non-settled elements of the population which would obscure the racial picture of the local population." Despite these restrictions, the Breslau research team managed to diagnose about a tenth of the Silesian rural population, and by 1940 they had registered 65,000 persons in thirty-seven districts and eight hundred villages. The anthropologists measured their skulls, noses, height, and body stature, and categorized the color of their hair and eyes. Next, the physical characteristics of each person were correlated, resulting in Eickstedt's "race formula" for each tested individual. According to its inventor, the "race formula" proved a great success because it allowed the quantification of the data that had been collected: "The approach of the Breslau school is the racial diagnosis on the basis of the race formula. The essence of this race formula lies in the summary of an individual racial appearance by means of a short and unambiguous equation (*Ausdruck*). Instead of vague guessing, there is now controlled measurement. Its basis is the registration of single traits, its ultimate goal the exact knowledge of a living type."[19]

"The Race Study of Silesia" received funding from the German Research Association (*Deutsche Forschungsgemeinschaft*); this can be seen as an indicator of the esteem in which the scientific community held Eickstedt's research.[20] At the same time, the study served a political purpose. Eickstedt and his team of researchers were encouraged and aided by the SS officer Fritz Arlt, the local representative of the "Reich's Commissar for the Stabilization of the German Nation" in Upper Silesia. Arlt had earned his doctorate with a study on race psychology, and had also co-edited the publications of Eickstedt's research on Silesia.[21] The reasons for a study on the racial makeup of Silesia originated in the ethnic-political struggles between Germany and Poland after the First World War. Eickstedt tried to provide scientific evidence for the notion that the majority of the Silesian population were of "Nordic stock," and hence German. In the light of this, Eickstedt maintained, Polish claims to Silesia were unjustified. According to Eickstedt, "The Race Study of Silesia" had been successful in proving this point: "In Silesia, we find Nordic people in great numbers."[22] After the beginning of the Second World War and the German occupation of Poland, the data collected by Eickstedt's team proved to be of yet greater use for German politicians and administrators, insofar as it was used to support the implementation of German resettlement policies.[23] Eickstedt's research team joined an army of experts involved in the policies of ethnic cleansing in Central Europe during the Second World War.

Another academic race psychologist whose work was even more closely connected to these policies, and the academic institutions of the SS that supported them, was the psychologist Rudolf Hippius (1906–1945). As an ethnic German from Estonia, Hippius was personally affected by German attempts to redraw the ethnic map of Central and Eastern Europe. After graduating from the University of Dorpat (Tartu) in 1929, Hippius had worked as a postgraduate student of Felix Krüger (1874–1948), professor at the prestigious Institute of Psychology at the University of Leipzig. In 1934, Hippius received his doctorate from the University of Dorpat for a study in experimental psychology.[24] He then taught at the University of Dorpat as lecturer in psychology until 1939. During this time, he conducted so-called "character and ability studies" on the ethnic German population in Estonia and Lithuania. These studies, which served as a blueprint for his later research at the Reich University of Posen, already attracted the attention of the SS in Germany, and were subsequently sponsored by the Office for the Support of Ethnic Germans (VoMi, *Volksdeutsche Mittelstelle*).[25] In 1939, Hippius responded to the "call back home" from the German Reich after the occupation of the Baltic States by the Soviet Union, in accordance with the German-Soviet non-aggression pact of 1939. After a short spell as a psychologist with the German army, in 1940 Hippius became lecturer in psychology at the recently established Reich University in Posen (Poznan). In 1942 he moved to

Prague, where he became professor of social and national psychology at the German Charles University, as well as deputy director of the Reinhard Heydrich Foundation.[26]

On his arrival in Germany, Hippius wasted no time in offering his services to the Nazi authorities. On 5 December 1939, he sent a letter to Professor Konrad Meyer (1901–1973), member of the SS and one of the authors of the *Generalplan Ost* (General Plan East), in which Hippius suggested conducting a psychological study to aid "demographic planning" in the Posen area. The letter included a draft proposal for a research project that would scrutinize the "human building material" in the annexed Polish territory, on the basis of its "ability to work, social attitudes, character structure, and suitability." The results of this study would provide the "raw material" for demographic policies "according to the principles of the National Socialist living order," and would make possible the "best exploitation of the human building material through adequate usage."[27] Hippius's draft proposal was forwarded to the office of Heinrich Himmler (1900–1945)—the "Reich's Commissar for the Stabilization of the German Nation" (*Reichssicherheitshauptamt*, RSHA)—where it caught the attention of the historian and SS *Obersturmführer*, Hans-Joachim Beyer (1908–1971). "The Race and Settlement Main Office" of the SS (RuSHA) agreed to fund Hippius's project with the sum of 2,500 Reichsmark so that he could test his methods. Beyer became, in due course, Hippius's closest collaborator and was responsible for his move to the University of Posen and, in 1942, to the Charles University and the Reinhard Heydrich Foundation in Prague. Hippius's proposed study was to provide much-welcomed expertise for local SS administrators in the Posen area.

Shortly after the occupation of the Western Polish provinces in September 1939, the German administrators were faced with a major obstacle to their plans for expelling the Polish population and replacing them with ethnic Germans from as yet unoccupied Eastern Europe. The administration of the annexed parts of Poland—especially Western Prussia (*Danzig*) and Posen (*Warthegau*)—encountered difficulties in distinguishing between ethnic Germans and Poles. In a detailed memorandum on the policies of ethnic cleansing in the area around Posen, the local representative of the Security Service (*Sicherheitsdienst*) of the SS, Herbert Strickner, described these difficulties: "After the introduction of a German administration (...) a number of difficulties arose, because no one was at all able to tell the difference between a German and a Pole."[28] There was a general lack of reliable census data; moreover, a number of organizations of ethnic Germans had, according to Strickner, indiscriminately given out certificates to people who wanted to claim German citizenship, regardless of their "ethnic" origin and without much testing. As a result of this, the Security Service, in cooperation with the office of the *Gauleiter* in the Posen area, Arthur Greiser (1897–1946), created a "List of Ethnic Germans" (later to become the "German People's List"), which would provide a register of all ethnic Germans in the *Warthegau* to whom German citizenship would be granted.

The first version of this register introduced two categories as means of identifying ethnic Germans. Category A included those who had been members of German political organizations or cultural associations before 1939, and Category B consisted of people who were undoubtedly of German stock (that is, those who spoke German and were Protestants), but had been prohibited by "Polish terror" from showing their allegiance to the German nation. Applicants for the "German People's List" had to fill in a detailed questionnaire and undergo testing by a commission of German administrators and members of the SD. According to Strickner, this procedure made it possible to identify the "core group of ethnic Germans" (*Kerntruppe des Deutschtums*) which would be granted German citizenship.[29] This original version of the "German People's List" did not, however, resolve all problems facing the German administration. Despite Strickner's insistence that no German-Polish "Zwischenschicht" (a mixed "ethnic layer in-between") existed in the Posen area, the large number of mixed German-Polish marriages posed a threat to his clear distinction between "Germans" and "Poles" that underpinned the utopian idea of ethnic cleansing (*völkische Flurbereinigung*). Hence, in May 1940, the new Category C was added to the "German People's List" in Posen. This introduced the inclusion of people who were of German origin but "had slithered into the Polish nation," especially those from mixed German-Polish families. This category included ethnic Germans who, for personal and material reasons, had renounced their German heritage in the interwar period. These people, Strickner claimed, had to be considered traitors to their nation and people (*Gesinnungslumpen im volkspolitischen Sinne*). Nevertheless, since they were not yet completely Polonized and carried "German blood," they could not be allowed to strengthen the Polish nation with their Germanic stock, and thus had to be re-Germanized. Finally, in January 1941, Category D was added to include "persons of German origin who have disappeared into the Polish nation but should be reclaimed for the German nation." This least favorable category would also include those Polish spouses of ethnic Germans who had been entered into Category C, to whom German citizenship would be granted on probation only. In March 1941, the procedure developed by the local administration in Posen (Poznan) for the "German People's List" provided the blueprint for a general law in the German Reich; categories A to D were simply renamed I to IV.[30]

Strickner's detailed report on the creation of the "German People's List" in the Posen area made explicit use of Rudolf Hippius's studies. Strickner's report referred to Hippius's work—and that of his colleague Hans-Joachim Beyer—as a "valuable contribution to the whole problem of Categories III and IV in the 'German People's List.'"[31] Hippius's work was especially helpful for Strickner and his colleagues, who had to rely on conventional, non-"racial" criteria, like language, religion, and national allegiance, in drawing up the "German People's List." Strickner was convinced that Hippius's study demonstrated that these cultural criteria were determined by racial factors, and consequently could be used as indicators for ethnicity. Strickner drew here upon a major empirical study on people of mixed Polish-German background that Hippius had conducted at the University of Posen (Poznan) in 1942. Shortly after his arrival in Posen (Poznan) in 1940, Hippius had drafted a memorandum on his proposed research project, which outlined the necessity and usefulness of such a study.[32] The main political purpose of the study was to provide greater knowledge of the least favourable and—in the eyes of the German occupiers—the most problematic categories, Categories III and IV, of the "German People's List." The studies were carried out with the aid of several teams of interviewers, who tested a total of 877 people. Of these, 262 persons belonged to Category III, 310 persons to Category IV, while 305 persons constituted Poles who had not been registered on the "German People's List."[33] Among the interviewers was the biologist Konrad Lorenz (1903–1989), who was later to win the Nobel Prize for his work in ethology.[34] The team concluded that there were "genetic values (*Erbwerte*) which are fixed according to peoples, and which undergo specific and regular changes when peoples interbreed." By testing and comparing the emotional behavior of separate groups, Hippius and his team tried to "shed light on the psychological background to national

character, namely as a hereditary condition as well as a *völkisch* sentiment."[35] While Hippius accepted that European nations were racially
mixed, he maintained that these mixtures had been "stabilized" and
could therefore be distinguished. Thus a "Polish genetic substance"
was distinguished from a "German" one. It followed that the findings
of the study were quite predictable. Hippius claimed to have proved
that a mixture of the basic "mental," or psychological, structure of
Germans and Poles would lead to negative results. The German basic
structure (*Grundstruktur*) was characterized by "persistence, dependence, energetic dynamism, and aggravated dynamism." The Polish
character, in contrast, showed "openness to the fullness of life, com-
pulsive dynamism, and a poverty of vital roots."[36] The analysis of the
productivity of people of German-Polish background concluded that
"the German aptitude for working ability was largely lost in interbreeding," and that "substantial damages in an interbred population mean
not only an irksome population difficult to guide, but a considerable
defect also in practical and civil life."[37]

Hippius's approach to "race psychology"—or, as he preferred to
call it, "ethnic (or national) psychology" (*Völkerpsychologie*)—was
a cross between traditional and modern racial studies. The methods
applied by Hippius and his team differed considerably from older
forms of anthropology, like those of the "Breslau school" or Eugen
Fischer's approach. Hippius, trained as a modern experimental psychologist, used association and aptitude tests, not unlike the American
race psychologists in which Eickstedt had shown so much interest. He
ignored the traditional approach of physical anthropology that measured skulls and categorized hair colors; similarly, he did not explicitly
rely on Günther's or other popular typologies of European races.
Implicitly, however, insofar as Hippius adopted the categories of the
"German People's List" derived from such racial typologies, his study
served not only to reaffirm these typologies, but also to establish them
as scientific facts.

Hippius was not greatly concerned with the distinctions between
"race" and "nation," or "people." He employed a range of extravagant
neologisms and avoided the established language of racial studies, so
that the racist nature of his approach emerges only upon close inspec-
tion and contextualization. On the one hand, the design and conduct of
the study of Germans and Poles in Posen (Poznan) resembled modern
empirical social-science research; but on the other, Hippius was engaged

in an already classical topic of scientific racism, *Mischlingsforschung*—that is, research on racially mixed people which had provided the impetus for Eugen Fischer's career. Hippius thus tried to prove, once and for all, the validity of the belief that interbreeding and the mixing of races had undesirable results and was to be avoided. It was not Hippius's methods that made his work racist, but rather the basic categories and assumptions that these methods were to prove, not least the political purposes of the "Posen study" and its ultimate implementation. Hippius worked at the heart of the scientific network that the SS had established in the occupied territories, and the data that his research team produced was immediately put to use by the German administration in occupied Poland in the service of the "Germanization of the land and the people."

Hippius's study of Germans and Poles in Posen (Poznan) remained his only major piece of research completed during the war. Plans were made for the continuation of this form of psychological research in the "Protectorate of Bohemia and Moravia" in order to complete the Germanization of the Czech lands. The situation here had, however, posed yet greater difficulties than encountered in the annexed parts of Poland. According to Karl Hermann Frank (1898–1946), the "key idea" of the policy was the "complete Germanization of space and people" by means of the "racial integration of suitable Czechs," the expulsion of "racially indigestible Czechs," and the expulsion or "special treatment" of the Czech intelligentsia and "all other destructive elements."[38] A precondition for this aim, as Reinhard Heydrich (1904–1942) reminded his colleague Frank, was the complete racial screening of the population in Bohemia and Moravia. In October 1940, Adolf Hitler (1889–1945) issued an order legitimizing Heydrich's ideas.[39] Due to the importance of the Czech military industry to the German war effort, however, German administrators were cautious not to stir up protest among the Czech population, and thus proceeded in a much less open and less brutal way than in the Posen and Danzig areas. The completion of the "Germanization" of Bohemia and Moravia—the resettlement of large parts of the Czech population—was postponed until after the war. Hippius and his team arrived too late in Prague to conduct another major research project in support of these plans; he was subsequently killed during the Red Army occupation of Prague in 1945.[40]

Both Egon von Eickstedt and Rudolf Hippius sought to apply the results of their research in the German borderlands of Central Europe: Eickstedt's major research project from the mid-1930s, the "Racial Study of Silesia," tried to show that, contrary to Polish claims, the Silesian population was of predominantly "Nordic" stock. The methods applied in this research project, Eickstedt claimed, had modernized older forms of physical anthropology and provided a basis by which the racial makeup of whole populations might be judged. Eickstedt's research, nevertheless, was more traditional than he pretended. He was aware of the shortcomings of popular typologies of race groups like Günther's, and wanted to turn them into proper scientific theories, but he remained wedded to traditional methods of physical anthropology, such as craniology and phrenology, that assumed that the physical appearance of people gave clues to their mentality and character.

The significance of the "Breslau school" lies less in its connection with Nazi policies during the Second World War than in the fact that Eickstedt and his team were able to survive the collapse of the Third Reich and re-establish themselves in the Federal Republic of Germany, at the newly founded University of Mainz. Here Eickstedt became professor of anthropology in 1947 and was able to continue his work in the Federal Republic of Germany. After the Second World War, he and his student Ilse Schwidetzky (1907–1997)—who had followed him to Mainz and would succeed him as professor of anthropology—made some semantic concessions to the new political circumstances.

Until the early 1960s, the term "race" was dropped and was replaced by less suspicious-sounding terminology. Schwidetzky, for instance, now wrote of *Völkerbiologie* (national biology) instead of racial studies; Eickstedt gave the completely revised and enlarged three-volume edition of his "Racial Study and Racial History of Humanity" the title *Forschung am Menschen* (Research on Man). *Zeitschrift für Rassenkunde* was renamed *Homo* and became the official journal of the German Association of Anthropology. Thus Eickstedt finally achieved his aim of establishing his "Breslau school" as the leading anthropological school in the Federal Republic of Germany, albeit only in a much-overlooked niche of the academic field.[41]

Compared to Eickstedt, Rudolf Hippius represented a particularly modern version of racial research in the Third Reich. He specialized in the "psychology of peoples" and developed his own method of "screening" populations and their racial composition. He used interviews and associations tests to study the mentality of racial groups.

Although Hippius avoided the terminology of traditional physical anthropology and racial theories (in fact he developed an inventive, if not esoteric, language of his own), and although he did not use the craniological and phrenological methods that Eickstedt had relied on, the purpose and the outcomes of his research proved to be no less racist than Eickstedt's more traditional approach: they helped to decide the national-ethnic classification of Poles according to the categories of the "German People's List," and hence were instrumental in the "resettlement" of large parts of the population in the occupied parts of Poland.

Endnotes:

[1] Erwin Baur, Eugen Fischer, Fritz Lenz, *Menschliche Erblehre*, vol. 1 (Munich: J. F. Lehmanns, 1936), 713. On Lenz, see Robert N. Proctor, *Racial Hygiene. Medicine under the Nazis* (Cambridge, Mass.: Cambridge University Press 1995), 46–63; Proctor relies heavily on Renate Rissom, *Fritz Lenz und die Rassenhygiene* (Husum: Matthiesen, 1983).

[2] See Mitchell G. Ash, *Gestaltpsychology in German Culture, 1890–1967: Holism and the Quest for Objectivity* (Cambridge, Mass.: Cambridge University Press, 1995), 356–360; on the Kaiser Wilhelm Institute in Berlin, see Peter Weingart, Kurt Bayertz, Jürgen Kroll, *Rasse, Blut und Gene. Geschichte der Eugenik und Rassenhygiene in Deutschland*, 2nd. ed. (Frankfurt-am-Main: Suhrkamp, 1996), 239–246, 396–424; Niels C. Lösch, *Rasse als Konstrukt. Leben und Werk Eugen Fischers* (Frankfurt-am-Main: Peter Lang, 1996); and Paul Weindling, *Health, Race and German Politics between National Unification and Nazism, 1870–1945* (Cambridge: Cambridge University Press, 1989), 430–439.

[3] On Günther, see the problematic monograph by Hans-Jürgen Lutzhöft, *Der Nordische Gedanke in Deutschland, 1920–1940* (Stuttgart: Klett, 1971); on Clauss, see Peter Weingart, *Doppel-Leben. Ludwig Ferdinand Clauss: Zwischen Rassenforschung und Widerstand* (Frankfurt-am-Main: Campus, 1995).

[4] Hans F. K. Günther, *Rassenkunde des deutschen Volkes*, 17th ed. (Munich: J. F. Lehmanns, 1933 [first ed. 1922]), 190. See also Hans F. K. Günther, *Rassenkunde Europas* (Munich: J. F. Lehmanns, 1925); and Hans F. K. Günther, *Kleine Rassenkunde des deutschen Volkes* (Munich: J. F. Lehmanns, 1929).

[5] A typical and similar example is Friedrich Keiter, *Rassenpsychologie. Einführung in eine werdende Wissenschaft* (Leipzig: Reclam, 1941), 14. See also Wilhelm E. Mühlmann, *Rassen- und Völkerkunde* (Braunschweig: Vieweg, 1936); Bruno Petermann, *Das Problem der Rassenseele. Vorlesungen zur Grundlegung einer allgemeinen Rassenpsychologie* (Leipzig: Barth, 1935); Eduard Ortner, *Biologische Typen des Menschen und ihr Verhalten zu Rasse und Wert. Zugleich ein Beitrag zur Clauss'schen Rassenpsychologie* (Leipzig: Thieme, 1937); Kurt Rau, *Untersuchungen zur Rassenpsychologie nach typologischer Methode* (Leipzig: Barth, 1936); and Ottmar Rutz, *Grundlagen einer psychologischen Rassenkunde* (Tübingen: Heine, 1934).

[6] Günther, *Rassenkunde des deutschen Volkes*; and Ludwig Ferdinand Clauss, *Rasse und Seele. Eine Einführung in den Sinn der leiblichen Gestalt*, 3rd ed. (Munich: J. F. Lehmanns, 1933).

[7] Günther, *Rassenkunde des deutschen Volkes*, 197.

[8] Egon Freiherr von Eickstedt, "Anthropologische Forschungen in Südindien," *Anthropologischer Anzeiger* 6, 1 (1929), 64–85; Egon Freiherr von Eickstedt, "Die Indien-Expedition des Staatlichen Forschungsinstituts für Völkerkunde zu Leipzig. Erster ethnographischer Bericht," *Ethnologischer Anzeiger* 1 (1927), 277–285; and Egon Freiherr von Eickstedt, *Untersuchungen an philippinischen Negrito-Skeletten. Ein Beitrag zum Pygmäenproblem und zur osteomorphologischen Methodik* (Stuttgart: Schweizerbart, 1931).

[9] Egon Freiherr von Eickstedt, *Rassenkunde und Rassengeschichte der Menschheit* (Stuttgart: Enke, 1934); Egon Freiherr von Eickstedt, *Die rassischen Grundlagen des deutschen Volkstums* (Cologne: Schaffstein, 1934); and Egon Freiherr von Eickstedt, *Grundlagen der Rassenpsychologie* (Stuttgart: Enke, 1936).

[10] Eickstedt, *Rassenpsychologie*, 11.

[11] Eickstedt, *Rassenpsychologie*, 12.

[12] Günther, *Rassenkunde*, 1–15.

[13] Isabel Heinemann, *'Rasse, Siedlung, deutsches Blut'. Das Rasse- und Siedlungshauptamt der SS un die rassenpolitische Neuordnung Europas* (Göttingen: Wallstein, 2003), 233.

[14] Eickstedt, *Rassenpsychologie*, 109.

[15] Egon Freiherr von Eickstedt, "Verfahren der Forschung am schlesischen Menschen," in Egon Freiherr von Eickstedt and Ilse Schwidetzky, *Die Rassenuntersuchung Schlesiens. Eine Einführung in ihre Aufgaben und Methoden* (Breslau: Priebatsch, 1940), 8–10. See the work by Eickstedt's student and research assistant Ilse Schwidetzky, "Methoden zur Kontrolle der v. Eickstedtschen Rasseformeln," *Zeitschrift für Rassenkunde* 2 (1935), 33–40.

[16] Thomas Russell Garth, *Race Psychology. A Study of Racial Mental Differences* (New York: McGraw-Hill, 1931).

[17] Eickstedt, *Rassenpsychologie*, 22, 108. On Garth, see Graham Richards, *'Race', Racism and Psychology. Towards a Reflexive History* (London: Routledge, 1997), 65; Graham Richards, "Reconceptualizing the History of Race Psychology: Thomas Russell Garth (1872–1939) and How He Changed His Mind," *Journal of the History of the Behavioral Sciences* 34, 1 (1998), 15–32; and Stephen J. Gould, *The Mismeasure of Man* (New York: Norton, 1981).

[18] Eickstedt and Schwidetzky, *Die Rassenuntersuchung Schlesiens*; Fritz Arlt, Egon von Eickstedt, eds., *Rasse, Volk, Erbgut in Schlesien*, 15 vols (Breslau: Priebatsch, 1939–1942).

[19] Eickstedt, *Verfahren der Forschung am schlesischen Menschen*, 16, 19. See Benoît Massin, "Anthropologie und Humangenetik im Nationalsozialismus oder: Wie schreiben deutsche Wissenschaftler ihre eigene Wissenschaftsgeschichte?" In Heidrun Kaupen-Haas and Christian Saller, eds., *Wissenschaftlicher Rassismus. Analysen einer Kontinuität in den Human- und Naturwissenschaften* (Frankfurt am Main: Campus, 1999), 23.

[20] Notker Hammerstein, *Die Deutsche Forschungsgemeinschaft in der Weimarer Republik und im Dritten Reich. Wissenschaftspolitik in Republik und Diktatur 1920–1945* (Munich: C. H. Beck, 1999), 353.

[21] Fritz Arlt, *Die Frauen der altisländischen Bauernsagen und die Frauen der vorexilischen Bücher des Alten Testaments, verglichen nach ihren Handlungswerten, ihrer Bewertung, ihrer Erscheinungsweise, ihrer Behandlung. Ein Beitrag zur Rassenpsychologie* (Leipzig: Jordan & Gramberg, 1936). On Arlt, see Götz Aly and Susanne Heim, *Vordenker der Vernichtung. Auschwitz und die deutschen Pläne für eine neue europäische Ordnung* (Frankfurt-am-Main: Fischer, 1993), 168–176, 207–220; and Michael Burleigh, *Germany Turns*

Eastwards. A Study of Ostforschung in the Third Reich (Cambridge: Cambridge University Press, 1988), 214, 266.

[22] Eickstedt, *Verfahren*, 12, 29.

[23] See Martin Broszat, *Nationalsozialistische Polenpolitik, 1939–1945* (Stuttgart: Deutsche Verlags-Anstalt, 1961); Heinemann, *'Rasse'*, 187–303; Mechtild Rössler and Sabine Schleiermacher, eds., *Der 'Generalplan Ost'. Hauptlinien der nationalsozialisitischen Planungs- und Vernichtungspolitik* (Berlin: Akademie, 1993); Götz Aly, *'Final Solution'. Nazi Population Policy and the Murder of the European Jews* (London: Arnold 1999).

[24] Rudolf Hippius, *Erkennendes Tasten als Wahrnehmung und Erkenntnisvorgang* (Munich: Beck, 1934).

[25] Subsequently published as Rudolf Hippius, *Die Umsiedlergruppe aus Estland. Ihre soziale, geistige und seelische Struktur* (Posen: Jacob, 1940).

[26] On Rudolf Hippius's biography, see Ulfried Geuter, *Die Professionalisierung der deutschen Psychologie im Nationalsozialismus* (Frankfurt-am-Main: Suhrkamp, 1988), 571; Burleigh, *Germany*, 291; Frank-Rutger Hausmann, *'Deutsche Geisteswissenschaft' im Zweiten Weltkrieg. Die Aktion Ritterbusch (1940–1945)* (Dresden: Dresden University Press, 1998), 267; Karl-Heinz Roth, "Heydrichs Professor. Historiographie des 'Volkstums' und der Massenvernichtungen: Der Fall Hans Joachim Beyer," in Peter Schöttler, ed., *Geschichtsschreibung als Legitimationswissenschaft 1918–1945* (Frankfurt-am-Main: Suhrkamp, 1997), 299, 304–307; Andreas Wiedemann, *Die Reinhard-Heydrich-Stiftung in Prag (1942–1945)* (Dresden: Hannah-Arendt-Institut für Totalitarismusforschung, 2000), 61.

[27] 'Entwurf über die Aufgaben und Wege einer Bevölkerungsplanung im Warthegau'. Letter from Hippius to Konrad Meyer, 5 December 1939, attachment, in Bundesarchiv Berlin, former Berlin Document Center (BDC), File "Rudolf Hippius".

[28] See the memorandum on the "German People's List" by Dr Herbert Strickner, a member of the *Einsatzgruppe* IV of the Sipo in Posen (Poznan), printed in Karol Mariam Pospieszalski, ed., *Niemiecka Lista Narodowa w 'Kraju Warty'. Documenta Occupationis Teutonicae*, vol. 4 (Poznan: Institutu Zachodniego, 1949), 19–130, here 36.

[29] Pospieszalski, ed., *Niemiecka Lista Narodowa w 'Kraju Warty'*, 55–65.

[30] Pospieszalski, ed., *Niemiecka Lista Narodowa w 'Kraju Warty'*, 66–70.

[31] Pospieszalski, ed., *Niemiecka Lista Narodowa w 'Kraju Warty'*, 124–125. See also Hans-Joachim Beyer, *Das Schicksal der Polen. Rasse, Volkscharakter, Stammesart* (Leipzig: Teubner, 1942).

[32] Rudolf Hippius, I.G. Feldmann, K. Jelinek, K. Leider, *Volkstum, Gesinnung und Charakter. Bericht über psychologische Untersuchungen an Posener deutsch-polnischen Mischlingen und Polen, Sommer 1942* (Stuttgart: Kohlhammer, 1943).

[33] The memorandum is printed in Rudi Goguel, *Über die Mitwirkung deutscher Wissenschaftler am Okkupationsregime in Polen im Zweiten Weltkrieg, untersucht an drei Institutionen der deutschen Ostforschung* (unpublished PhD dissertation: University of Berlin, 1964), Appendix, 46–49.

[34] See the internal report on the research project by Hippius, Bundesarchiv Berlin, File "Rudolf Hippius".

[35] Theodora Kalikow, "Die ethologische Theorie von Konrad Lorenz: Erklärung und Ideologie, 1938 bis 1943," in Herbert Mehrtens and Stefan Richter, eds., *Naturwissenschaft, Technik und NS-Ideologie. Beiträge zur Wissenschaftsgeschichte des Dritten Reiches* (Frankfurt-am-Main: Suhrkamp, 1980), 189–214; Ute Deichmann, *Biologists under Hitler* (Cambridge, Mass.: Harvard University Press, 1996), 179–205; and Benedikt Föger, Klaus Taschwer, *Die andere Seite des Spiegels. Konrad Lorenz und der Nationalsozialismus* (Vienna: Czernin, 2001), 148–153.

[36] Hippius et al., *Volkstum, Gesinnung und Charakter*, 16.

[37] Hippius et al., *Volkstum, Gesinnung und Charakter,* 64.

[38] Hippius et al., *Volkstum, Gesinnung und Charakter,* 112, 114. See Deichmann, *Biologists*, 193–196.

[39] "Memorandum by Karl Hermann Frank, 28 August 1940," in Václav Král, ed., *Die Deutschen in der Tschechoslowakei 1933–1947. Acta Occupationis Bohemiae et Moraviae* (Prague: Nak. Československé akademie, 1964), 417–421, here 419.

[40] Král, ed., *Die Deutschen in der Tschechoslowakei 1933–1947.* Document 318, letter from Heydrich to Frank, 14 September 1940; document 320, order by Hitler, 5 October 1940.

[41] On the German occupation policies in the "Protectorate," see Heinemann, *Rasse*, 127–186, Detlef Brandes, *Die Tschechen unter deutschem Protektorat*, vol. 1 (*Besatzungspolitik, Kollaboration und Widerstand im Protektorat Böhmen und Mähren bis Heydrichs Tod, 1939–1942*) (Munich: Oldenbourg, 1969); and Vojtech Mastny, *The Czechs under Nazi Rule: The Failure of National Resistance, 1939–1942* (New York: Columbia University Press, 1971).

[42] Egon Freiherr von Eickstedt, *Die Forschung am Menschen*, 3 vols. (Stuttgart: Enke, 1963); and Ilse Schwidetzky, *Einführung in die Völkerbiologie* (Stuttgart: Gustav Fischer, 1950).

From "Prisoner of War Studies" to Proof of Paternity: Racial Anthropologists and the Measuring of "Others" in Austria

Margit Berner

From the beginning of the twentieth century, the separation of physical and cultural anthropology occurred differently in English-speaking and German-speaking countries. Traditional academic seats of learning in Germany, and the names of the oldest learned societies, such as the German Society for Anthropology, Ethnology and Prehistory (*Berliner Gesellschaft für Anthropologie, Ethnologie und Urgeschchte*), or the Viennese Anthropological Society (*Anthropologische Gesellschaft in Wien*), reflected distinct branches of anthropology. Accordingly, separate faculties were created in acknowledgement of the different strands within the discipline. In the textbook *Lehrbuch für Anthropologie* (Textbook of Anthropology), first published in 1914, Rudolf Martin (1864–1925) urged German-speaking anthropologists to follow his methods of classifying observations, morphognosis and the verifying of hypotheses through measurements. The shortcoming of this methodology is that it did not allow for a broader biological and evolutionary context.[1] In 1908, the racial anthropologist Eugen Fischer (1874–1967) traveled to the German protectorate of South-West Africa (*Deutsch-Südwestafrika*) to investigate the anthropological traits of "mixed-race" inhabitants, the offspring of "Boers and Hottentots." He published the results of his work in 1913 as *Die Rehobother Bastards und das Bastardisierungsproblem beim Menschen* (The Bastards of Rehoboth and the Problem of Miscegenation in Man). This study led to the assumption that complex racial traits segregate in Mendelian fashion. The consequences of free combination of genes for the development of races were not taken into account. The concept of race therefore remained static.[2]

In 1913, the first Chair in Anthropology and Ethnography was established at the University of Vienna; Rudolf Pöch (1870–1921) was its first recipient. Prior to his appointment Pöch had trained as a physician, later joining an expedition in 1897 to study the plague in Bombay. One year later, he became renowned for treating cases of plague in Vienna, which had been caused by imported serum from the same expedition. In 1900/1901, Pöch studied anthropology in Berlin with the Austrian anthropologist Felix von Luschan (1854–1924). Between 1904 and 1906, Pöch undertook an anthropological-ethnographical expedition to Australia and New Guinea in order to study native populations. Following the war in the German colony of South-West Africa (1904–1907) in which 80 per cent of the Herero people—half of the Nama population including many Dama and San—were killed, or perished (after being driven into the desert by colonial German troops), Pöch joined another expedition to Namibia between 1907 and 1909. Once there, Pöch's research focused on the observation of survivors whom he considered as prime examples of a biologically primitive race of "Bushmen," soon to become extinct. He regarded them as belonging to a lower stage of cultural evolution.[3]

Pöch further developed these ideas in his doctoral dissertation on the racial characteristics of Australian aborigines, completed in 1913 under the supervision of the Munich anthropologist Johannes Ranke (1836–1916). Of the Australian groups, only individuals regarded as belonging to a "pure race" were of interest, as defined by the colonial authorities. Pöch believed that it was unlikely that Australian aborigines, for instance, constituted a source of degeneration for the simple reason that they were not "evolutionarily adapted" for agricultural labor. Pöch regarded Europeans as superior and ranked them above the indigenous populations of Australia.[4]

The First World War played an important role in the development of physical anthropology as a discipline in Vienna. Austrian and German anthropologists regarded the soldiers held in prisoner of war camps as prime "material" for scientific research. The racial study of POWs had been initiated and financially supported by the Viennese Anthropological Society from 1915. Between 1915 and 1918, the anthropometric studies conducted on several thousand captured soldiers in camps located in Austria-Hungary and Germany were carried out by a team led by Pöch and his assistant Josef Weninger (1886–1959). Further support was provided by the Austrian Academy of Sciences and the Austro-Hungarian Imperial War Ministry. At the beginning, Pöch also made phonographic recordings of songs and filmed various ethnographic scenes including dances. Later, he initiated further phonographic research in the camps, undertaken by the Phonographic Commission of the academy of Sciences. This allowed Pöch to concentrate on his racial research.[5]

Pöch prompted further anthropometric studies between 1914 and 1918. The anthropologist Viktor Lebzelter (1889–1936) measured Serbs, Roma and Sinti during his stationing in Krakow; the ethnologist Arthur Haberlandt (1889–1964) undertook a similar study of Albanians in Montenegro, which was supported by the Austrian Ministry of Education and the Austrian Academy of Sciences. Georg Kyrle (1887–1937), a geologist and pharmacist who participated in Pöch's 1915 project, undertook family studies of Wolhynian people in the course of his work as commander of an epidemic laboratory. Kyrle measured families, especially naked women and men, while they were being taken for delousing. Finally, Pöch initiated heritage studies on Wolhynian refugees in an Austrian camp, carried out by his fellow scholar and future wife Hella Pöch (née Helene Schürer von Waldheim) (1893– 1976), a study also supported by the Austrian Academy of Sciences.[6]

Several thousand data sheets, hundreds of photographs and plaster casts of heads, as well as hair samples, were collected for further statistical and racial analysis. Pöch and Weninger compiled initial statistical evaluations, published comprehensive reports on methodology, and delivered lectures. Samples of photographs and colored plaster busts considered characteristic of certain human "types" were selected from the collected "human materials" on display at the War Exhibition in Vienna between 1916 and 1917.[7]

Stressing the singular opportunity given by war, the Austrian "prisoner of war project" focused on investigating anthropologically lesser-known ethnic groups within the Russian Empire, particularly those considered close to extinction. Later the project was expanded to include more inhabitants of the Russian Empire, as well as Africans. In comparison to previous field surveys the "prisoner of war project" presented Pöch with ideal conditions for scientific research and comparative racial studies, particularly the dependence of captured soldiers upon their captors. The situation inside the camps was exacerbated by the surge of national feeling among the inmates, thereby strengthening the tendency to portray the prisoners as "racial" as opposed to "political" enemies. The studying of people in the context of camps and prisons had already been established by colonial administrations. However, the prisoner of war camps in the First World War allowed for an immediate comparison between different ethnic groups able to be studied in one place, a special situation unlike any other colonial or civilian context. In his study, Pöch avoided investigations of ethnic groups that were part of the Austro-Hungarian Empire.[8]

Austrians were not the only anthropologists enthusiastic about the possibility of studying POWs. At the same time as Pöch started his project a commission was set up in Germany to collect sound documents of languages and songs of different nationalities in the camps. Felix von Luschan, the director of the Museum of Ethnology in Berlin, introduced physical anthropology as part of the scientific method used in the study of POWs in German camps. He collaborated with Pöch in standardizing anthropometric methods, supported Pöch's studies, and invited his student, Egon von Eickstedt (1892–1965), and the anthropologist Otto Reche (1879–1966) to join the "prisoner of war project." Reche joined the project in 1917 after being injured at the front.

Reche commenced his research with a study on ethnic groups from Central Asia, but shifted his attention to Western Europeans after several weeks.[9] In order to legitimize his own political agenda and support for German expansionism, Reche portrayed certain European groups as being racially related to the Germans. For instance, he reported that Estonians, Latvians and Lithuanians had an extraordinarily strong element of "Northern European blood," and that the Flemish were "anthropologically German."[10] In contrast, Eickstedt chose a selection of ethnic groups within nations fighting against Germany as racial case studies. Following Luschan's suggestion, Eickstedt started with "anthropologically interesting" groups like Indians, Turks and Asians, but several months later changed his focus to Scots, Irish, English, Ukrainians, Poles and Russians. The racial characteristics of Sikhs became the topic of his doctoral dissertation. Similarly, Pöch considered Asian Russians, Austria-Hungary's eastern war enemies, to be non-European "others" in terms of race.[11]

Apart from writing reports and lecturing on racial studies, Pöch never published the results of his "prisoner of war project." On his death in 1921, a part of Pöch's estate was donated to the Austrian Academy of Sciences for the twofold purpose of further scientific research into race and the posthumous publication of his work.[12] Some members of the Viennese scientific community were eager for Eugen Fischer to succeed Pöch, but he refused. Eventually, Otto Reche assumed the position in 1924.[13] His successor was Pöch's assistant, Josef Weninger, who was appointed professor of anthropology in 1927. Eventually, Weninger evaluated and published the data collected by Pöch.

By the mid-1920s, race had become the single most important concept in German anthropology. In his studies, Pöch, for example, differentiated between people (*Volk*) and race, and focused on methodology and racial classifications, as well as racial mixing between people of European and Asian descent.[14] A multitude of then recent "ethnic groups" were formed from a few originally widespread races; the differences were thought to derive from the degree of mixture. Pöch emphasized that the proportion of various psychological qualities, such as character traits and mental qualities, determined the particular character of any *Volk*, which in turn characterized hereditary racial peculiarities. Based on Eugen Fischer's *Die Rehobother Bastards*, German and Austrian anthropologists were convinced that race was not the product of genetic inheritance in general, but that only specific traits were hereditary. The mixing of races would thus result in hybrids and bastards, not the emergence of new races. However, many believed it was still possible that some individuals belonging to particular ethnic groups could be representatives of former racial "pure types," following Mendel's theory of backcrossing.[15] German and Austrian anthropologists therefore attempted to identify individuals or groups that corresponded to "pure racial types."

One of the first publications written on this subject was Josef Weninger's postdoctoral work (*Habilitation*), *Eine morphologisch-anthropologische Studie. Durchgeführt an 100 westafrikanischen Negern, als Beitrag zur Anthropologie von Afrika* (A Morphological-Anthropological Study. Carried out on 100 West African Negroes, as a Contribution to African Anthropology), published in 1927.[16] This study was based on measurements taken from Africans in Wünsdorf, a prisoner of war camp near Berlin, in 1917, and from Turnu Măgurele in Romania in 1918. Weninger regarded his work as an attempt to develop a new morphological method, one possibly leading to more detailed racial differentiation. For instance, he grouped morphological traits into series, thereafter constructing racial typologies. An example of his qualitative approach, based on photographic evidence, was the classification of nose types: the primitive button-nose (*Knopfnase*) of "pygmies and Bushmen" was located at the bottom end of his typology, while the "highly specialized European" nasal form was listed at the top end. The nasal form, again, was not inherited as a whole, but as the sum of various traits. Each trait would follow Mendelian laws, and the observed traits were interpreted as belonging to pure or mixed races.[17] The Austrian anthropologist Viktor Lebzelter developed another form of race typology based on his "prisoner of war studies," consisting of a combination of morphological and statistical data.[18]

Others, too, were concerned about the "problem of the Eastern race" (*Problem der Ostrasse*), a theme engaging Pöch during his study of Baltic POWs. Joseph Deniker (1852–1918) had first defined the "Eastern race" in 1900, as a blond, grey-eyed, and short-headed race.[19] Later contributions—like those by Hella Pöch (1893–1976) on Wolhynians, and Michael Hesch (1896–1971) on Latvians, Lithuanians and Belarusians—added many traits to be held typical of the "Eastern race." Pöch and Hesch distinguished two types within this race: a light and a dark variant, both of which were considered to be Asian in origin. In her work, Hella Pöch classified the population as a "Mendelian F2" mixed generation, and attempted to identify inheritance of racial characteristics and racial mixture in the theoretical and actual distributions of traits.[20] Others came to the conclusion that the Bashkirian and Turkish inhabitants of the Russian Empire did not represent a unitary racial type and were thus characterized as representative of a disharmonic racial mixture of European and Asian types.[21] Among anthropologists at the time, it was quite common to assign to people of mixed origins negative aesthetical criteria like disharmonic and ugly.[22]

In their work on prisoners of war, the anthropologists discussed here presented case studies in relation to varying taxonomies, such as ethnographic, historic and prehistoric factors. In particular they stressed that further studies on the principle of heredity, and the passing on of racial traits among and between races, were necessary. Yet biographical and demographical data, as well as data concerning general health, was not analyzed from a eugenic perspective. Such questions were only addressed in the publications appearing after 1938. Prior to this date, all publications simply described human types. They did not explicitly stress the superiority of the Nordic race; however, from today's perspective, the personal prejudices and racist thinking of the authors can be clearly discerned. These authors essentially assumed that racial variety implied racial inequality.[23]

Egon von Eickstedt developed a similar method of analysis in his study on prisoners of Sikh origin.[24] The Nazis extensively employed his formulae on race, and the application of this method of classification continued after the Second World War. As the German anthropologist Holger Preuschoft argued: "Germans were for some time disabled intellectually by pure race-systematic in the sense of von Eickstedt, which became quite sterile in the course of time."[25] Working on methods applied in the "prisoner of war studies" Viennese anthropologists developed an extensive data sheet for the purpose of detailed morphological studies, best known for an in-depth analysis of facial "racial" traits.[26]

The Austrian "prisoner of war studies" also informed the work of Hans F. K. Günther (1891–1968), particularly *Rassenkunde des deutschen Volkes* (Racial Science of the German People) and his teaching materials.[27] Günther had initially studied linguistics and German philology, but he gained his anthropological knowledge during study trips to Vienna and Dresden. In the preface of *Rassenkunde des deutschen Volkes* Günther thanked the Institute of Anthropology and the Museum of Natural History in Vienna for their support, and for photographs taken during the "prisoner of war project, which were reproduced in this book." In a discussion of certain traits of the eye due to Asian-European mixture in the ethnic groups of Eastern Europe, Günther referred directly to one of Pöch's schemes.[28]

Günther distinguished four "indigenous" races as European: Nordic (*nordisch*), Alpine (*ostisch*), Mediterranean (*westisch*), and Dinaric (*dinarisch*), to which he added two more races in later works: East Baltic (*ostbaltisch*), and a new racial category—Sudetic or Phalian (*sudetische/fälische*)—previously described by Otto Reche.[29] The Austrian school teacher Gustav Kraitschek had published a similar book, under the title *Rassenkunde mit besonderer Berücksichtigung des deutschen Volkes, vor allem der Ostalpenländer* (Racial Science with Particular Consideration of the German People, mainly the East Alpine countries).[30] In this monograph he attempted to demonstrate that Austrians were of Nordic/Germanic descent, and that, countering Günther's argument, the Alpine race was not inferior to the Nordic race.[31] Many anthropologists rejected Günther's arguments, among them Viktor Lebzelter, who, for instance, criticized Günther's psychological description of races. However, the scientific community as a whole welcomed the popularization of "racial science" through Günther's work, making it readily available to the general public.[32]

The conclusions of the "prisoner of war project," and the newly created standardized research methods, encouraged scientists to undertake heritage studies on families and twins, as well as extensive research on Austria's population.[33] Hella Pöch's studies on Wolhynian refugees between 1925 and 1926 were the first in a series of such studies, where metrical and morphological traits, as wells as palmar lines, were studied within families. Soon thereafter, questions of nutrition, health and lifestyle were considered as matters related to racial hygiene.[34] During the 1920s, Otto Reche devised a certificate that confirmed proof of paternity for Viennese courts. This was later extended to a genetic certificate encompassing race and family origin under the Nazis.[35]

In conclusion, the interwar period witnessed a methodological shift from theory-based to applied anthropology, resulting in close co-operation between medical institutions, judicial courts and law enforcement agencies.[36] This offered new academic positions and income possibilities for racial anthropologists at a time when financial support and institutional funding were difficult to obtain. For instance, Eberhard Geyer (1899–1942), head research assistant at the Institute of Anthropology in Vienna, planned to create an institute separate from the university for the specific purposes of verifying paternity.[37] At the same time, racial scientists became increasingly engaged in the eugenic movement, as evidenced by the fact that the Viennese Society of Racial Hygiene held their meetings at the Institute of Anthropology.[38]

Robert Proctor considered the 1920s as the beginning of the end of physical anthropology, at least as a discipline practiced in the tradition of Rudolf Martin. Proctor argued that the focus of research had shifted to explain how physical and cultural qualities might be combined to fit the rubric of human genetics. Anthropologists also gained an interest in how social, and not just psychological, problems might be solved by "racial hygiene."[39]

During the Second World War anthropologists working at the Natural History Museum in Vienna applied a method almost identical to that of their predecessors in investigating around 7,000 people, among them Jews and other prisoners of war.[40] Most of those measured were not volunteers, despite Rudolf Pöch's insistence otherwise. In truth, some of the prisoners sought to avoid or undermine the studies through resistance. In order to surmount these obstacles, Pöch asked a camp commander for assistance. The commander allowed a molding of his own head to be taken, responding to Pöch's remark about the "persistent shyness of the people towards this procedure."[41] Egon von Eickstedt also complained about the unwillingness of some prisoners to participate in the investigation.

Racial studies conducted on prisoners of war may be regarded as the main project of the newly founded Institute of Anthropology in Vienna. The anthropologists associated with the project pioneered a new method in the collection of human data: matter-of-fact presentation of gathered material (photographs, plaster casts, statistics and so on) to the scientific community, and to the general public, further affirmed a racist attitude of biological determinism.

These racial studies on prisoners of war demonstrate that physical anthropology as a distinct scientific discipline was not based on epistemological claims, and that results obtained during the research were not just objective facts. They serve as an example of how cultural and political contexts could shape scientific research. For the interwar period in particular, these studies show how a combination of ideology, personal career planning, and the special situation of the war, all influenced the Viennese physical anthropologists in their scientific motivations and undertakings. The testimony of this research, the photographs and data sheets, plaster masks and data sheets are still kept at the University of Vienna and the Natural History Museum in Vienna. It is time to ask entirely different questions about ethics in anthropology—such as, for instance, how to handle previously collected "human material"—by taking into account the transgressed dignity of the often involuntarily measured individuals.[42]

Endnotes:

[1] Holger Preuschoft, "Physical Anthropology in German-Speaking Europe," *Yearbook of Physical Anthropology* 16 (1972), 122–140; and Rudolf Martin, *Lehrbuch der Anthropologie in systematischer Darstellung mit besonderer Berücksichtigung der anthropologischen Methoden* (Jena: Gustav Fischer, 1914).
[2] Eugen Fischer, *Die Rehobother Bastards und das Bastardisierungsproblem beim Menschen* (Jena: Gustav Fischer, 1913). See also Robert Proctor, "From Anthropology to *Rassenkunde* in the German Anthropological Tradition," in George W. Stocking, ed., *Bones, Bodies, Behavior. Essays on Biological Anthropology* (Madison: University of Wisconsin Press, 1988), 138–179; and Niels C. Lösch, "Rasse als Konstrukt. Leben und Werk Eugen Fischers," *Europäische Hochschulschriften* 737 (1997), 53–81.
[3] Brigitte Fuchs, *"Rasse," "Volk," "Geschlecht." Anthropologische Diskurse in Österreich 1850–1960* (Frankfurt-am-Main: Campus, 2003), 190–211; and Eigen Oberhummer, "Rudolf Pöch (gestorben am 4. März 1921)," *Mitteilungen der Anthropologischen Gesellschaft Wien* 51 (1921), 95–104.
[4] Rudolf Pöch, *Studien an Eingeborenen von Neu-Südwales und an australischen Schädeln* (Munich–Vienna: Hamburger, 1913), 12–16.
[5] See Margit Berner, "Die 'rassenkundlichen' Untersuchungen der Wiener Anthropologen in Kriegsgefangenenlagern 1915–1918," *Zeitgeschichte* 30, 3 (2003), 124–136; Margit Berner, "Forschungs-'Material' Kriegsgefangene: Die Massenuntersuchungen der Wiener Anthropologen an gefangenen Soldaten 1915–1918," in Heinz Eberhard Gabriel and Wolfgang Neugebauer, eds., *Vorreiter der Vernichtung? Eugenik, Rassenhygiene und Euthanasie in der österreichischen Diskussion vor 1938*, (Vienna: Böhlau, 2005), 171–173; Andrea Gschwendtner, "Als Anthropologe im Kriegsgefangenenlager—Rudolf Pöchs Filmaufnahmen im Jahre 1915. Film P2208 des ÖWF 1991," *Wissenschaftlicher Film* 42, April (1991), 105–118; and Andrea Gschwendtner, "Frühe Wurzeln für Rassismus und Ideologie in der Anthropologie der Jahrhundertwende am Beispiel des wissenschaftlichen Werkes des Anthropologen und Ethnographen Rudolf Pöchs," in Claudia Lepp and Barbara Danckwortt, eds., *Von Grenzen und Ausgrenzung* (Marburg: Schüren, 1997), 136–158.
[6] Berner, "Die 'rassenkundlichen' Untersuchungen der Wiener Anthropologen," 126; Berner, "Forschungs-'Material' Kriegsgefangene," 175–177; Hella Pöch, "Beiträge zur Anthropologie der ukrainischen Wolhynier," *Mitteilungen der Anthropologischen Gesellschaft Wien* 55 (1925), 289–321; and Hella Pöch, "Beiträge zur Anthropologie der ukrainischen Wolhynier," *Mitteilungen der Anthropologischen Gesellschaft Wien* 56 (1926), 16–47.
[7] Berner, "Die 'rassenkundlichen' Untersuchungen der Wiener Anthropologen ," 127; Berner, "Forschungs-'Material' Kriegsgefangene," 177–178.
[8] Berner, "Die 'rassenkundlichen' Untersuchungen der Wiener Anthropologen," 131–132; and Andrew D. Evans, "Anthropology at War: Racial Studies of POWs during World War I," in Glenn H. Penny and Matti Bunzl, eds., *Worldly Provincialism: German Anthropology in the Age of Empire* (Michigan: University of Michigan, 2003), 198–228.

[9] Evans, "Anthropology at War," 198–228.

[10] Evans, "Anthropology at War," 220–222.

[11] Evans, "Anthropology at War," 207–226.

[12] Berner, "Die 'rassenkundlichen' Untersuchungen der Wiener Anthropologen in Kriegsgefangenenlagern 1915–1918," 127; Berner, "Forschungs-'Material' Kriegsgefangene," 177–181.

[13] Lösch, "Rasse als Konstrukt," 116–118; and Katja Geisenhainer, *"Rasse ist Schicksal": Otto Reche (1879–1966)—ein Leben als Anthropologe und Völkerkundler* (Leipzig: Evangelische Verlagsanstalt, 2002), 108–113.

[14] Berner, "Forschungs-'Material' Kriegsgefangene," 178–181.

[15] Rudolf Pöch, "Neue anthropologische Fragestellungen," *Mitteilungen der Geographischen Gesellschaft in Wien* 62 (1919), 193–209.

[16] Josef Weninger, *Eine morphologisch-anthropologische Studie. Durchgeführt an 100 westafrikanischen Negern, als Beitrag zur Anthropologie von Afrika*, Rudolf Pöch's Nachlass, A, vol. 1 (Vienna: Anthropologische Gesellschaft, 1927).

[17] Berner, "Die 'rassenkundlichen' Untersuchungen der Wiener Anthropologen," 128–129; Berner, "Forschungs-'Material' Kriegsgefangene," 183; Margit Berner, "Die Forschungen der Wiener Anthropologen an schwarzen Kriegsgefangenen im Ersten und Zweiten Weltkrieg," in Peter Martin and Christine Alonzo, eds., *Zwischen Charleston und Stechschritt. Schwarze im Nationalsozialismus* (Hamburg: Dölling und Galitz, 2004), 605–613.

[18] Viktor Lebzelter, "Beiträge zur physischen Anthropologie der Balkanhalbinsel. I Teil Zur physischen Anthropologie der Südslawen," *Mitteilungen der Anthropologischen Gesellschaft Wien* 53 (1923), 1–48; and Viktor Lebzelter, "Beiträge zur physischen Anthropologie der Balkanhalbinsel II," *Mitteilungen der Anthropologischen Gesellschaft Wien* 63 (1933), 233–247.

[19] Joseph Deniker, *The Races of Man* (London: Walter Scott, 1900), 286.

[20] Hella Pöch, "Beiträge zur Anthropologie der ukrainischen Wolhynier," 16–47, 289–321; Michael Hesch, *Letten, Litauer, Weißrussen. Ein Beitrag zur Anthropologie des Ost-Baltikums mit Berücksichtigung der siedlungs- und stammesgeschichtlichen Grundlagen*, Rudolf Pöch's Nachlass, A, vol. 3 (Vienna: Anthropologische Gesellschaft, 1933).

[21] Josef Wastl, *Baschkiren: Ein Beitrag zur Klärung der Rassenprobleme Osteuropas*, Rudolf Pöch's Nachlass A, vol. V (Vienna: Anthropologische Gesellschaft, 1938); and Karl Tuppa, *Mischeren und Tipteren. Ein Beitrag zur Anthropologie der Türkvölker in Rußland*, Rudolf Pöch's Nachlass, A, vol. 6 (Berlin: Ahnenerbe-Stiftung 1941).

[22] Pöch, "Neue anthropologische Fragestellungen," 197.

[23] Berner, "Forschungs 'Material' Kriegsgefangene," 186–187; Berner, "Die Forschungen der Wiener Anthropologen an schwarzen Kriegsgefangenen," 607–609.

[24] Egon von Eickstedt, "Rassenelemente der Sikh," *Zeitschrift für Ethnologie* 52 (1920–21), 317–80.

[25] Preuschoft, "Physical Anthropology in German-Speaking Europe," 133.

[26] Josef Weninger (mit einem Beitrag von Hella Pöch), "Leitlinien zur Be-

52 *"Blood and Homeland"*

obachtung der somatischen Merkmale des Kopfes und Gesichtes am Menschen," *Mitteilungen der Anthropologischen Gesellschaft Wien* 54 (1924), 232–261.

[27] Hans F. K. Günther, *Rassenkunde des deutschen Volkes*, 3d. ed. (Munich: Lehmann, 1923).

[28] Günther, *Rassenkunde des deutschen Volkes*, 125, 129.

[29] Proctor, "From Anthropology to *Rassenkunde* in the German Anthropological Tradition," 149.

[30] Gustav Kraitschek, *Rassenkunde mit besonderer Berücksichtigung des deutschen Volkes, vor allem der Ostalpenländer* (Vienna: Burg, 1924).

[31] Fuchs, *"Rasse," "Volk," "Geschlecht,"* 250–260.

[32] Viktor Lebzelter, "Günther, Dr. Hans K.F.: Kleine Rassenkunde Europas," *Mitteilungen der Anthropologischen Gesellschaft Wien* 56 (1926), 128–129; Benoît Massin, "Rasse und Vererbung als Beruf," in Hans-Walter Schmuhl, ed., *Rassenforschung an Kaiser-Wilhelm-Instituten vor und nach 1933* (Göttingen: Wallstein, 2003), 190–244, especially 193–194.

[33] Josef Weninger, "25 Jahre Anthropologisches Institut an der Universität Wien," *Mitteilungen der Anthropologischen Gesellschaft Wien* 68 (1938), 191–205.

[34] Fuchs, *"Rasse," "Volk," "Geschlecht,"* 241–44; Berner, "Forschungs-'Material' Kriegsgefangene," 184.

[35] Georg Lilienthal, "Arier oder Jude? Die Geschichte des erb- und rassenkundlichen Abstammungsgutachtens," in Peter Propping and Heinz Schott, eds., *Wissenschaft auf Irrwegen. Biologismus—Rassenhygiene—Eugenik* (Bonn: Bouvier, 1992), 66–84; Geisenhainer, *"Rasse ist Schicksal,"* 125–127; Maria Teschler-Nicola, "Der diagnostische Blick—Zur Geschichte der erbbiologischen und rassenkundlichen Gutachtertätigkeit in Österreich vor 1938," *Zeitgeschichte* 30, 3 (2003), 137–149. Testing for paternity became a fairly profitable job after World War II; many anthropologists that were dismissed from their positions later became anthropological experts for the courts.

[36] Weninger, "25 Jahre Anthropologisches Institut," 200.

[37] Berner, "Forschungs-'Material' Kriegsgefangene," 190.

[38] Weninger, "25 Jahre Anthropologisches Institut an der Universität Wien," 204; see also Thomas Mayer, *Akademische Netzwerke um die "Wiener Gesellschaft für Rassenpflege (Rassenhygiene)" von 1924–1948* (unpublished PhD dissertation: University of Vienna, 2004), 165.

[39] Proctor, "From Anthropology to *Rassenkunde* in the German Anthropological Tradition," 148–156.

[40] Verena Pawlowsky, "Erweiterung der Bestände. Die Anthropologische Abteilung des Naturhistorischen Museums 1938–1945," *Zeitgeschichte* 32, 2 (2005), 69–90; Claudia Spring, "Vermessen, deklassiert und deportiert. Dokumentation zur Anthropologischen Untersuchung an 440 Juden im Wiener Stadion im September 1939 unter der Leitung von Josef Wastl vom Naturhistorischen Museum Wien," *Zeitgeschichte* 32, 2 (2005), 91–110; Margit Berner and Claudia Spring, "Gipsmasken und Messblätter. Relikte in den Anthropologischen Sammlungen des Wiener Naturhistorischen Museums," *Jüdisches Echo* 53 (2004), 222–226; and Maria Teschler-Nicola and Margit Berner, "Die

Anthropologische Abteilung des Naturhistorischen Museums in der NS-Zeit; Berichte und Dokumentation von Forschungs- und Sammlungsaktivitäten 1938–1945," in *Senatsprojekt der Universität Wien. Untersuchungen zur Anatomischen Wissenschaft in Wien 1938–1945* (Vienna: Akademischer Senat der Universität Wien, 1998), 333–358.

[41] Evans, "Anthropology at War," 216.

[42] Margit Berner, "Macht und Ohnmacht vor dem musealen Bestand: Eine anthropologische Untersuchung an Juden im September 1939 in Wien. Anmerkungen und Annäherungen einer Kuratorin," in Ingrid Bauer et. al., eds., *Kunst->Kommunikation>Macht, Sechster Österreichischer Zeitgeschichtetag 2003* (Innsbruck: Studien Verlag, 2004), 261–265; and Claudia Spring, "Macht und Ohnmacht vor dem musealen Bestand: Eine anthropologische Untersuchung an Juden im September 1939 in Wien. Anmerkungen und Annäherungen einer Historikerin," in Bauer et. al., eds., *Kunst>Kommunikation>Macht*, 266–270.

Volksdeutsche and Racial Anthropology in Interwar Vienna: The "Marienfeld Project"

Maria Teschler-Nicola

In the second half of the nineteenth century, racial anthropology was shaped by positivist and materialist thinking, initially aiming at a quantitative assessment of physical traits and comparative anatomical "studies of race" in order to identify "ideal racial types." But in contrast to the descent-based anthropological orientation, this branch of physical anthropology soon arrived at a deadlock. In *Geschichte der Anthropologie* (History of Anthropology), Wolfgang Mühlmann described this phenomenon as an "accumulation of a large number of facts whose interpretative value to biology has remained questionable."[1] One of the most prominent exponents of this previously static approach was Augustin Weisbach (1837–1914), a Viennese anatomist whose scientific "work and life program" aimed to identify, the "racial differences" among the populations of the Habsburg Empire.[2]

Rudolf Pöch (1870–1921), recipient of the first Chair of Anthropology at the University of Vienna, conceded that Weisbach had "dealt with an enormous volume of material" and stressed his achievements in the field of anthropology in Austria.[3] Pöch also raised subtle criticism of Weisbach's work which, in his opinion, failed to make use of the available "resources and methods of modern anthropology" (implying not only technique, but also the genetic-biological approach). The rediscovery of Mendel's laws of inheritance at the beginning of the twentieth century ushered in a paradigm shift in anthropology, which placed in question prior comparative anatomical studies of race. Pöch belonged to the generation of physical anthropologists that reopened the discussion of racial anthropology. This new approach evolved against the background of complex developments in society, politics, and the humanities in Europe towards the end of the nineteenth century, shaped by nationalist movements and debates on the nation-state, national character, and "racial theories." In Central Europe the debate about race was to a large extent initiated by Houston Stewart Chamberlain (1855–1927), a Vienna-based advocate of "racial purity" who, as the Nazi racial theorist Hans F. G. Günther (1891–1968) later attested, "introduced racial thought to broad sections of the public for the first time."[4]

The First "Racial" Anthropological Project in Vienna

At the turn of the twentieth century, "modern biology" increasingly focused on the theory of inheritance,[5] while anthropology emphasized the "research of cause" rather than "research of facts."[6] Leading theoreticians of the discipline, including Eugen Fischer (1874–1967) and Erwin Baur (1875–1933), defined anthropology as the science of genetic differences in man.[7] The concept of race became inheritance-orientated and it was assumed that "physical and psychological racial traits" were due to genetic factors, while "racial formation," "racial reshaping," "bastardization," and the identification of the different components comprising the mix of "hypothetically pure races" became major research topics.[8]

The physician, anthropologist and ethnographer Rudolf Pöch was an early supporter of the genetic approach. The anthropologist Josef Weninger (1886–1959), one of Pöch's students and associates, supported Pöch's biological approach at a time when anthropology provided little information about genetics. According to Weninger, Pöch delivered a "rather portentous lecture" on the biology of the human race in 1912, which focused on questions of racial hygiene.[9] In this respect, and in the light of recently discovered documentary evidence, a new investigation is required to explore whether Pöch, a founding member of the *Deutsche Gesellschaft für Rassenhygiene* (German Society for Racial Hygiene), paved the way for the development of racial hygiene in Austria. Evidence suggests, *inter alia*, that Pöch was part of a national and international network of anthropologists, anatomists, ethnologists and politicians that included Felix von Luschan (1854–1924); Rudolf Martin (1864–1925); Viktor Adler (1852–1918); Julius Tandler (1869–1936); Emil Zuckerkandl (1849–1910); Carl Toldt (1840–1920); and in particular Richard Thurnwald (1869–1954), the co-editor, with Alfred Ploetz (1860–1940), of the *Archiv für Rassen und Gesellschaftsbiolo-*

gie (Journal for Racial and Social Biology).[10] Pöch's circle of friends included Richard Fröhlich (1864–1926) and Rudolf Wlassak (1865–1930), as well as notable exponents of the "anti-alcohol" movement. Following the publication of Eugen Fischer's *Die Rehobother Bastards und das Bastardierungsproblem beim Menschen* (The Rehoboth Bastards and the Problem of Bastardization in Mankind) in 1913, the German anthropological community focused on questions of "racial biology" with a view towards creating a genetically orientated "racial science." Viennese anthropologists subsequently extracted enormous sets of metric and morphological data obtained from non-European ethnicities during expeditions and military missions, and from prisoners detained in Austro-Hungarian and German camps during the First World War.[11] Racial investigators—Pöch in particular—were well aware of the time, effort, and cost involved in organizing such an inquiry and thus welcomed this "unique, never recurring opportunity," providing food for thought for, and ensuring the careers of, generations of Viennese anthropologists.[12]

According to Weninger's retrospective interpretation, Pöch "progressed as scheduled on all main roads, in particular as far as his biological attitude was concerned."[13] The first "family projects," consisting of two parts, originated during the war: while stationed as a commander of an epidemiological laboratory in Western Wolhynia in 1916, Georg Kyrle (1887–1937), pharmacist, speleologist, and Rudolf Pöch's research assistant during the First World War, collected various bits of anthropological data.[14] Arguing that "it was possible to observe the people completely undressed," Kyrle sought to justify the reliability of his data. Similarly, Helene Schürer (1893–1976) conducted her family research project according to the precepts of "racial and genetic biology" by studying Ukrainian Wolhynians detained in a camp at Niederalm, Grödig (near Salzburg) from 1917 to 1918.[15] In principle, the aim was to collect metric and morphological data on women. On the one hand, Schürer focused on the racial character of her cases in order to identify the "most important racial types" of Ukrainian Wolhynians ("people of non-Ukrainian ethnicity," such as "Germans, Poles, Russians and Jews" were excluded from the study).[16] On the other hand, her genetic approach centered on subdividing morphological features into the smallest constituents that could be ascribed to individual inherited elements; she classified her concepts as "race admixture" and "bastardization." According to Schürer, "This way of inspecting different types, which is difficult in the beginning and often successful only if we allowed for a certain interval from one anthropological survey to the next, eventually enabled us to identify the common constituents that belong together, and to attempt racial typing."[17] But apart from these "attempts at racial typing," Schürer also discovered the "scientific potential" of this research. The family thus became "ideal" for studies in racial hygiene.[18]

During the interwar period, racial research conducted at Pöch's Institute of Anthropology essentially focused on two areas: racial anthropology, which had undergone a major transformation at the beginning of the twentieth century to become racial biology; and genetic research, which aimed to identify hereditary patterns as well as the "regular anthropological relationships existing between blood relatives."[19] The "twin method" mainly addressed general issues relating to inheritance. It was by adding the biological dimension that family research (for some time equivalent to "genealogy," and representing a narrow branch of historical research) became the "science of immediate importance for life" which, according to anthropologist Egon Freiherr von Eickstedt (1892–1965), "could also be subjected to the goals of racial hygiene."[20]

Otto Reche (1879–1966), holder of the Chair of Anthropology at the University of Vienna, was the first to prepare genetic assessments during paternity trials in the mid-1920s.[21] His successor, Josef Weninger, argued that the theoretical basis of these assessments, based on a comparative study of similar morphological traits, was too narrow. Accordingly, he adopted a critical stance towards this "unscholarly" procedure. Weninger's attitude changed later after a ruling of the Austrian Supreme Court that the lack of a biological examination in a paternity case constituted a procedural deficiency. In the early 1930s, after founding *Erbbiologische Arbeitsgemeinschaft* (The Working Group on Genetic Biology) and launching the "Marienfeld project," he eventually established the organizational basis for improving the reliability of paternity diagnosis.

The Working Group on Genetic Biology

Realizing that the practical scope of a family-based approach to hereditary research and paternity assessment would exceed the physical capacity of any individual, efforts were made to create an appropriate organizational structure for the new discipline. In 1932, these efforts culminated in the foundation of the Working Group on Genetic Biology under Josef Weninger at the Anthropological Institute of Vienna University, whose purpose was to investigate the frequency distribution and hereditary pattern of morphological traits through the anthropological examination of families.[22]

The Working Group on Genetic Biology consisted of an inner circle of associates including the anthropologist Eberhard Geyer (1899–1943) (an assistant at the Anthropological Institute from 1927); Robert Routil (1893–1955) (a biometrician and assistant at the Anthropological Institute from 1931); other non-civil-servant anthropologists, as well as members of the medical and legal professions specializing in anthropology, including Dora Maria Könner (1905–1970); Margarete Weninger (1896–1987); Karl Tuppa (1899–1981); Karl Thums (1904–1976); Wolfgang Müller; Albert Harrasser; and Friedrich Stumpfl (1902–1997).[23]

The method adopted was "metric measurement" involving documentation of the morphological features of the head, face, and body through the collection of drawings and photographs (as a basis for "classification"). Weninger referred to this method as the "Viennese School of Anthropology."[24] His associates were responsible for collecting and publishing material in their specific fields, conducting relevant research, and disseminating their findings to other members of the group. Due to this well-structured division of tasks, by 1932 the Working Group on Genetic Biology had amassed more than 100 expert opinions in Viennese legal paternity disputes.[25] Weninger's research assistant until the end of 1934, Albert Harrasser, observed an increase in forensic work and research activities at the Anthropological Institute.[26] The general consensus was that anthropology could, from that point forth, serve the larger interests of society.[27] Moreover, these new applications started to have an impact on the number and orientation of scientific projects and publications hosted by the Vienna Institute of Anthropology.

Selected Aspects of German Anthropology

During the 1920s and 1930s, there was an increase in the circulation of popular writings on the subject of race, particularly through the work of Hans F. G. Günther,[28] which aimed at raising the awareness of "racial science" and "racial theory."[29] However, Eugen Fischer, the most renowned anthropologist of the time, insisted that "the German people" (*Deutsche Volk*) remained an under-researched topic.[30] Questioning all previous efforts by anthropologists, Fischer was especially critical of the attempt by Gustav Schwalbe (1844–1916) to establish a commission for anthropological investigations in Germany, Austria, and Switzerland.[31] In the wake of political developments in the late 1920s, Fischer considered a comprehensive survey of German anthropology "an urgent matter not to be postponed," especially the study of "facial structures and nose shapes," which he considered the most important of all "racial traits."

For the first time, the focus of racial research turned to regional populations. This was accompanied by the use of new historical sources, including parish registers and local village records, in order to survey a population within its genealogical sources. According to Fischer, the exercise was "truly anthropological."[32] Other anthropologists, including Karl Saller, Friedrich Keiter and Walter Scheidt, adopted this new anthropological method. Their findings were published in several volumes of Eugen Fischer's *Deutsche Rassenkunde* (German Racial Studies).[33] Scholarly reviews of *Deutsche Rassenkunde* highlighted the fact that Viennese anthropologists knew these approaches.[34] It was obvious that this form of racial anthropology, which included family surveying from a biological point of view, gradually acquired eugenic undertones. Fischer attached "enormous importance" to obtaining concrete material "concerning the question of numbers and distribution of several significant pathological genetic lineages among our people."[35]

According to the fashionable conception of *Volksdeutsche* (German people), German scientists believed that it was possible to ascertain whether descendants of German immigrants had evolved biologically or psychologically, in the absence of crossbreeding, admixture, potential inbreeding, or environmental impact. Consequently, it would be possible to identify the "final product" of ethnic mixing. Here the aim was to examine "Germans abroad in those places" where they had settled "massively" and remained "relatively pure."[36] Fischer's hopes for the adoption of similar attitudes by Austrian anthropologists were to be fulfilled with the "Marienfeld project."

The Anthropological Project in Marienfeld

The anthropological project carried out during the winter of 1933–34 in the village of Marienfeld (Teremia-Mare) in the Banat region of Romania endeavored to shed light on racial anthropology, genetic biology, and ontogenetics. The racial side of the project centered on studying the "population of Marienfeld, a German linguistic enclave in Romania," including the history and anthropology of the local population, in addition to the "possibilities and prospects for the development of a German ethnic group abroad."[37] The second aspect of the "Marienfeld project" aimed at surveying families from an anthropological perspective; namely, the "examination and description of the genetic stock based on new typological methods," according to the principles of the Viennese School of Anthropology.[38] Weninger emphasized that this approach was relevant not only for proof of descent, but also for the purposes of racial analysis. Finally, the ontogenetic aspect of the project dealt with the question of age modification, a problem hitherto barely investigated.

Simply put, the "Marienfeld project" combined three distinct areas of research under the umbrella of one scientific approach. The project followed the main thrust of biological research at the time, and was viewed by Weninger as a "genetic-biological" project aimed at addressing the prevailing shortcomings of genetic theory. The project also provided a genetic-biological basis for addressing "racial issues," which involved viewing the village of Marienfeld—in line with their ideological beliefs and *Weltanschauung*—as a nationality study of a German minority living in the Banat region of Romania. In Eberhard Geyer's view, problems arising from nationality research could only be addressed in racial-biological and anthropological terms. Geyer believed that the time had come for nationality research to "become a conservator and guardian of national values," and that Hans F. G. Günther's popular 1922 *Rassenkunde des deutschen Volkes* (Racial Study of the German People) outlined the transformation of physical into racial anthropology.[39]

At the time, it was assumed that most European ethnicities were characterized by a number of identical racial traits, and that both the racial makeup and the extent and type of ethnic admixture were *völkisch* in nature. As racial science had developed towards an even more sophisticated classification of ethnicity, anthropological measures were introduced to categorize small groups. This was an initial step towards the classification of regional *Gaue* (districts) and family typologies beyond racial boundaries, which was practiced, in particular, in the course of anthropological examinations of national minorities.[40] The linking of empirical data with research on the nationality question implied the following: in the context of "nationality research," anthropology was required to define the basic racial ingredients alleged to have shaped the so-called *völkisch* character of an ethnicity (in Marienfeld these examinations were undertaken by Friedrich Stumpfl).[41] Further issues were the "racial gap" between *Gast- und Wirtsvolk* (Guest and Host Peoples) and whether, and to what extent, "seclusion from the outside world" accounted for the biological preservation of a minority group. For Geyer, deeply influenced by Nazi ideology, the preservation of the so-called "racial divide" seemed to have been particularly important. He envisaged the "danger" of further admixture for many European minorities, including the Swabs in the Banat, because the two ethnic groups in question were racially similar, which implied the onset of "cultural assimilation."

Geyer considered the Swabs in the Banat to be "one of the best-founded pillars of Germanness abroad in Southeast Europe, biologically speaking."[42] Moreover, due to the occurrence of intermarriage between close relatives on the one hand, and religious and linguistic barriers to neighboring communities on the other, the inhabitants of Marienfeld were characterized as a "narrowly defined biological community."[43] Such conditions provided the ideal testing ground for the research project outlined above.[44]

The "Technique" of the Viennese School of Anthropology

The "Marienfeld project" consisted of an anthropological examination of 1,081 people from 251 families. The novel aspects of this "method," according to Geyer, were "its coverage of individuals of all ages including very young children," as well as specialized recording techniques. Eight workstations operated simultaneously, all focusing on different tasks: the first station compiled personal and genealogical data; the second took stereo-photographic and schematic images of the ear; the third made photographic and schematic images of the hand and the foot; the fourth assembled three-part photographic images of the head and the ocular region; the fifth measured the head and body as well as

taking hair samples; the sixth took fingerprints; the seventh made drawings of the iris and recorded eye color; and, finally, the eighth made
drawings of the nose.[45] (See figs. 1–6.)

Geyer recorded an average "daily output" of forty persons, or 280
to 300 photographic views (and approximately the same number of
schematic views), as well as forty data sheets and fingerprints.[46] The
traditional method employed for this allegedly "highly effective"
examination, which ensured fast collection of data, dates back to the
examinations of prisoners during the First World War and was applied
in a similar manner during anthropological examinations in prisoner of
war camps during the Second World War.[47]

Another relatively innovative methodological approach was to
extend these methods to the entire spectrum of morphological traits,
one of Weninger's "hobbies" developed in Vienna. According to him,
the conclusions drawn by the Viennese anthropological community
were "far too broad and came from ill-observed material"—a consequence of employing metric data.[48] In line with genetic research of the
time, whereby hereditary patterns are transmitted in the form of individual rather than combined traits, Weninger developed discrete schemes
for observing morphological features of the head and face, characteristics subsequently compared to existing metric findings. This meant that
separate assessments were made of minute details; for instance, of the
soft tissue around the eye, including the eyelid crease; the upper lid
region; the lid wrinkles; the color and structure of the iris; the outer
nose; the mouth and chin regions; the exterior ear; both hands and feet;
as well as hair and skin ridge patterns. For Weninger, this procedure
provided not only the basis for human *Erbnormalbiologie* (Hereditary
Biology of Normal Development); it was also able to provide scientific
proof of paternity. Thus, "far from serving science alone, our work also
pursues a practical purpose for the collective good of the people."[49]

Even so family surveys were based on voluntary participation, and
researchers were not allowed to conduct embarrassing full-body examinations. Yet Weninger deplored the fact that Viennese surveyors were
prevented full-body examinations: "Bodily examinations under an
extended program could probably only be conducted on primitives living without clothes; in this country, on a large scale, it would only be
feasible for athletes or in the course of clinical (constitution) studies.
Observing the head and face is arduous, but generally encounters much
less resistance."[50]

Fig. 1. *Volksdeutsche* – The "Marienfeld project" of the Anthropologisches Institut, Universität Wien, winter 1933–34, working area (Photo album Marienfeld, Department of Anthropology, University of Vienna)

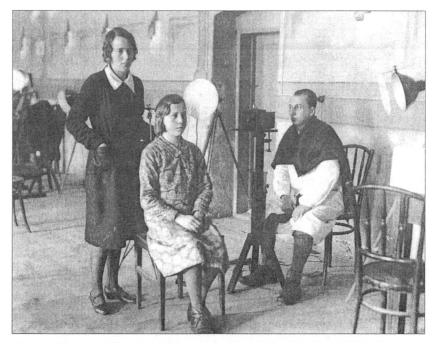

Fig. 2. *Volksdeutsche* – The "Marienfeld project," working station II, Eberhard Geyer takes photographic views of the ear (Photo album Marienfeld, Department of Anthropology, University of Vienna).

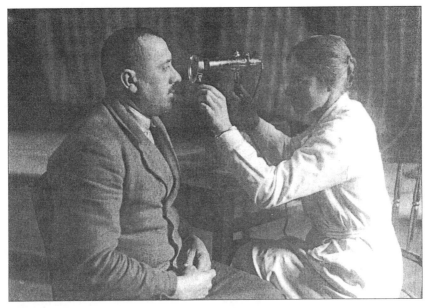

Fig. 3. *Volksdeutsche* – The "Marienfeld project," working station VII, an assistant takes drawings of the structure of the iris and eye colour (Photo album Marienfeld, Department of Anthropology, University of Vienna).

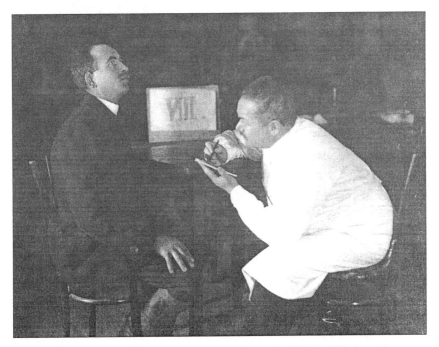

Fig. 4. *Volksdeutsche* – The "Marienfeld project," working station VIII, Josef Weninger takes drawings of the nasal features (Photo album Marienfeld, Department of Anthropology, University of Vienna).

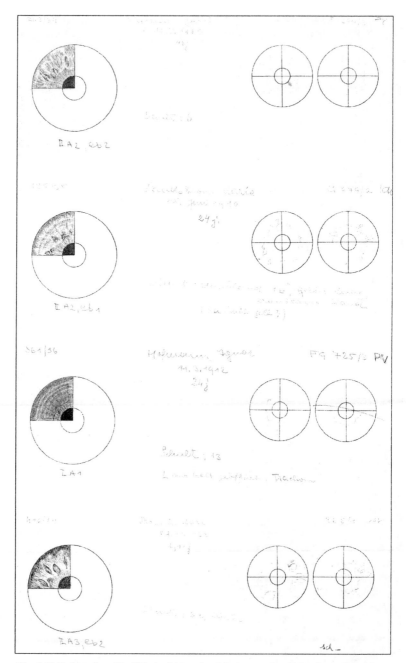

Fig. 5. *Volksdeutsche* – The "Marienfeld project," four examples of drawings of the iris structure (Marienfeld archive material, Department of Anthropology, University of Vienna).

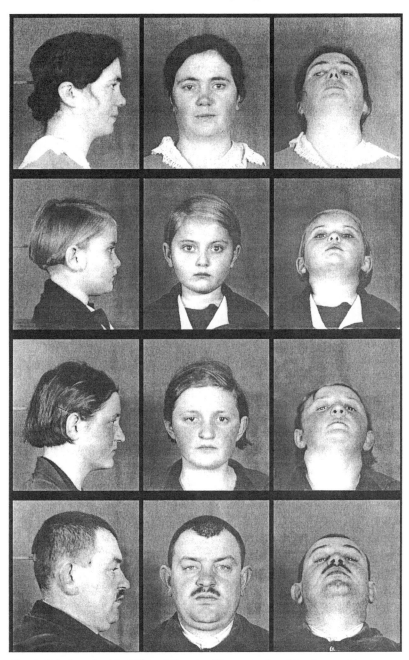

Fig. 6. *Volksdeutsche* - The "Marienfeld project," four examples of the photographic views
(facial form and features) of family members (Marienfeld archive material,
Department of Anthropology, University of Vienna).

Weninger stressed the novel character of his morphological approach, a formal analysis of particular characteristics, which provided the "basic instrument of anthropological research" in Austria for many years.[51] Anthropologists associated with the Viennese School of Anthropology also applied this method to the study of the "racial components [...] extracted from large series of various foreign people." At a later stage, researchers hoped that this method could also be applied to "our own *volk*, though not only to randomly sampled individuals, but also to narrower or extended family groups as well. In particular, this method allowed us to approach the domain of family anthropology, the source of human genetics."[52]

Theoretical Concepts in "Hereditary Biology"

The formulation of genetic-biological questions also inspired new thinking about the scientific study of genetic variation. Since the characteristics of the *Systemrassen* (The System of Races) were considered genetic, it was first necessary to ask whether this approach was scientifically appropriate: this was done through the "twin method," and "research of family and bastards."[53]

Research undertaken on twins revealed that if a genetic trait is inherited, then the question of how this occurred had to be addressed by analyzing families or family genealogies. Like many other anthropologists of the time, Weninger considered hereditary patterns in humans to follow the same Mendelian laws were the basis of the science of genetics. However, the enormous amount of resources and expense involved were beset by serious methodological problems. Measurements of children, for instance, had to be modified to allow comparison with adult data, while the idea of "sameness," "similarity" and "dissimilarity" had to be expressed numerically. For this purpose, both identical and fraternal twins were measured; these data were used to compute "percentage" and "mean percentage" deviations for individual features. Yet it remained unclear at the time how to deal with the morphological traits recorded by numerous photographic data. Finally, for the purposes of illustration, graphs showed the percentage deviation of two features with each family; it was an extremely complex method considering the scarcity of analytical results.[54]

Already in 1936, Weninger probably realized that the genetic patterns of most human traits did not follow the Mendelian laws; experimental studies had shown the limitations of dominant and recessive traits. Consequently, phenomena like gene modification and gene setting were able to modify the phonotypical manifestation of the individual gene. Weninger referred, somewhat cryptically, to "superior Mendelism," and considered himself "at the very beginning of a difficult field of research,"[55] that of human *Erbnormalbiologie*.[56] Had Weninger realized, at that point, that the racial research in Marienfeld might never satisfy its original goal?

Funding for the Project and Associate Workers

The participants in the "Marienfeld project" were all members of the Working Group on Genetic Biology, and were in charge of overseeing different aspects of the project: Josef Weninger concentrated on identifying traits of the eye; Eberhard Geyer focused on the ear; Dora Maria Könner was responsible for taking photographs of the hand and foot; Robert Routil was in charge of taking hair samples and analysing metric dimensions; Margarete Weninger's specialty were the papillary lines of the hand; Albert Harrasser was in charge of developing photographs; and Karl Stumpfl conducted psychological examinations. The project was funded from three sources: the Anthropological Institute provided the equipment, part of which was funded from the proceeds of genetic paternity assessments; in January and November 1933, and in May 1934, the *Deutsche Forschungsgemeinschaft* (German Research Foundation) approved funds to cover the travel costs of nine researchers (including material to make prints for genealogical diagrams from 5,800 photographic plates), and the statistical processing of measurements; and, finally, the community of Marienfeld provided food and lodging for eight project workers over two months.[57]

Despite all this, in 1936, three years after commencing the project, Weninger concluded that the enormous amount of material compiled, and the ideal situation prevailing for "scientific exploitation," nevertheless meant that the project was proceeding at a frustratingly slow pace, resulting in nothing but fragmentary pieces of analysis.[58] With a long list of open questions, Weninger approached the German Research

Foundation for further funding. The new request for the "counting" of traits derived from approximately 9,000 data sheets and 8,000 individual photographs. By that point, Weninger realized that the task at hand was no longer manageable without obtaining additional technical and financial support, as well as additional personnel.

In this request from 1936, Weninger assigned priority to the racial aspect although, originally, the genetic issue and the potential for paternity diagnosis by the "Marienfeld project" had been more important to him. It was only at a later stage that he referred to the "completely new and exact genetic findings, including their far-reaching theoretical and practical consequences for genealogy [and] verification of descent."[59] Weninger applied for additional funds to support another 779 working weeks, for which he had calculated a sum of 4,674 German Reichmarks. The application was passed to Eugen Fischer for consideration.[60] Fischer praised the application: "Apart from the findings about the German population in Romania, which are welcome from the scientific point of view, the present plan [will] most certainly render results that are important for paternity assessments. The more of these findings that are furnished, the greater our certainty in preparing paternity opinions not only for civil cases [alimony procedures] but upon requests of the Reich Sippenamt; the latter serving as a basis for decisions on Aryan or non-Aryan descent of extra-marital children, children of adultery, foundlings and so on."[61]

From this it is clear that the "Marienfeld project," originally drafted in the context of "genetic biology," was used in conjunction with Nazi ideology. Nevertheless, Fischer's positive opinion met with disapproval elsewhere, specifically from an "expert opinion" regarding Weninger's political reliability. The opinion belonged to Karl Thums. Whereas Weninger's national attitude was "indeed irreproachable," considered Thums, there was "one critical point" in Weninger's biography: he was married to a Jewess. Thums added, however, that the scientific work at the Institute of Anthropology in Vienna "was not without a positive political note in the National Socialist sense."[62] Weninger had extended the scope of the Institute's scientific aspirations for the benefit of "genetic care and racial hygiene."[63] It was unusual that Weninger's application was passed on to the Reich's Interior Ministry in Berlin, which considered that the project could just as well be performed on "German material from the Reich," meaning that a grant to a foreign institute was not justified. [64] Weninger's application was rejected in

September 1936 because of "excessive demands for funds, for racial-biological examinations in particular."[65]

In December, Weninger re-submitted an application drafted in an identical manner, except for a supplementary justification. Weninger argued that in contrast to his rejection, Fritz Stumpfl had received a grant for the psychiatric assessment of the Marienfeld population.[66] This time Weninger's political reliability was considered by Ernst Rüdin (1874–1952), director of the *Kaiser-Wilhelm-Instituts für Genealogie und Demographie der Deutschen Forschungsanstalt für Psychiatrie* (Kaiser Wilhelm Institute for Genealogy and Demography of the German Research Institute for Psychiatry), who was aware of Weninger's intentions regarding paternity assessments.[67] He argued that Weninger's Institute was "inspired by the spirit of family research in anthropology as a discipline."[68] Nevertheless, Weninger's application was rejected yet again.[69] Under normal circumstances, Weninger should have been skeptical of the wording of the second refusal, knowing that other Austrian funding requests had received favorable treatment. In principle, the project was to be supported, but rather than providing direct funds to Weninger, they were to be directed to "politically reliable" associates and support staff to be selected by Geyer.[70] Subsequently, Geyer and Dora Könner introduced their own requests.[71] In view of ongoing political and scientific developments, it almost seems as if the research in Marienfeld was not sufficiently "in the Nazi Party line." But as the *Deutsche Forschungsgemeinschaft* preferred to grant assistance to ongoing projects, Geyer reformulated his request to include the "Marienfeld project." The *Volksdeutsche Wissenschaftshilfe* (German Scientific Emergency Fund), in charge of processing this matter, finally granted 600 German Marks, emphasizing that this Austrian project was also of interest to the German *volk*.[72]

Scientific Results of the "Marienfeld Project"

According to Weninger, the "Marienfeld project" "delivered racial and genetic-biological material of a completeness and diversity hitherto unrivalled."[73] The material that Weninger and his co-workers collected is preserved to this day at the Institute of Anthropology of the University of Vienna. The collection contains the following material:

Genealogical and statistical material
- Approximately 1,000 extracts from parish registers (1928–1933) (30,000 from earlier years had been processed before the expedition)
- Approximately 100 drafts of family trees
- 1,000 sheets relating to the alphabetical index of registered individuals
- Statistics on population movement

Anthropological records
- 1,080 sheets with relevant personal data
- 1,080 anthropological data sheets with fifteen measurements
- 1,080 sheets with ten fingertip prints
- 1,080 sheets with two hand prints
- 1,080 colored drawings of the iris
- 1,080 mapping sheets of the soft tissue of the eye
- 1,080 mapping sheets of the bottom of the nose
- 1,080 mapping sheets of the mouth
- 1,080 mapping sheets of the shape of the hand and the foot
- 1,080 mapping sheets of the ear

In total there are 10,800 sheets. There are also 800 hair samples and photographs arranged as follows:
- 1,060 three-part photographic plates 13×18
- 250 photographic plates 6.5×9 of children under 14
- 1,080 photographic plates 6.5×9 of 2 hands each
- 1,080 photographic plates 6.5×9 of 2 feet each
- 2,300 photographic plates 9×12 (stereo) of ears

In total there are 5,770 photographic plates, and films with 7,890 single photos.[74]

By 1936, comprehensive preliminary work and the archiving of material had been completed in the field of genealogy and statistics, including: a cadastre of all inhabitants of Marienfeld derived from parish register lists and personal data sheets; a directory of all registered individuals including age groups and ancestors from seventy-six family trees; a compilation of seventy-six family trees and reprography; a compilation of seventy-five genealogical tables of surveyed individuals, and the identification of all cases of racial intermarriage; statistics on the number of children and births, alongside the age structure of the Marienfeld population; detailed analyses of different groups of physical traits; an examination of the age modification of the above measurements and indices; determination of the sexual dimorphism in relation to all measurements and indices; dermatoglyphic typing; statistics on the patterns of the ball of the thumb and their hereditary pattern; determination of the pattern of the little finger and the axial triradius; analysis of the iris pigmentation including reduction of the boundary layer, and change of pigmentation over age; and, finally, counting the morphological traits of hands, feet and the ear, and their modification over time.[75]

One cannot discuss all the publications resulting from the "Marienfeld project" in detail; a discussion of a few is in any case illustrative of the type of findings obtained. The work undertaken by Josef and Margarete Weninger essentially focused on the genetic pattern of eye color, the eye structure and the dermatoglyphic pattern of various regions of the hand. In the late 1920s and early 1930s, there was little knowledge of the hereditary pattern of the dermatoglyphic system and embryo development. Accordingly, Margarete Weninger's treatise on the dermatoglyphics of the thenar (thumb) and interdigital (between thumb and index finger) pads later became a standard work of reference.[76]

In 1942, Robert Routil completed what Geyer had started in the area of racial analysis. Incorporated into the team as a biostatistician, Routil's responsibility was to analyze the previously compiled measured data. His work covered the "position of the Marienfeld population within the German *volk*," determined the frequency of "types" among the Marienfeld population, analyzed age and sex variability, and developed a mathematical approach to genetic-biometric analysis.[77] Routil concluded that Southern German groups, as well as Romanians from Bucharest, were "close to the Marienfeld population in terms of their appearance and probably also in their genetic constitution"; furthermore, "the inhabitants of Marienfeld have preserved the physical peculiarities of their Southern German ancestors to this day; the genealogical identification of descent applied to our study material from the Saarland region, Swabia and Wurttemberg seems to be confirmed also from the anthropological point of view."[78] Routil welcomed "racial-hygienic protective measures as a preventive method to eliminate any defective genetic sequence particularly harmful to the *Volksgemeinschaft*."[79]

The results of the "Marienfeld project" were shown to the general public as part of an exhibit on family biology, family hygiene and family protection at the Natural History Museum in Vienna in June 1935. Moreover, in that year the former *Garnisonspital* building was remodeled, and the Institute in Vienna procured a hall, finally providing the required space "desperately needed and strongly desired to foster our method and follow the course of our work."[80] The new hall was used for exhibiting genetic-biological specimens and for the temporary installation of a new genetic-biological series.[81]

Conclusions

With the rediscovery of Mendel's laws of inheritance around 1900, anthropological attention was increasingly drawn to genetic-biological issues. The focus was on examining "bastards," twins and families to furnish and establish a basis for both "racial" assessments and paternity diagnoses. The first study in this vein was conducted at a time when attempts were being made to identify the "racial" peculiarities of European peoples.[82] In the late 1920s, researchers focused on communities located in small, limited geographical areas, later moving on to the German *Völker*.

In relation to the volumes of data collected, and in view of the complexity of its approach, the "Marienfeld project" constitutes a "special case" for anthropology in Austria and the international community as a whole. In design it followed the mainstream of German anthropology, which focused on "hereditary biology" and "racial questions of the German *volk*." Among other factors, Geyer attributed this development to the "forced abandonment of the colonies"[83] caused by the First World War, which had triggered a form of *Selbstbesinnung* (self-reflection). Hans Günther's *Rassenkunde des deutschen Volkes* and *Deutsche Rassenkunde*, published under the editorship of Eugen Fischer between 1929 and 1936, may also be considered as symptomatic products of this form of "self-reflection."

It was not least due to these popularization efforts that *Rassenkunde und Rassenpflege* (Racial Studies and Racial Care) constituted the material for subjects taught in schools in Germany from 1933 onwards.[84] This emphasis on "racial questions" of the German *volk* now offered anthropologists a plethora of new tasks and opportunities—though questionable—to make a name for themselves by starting centering upon issues related to "social hygiene," "racial hygiene" and "genetic pathology." Austrian anthropology followed its German counterpart "gradually."[85]

Eberhard Geyer, whose philosophical conclusions were published posthumously in the *Archiv für Rassen- und Gesellschaftsbiologie*, personified the development of anthropology in German-speaking areas.[86] Given his political allegiance (an illegal member of the NSDAP since 1933, he headed the Institute's operational cell and was also *Hauptstellenleiter* of the *Rassenpolitische Amt* in the Nieder-Donau region), Geyer approached the "Marienfeld project" from a totally different perspective than Josef Weninger, whose primary intention was to overcome the theoretical shortcomings of paternity assessments. In this regard it was no coincidence that the "Marienfeld project" took place in the winter months of 1933/34. Both protagonists assumed key roles in the project, yet both were largely unable to achieve their goals.

While initially skeptical of paternity diagnosis, Weninger most likely realized at the beginning of the 1930s that the findings of a discipline hitherto withdrawn from the realms of academia would now, at a time of economic hardship, acquire a new relevance. Paternity diagnoses had turned into a prosperous enterprise, and most members of the legal profession were relieved to be able to rely on expert opinions in support of their decisions. On an international scale, the Institute of Anthropology had the potential of rising to fame. Weninger had no clue about the direction anthropology would take, including paternity diagnosis.[87]

Eberhard Geyer considered the "Marienfeld project" a "matter of the heart,"[88] and a contribution to the study of nationalities and the minority problem.[89] Geyer rejected a research approach aiming primarily at "empirical assessment and theoretical treatment" without concurrent goals "of immediate applicability."[90] His objective, heavily influenced by Nazi ideology, was to identify the "racial differences between *Gast- und Wirtsvolk*," as well as the framework of the evolution and preservation of the German minority in Southeast Europe.[91]

Robert Routil's role in the project seems ambivalent, and was possibly opportunistic. He was rated a "heretic of the fatherland" by some, "cut off" from the scheming intrigues surrounding the "rights of processing" of the Marienfeld material,[92] though he clearly took a Nazi conformist, racialist stance in his study of the project published in 1942.[93]

The "Marienfeld project," originally linked to paternity assessment activities and the analysis of "classified" morphological traits, soon lost its importance. The large amounts of data, still available in the Institute of Anthropology, are symptomatic of the scholarly hypertrophy of the Viennese School of Anthropology during the interwar period.[94] The connection between "family projects" and Nazi doctrine was not immediately obvious, since the village of Marienfeld itself initiated the research. It became clear only afterwards that the members of the Working Group on Genetic Biology pursued rather distinct political, commercial, and personal interests, and that the impact of their diverging ideological positions proved crucial in the course the project would take, both in a positive and a negative direction.

Endnotes:

[1] Wilhelm E. Mühlmann, *Geschichte der Anthropologie* (Frankfurt-am-Main: Athenäum Verlag, 1968), 100.

[2] Rudolf Pöch, "Dr. Augustin Weisbach (1837–1914)," *Wiener Prähistorische Zeitschrift* 1 (1914), 1–6.

[3] Pöch, "Dr. Augustin Weisbach," 6.

[4] Hans F. K. Günther, *Rassenkunde des deutschen Volkes* (Munich: J. F. Lehmanns Verlag, 1942), 20.

[5] Rudolf Pöch, "Neue anthropologische Fragestellungen," *Mitteilungen der Anthropologischen Gesellschaft Wien* 62 (1919), 193–210; Margarete Weninger, "Rudolf Pöch zum 40. Jahrestag seines Todes (1870–1921)," *Mitteilungen der Anthropologischen Gesellschaft Wien* 91 (1961), 142–143; Josef Weninger, "25 Jahre Anthropologisches Institut an der Universität in Wien," *Mitteilungen der Anthropologischen Gesellschaft Wien* 68 (1938), 191–205.

[6] Otto Reche, "Die Anthropologie als biologische Wissenschaft," *Der Biologe* 8 (1939), 317–323, here 318.

[7] Niels C. Lösch, *Rasse als Konstrukt—Leben und Werk Eugen Fischers* (Berlin: Peter Lang, 1997).

[8] See Fritz Lenz, "Die Erblichkeit der geistigen Begabung," in Erwin Baur, Eugen Fischer and Fritz Lenz, eds., *Menschliche Erblichkeitslehre und Rassenhygiene* (Munich: J. F. Lehmanns Verlag, 1927), 471–507. See also Reche, "Die Anthropologie als biologische Wissenschaft," 318. For example, the anthropologist Viktor Lebzelter was one of the few that challenged the static concept of race underlying Nazi Party dogma. On the "static" concept of race see also Karl Saller, *Die Rassenlehre des Nationalsozialismus in Wissenschaft und Propaganda* (Darmstadt: Progreß, 1961).

[9] Josef Weninger, "Das Denkmal für Rudolf Pöch an der Wiener Universität," *Mitteilungen der Anthropologischen Gesellschaft Wien* 63 (1933), 255–256.

[10] Rudolf Pöch to Alfred Ploetz, 6 September 1906, Naturhistorisches Museum – Anthropologische Abteilung, Somatological Collection, New Guinea Correspondence, no. n.d., 83–84.

[11] Georg Kyrle, "Siedlungs- und Volkskundliches aus dem wolhynischen Poljesje," *Mitteilungen der Anthropologischen Gesellschaft Wien* 48 (1918), 118–45; Andrea Gschwendtner, "Frühe Wurzeln für Rassismus und Ideologie in der Anthropologie der Jahrhundertwende am Beispiel des wissenschaftlichen Werkes des Anthropologen und Ethnographen Rudolf Pöch," in Barbara Danckwortt and Claudia Lepp, eds., *Von Grenzen und Ausgrenzung* (Marburg: Schüren, 1997), 136–158; and Margit Berner, "Forschungs-'Material' Kriegsgefangene: Die Massenuntersuchungen der Wiener Anthropologen an gefangenen Soldaten 1915–1918," in Heinz E. Gabriel and Wolfgang Neugebauer, eds., *Vorreiter der Vernichtung? Eugenik, Rassenhygiene und Euthanasie in der österreichischen Diskussion vor 1938* (Vienna: Böhlau, 2005), 167–198.

[12] Rudolf Pöch to Academy of Sciences, 18 May 1916, Archiv der Österreichischen Akademie der Wissenschaften, request for financial support, 367/1916.

[13] Weninger, "Das Denkmal für Rudolf Pöch," 256.

[14] See Hella Pöch, "Beiträge zur Anthropologie der ukrainischen Wolhynier," *Mitteilungen der Anthropologischen Gesellschaft Wien* 55 (1925), 289–333, especially 321; see also Hella Pöch, "Beiträge zur Anthropologie der ukrainischen Wolhynier," *Mitteilungen der Anthropologischen Gesellschaft Wien* 56 (1926), 16–52.

[15] Pöch, "Anthropologie der ukrainischen Wolhynier," 290.

[16] Pöch, "Anthropologie der ukrainischen Wolhynier," 292.

[17] Pöch, "Anthropologie der ukrainischen Wolhynier," 290.

[18] Hella v. Schürer (Pöch) to Academy of Sciences, 24 March 1918, Archiv der Österreichischen Akademie der Wissenschaften, request for financial support, 250/1918.

[19] Eberhard Geyer, "Probleme der Familienanthropologie," *Mitteilungen der Anthropologischen Gesellschaft Wien* 64 (1934), 295–326.

[20] Egon Frhr. v. Eickstedt, *Rassenkunde und Rassengeschichte der Menschheit*, vol. 1 (Stuttgart: Ferdinand Enke, 1940), 584.

[21] Katja Geisenhainer, *"Rasse ist Schicksal": Otto Reche (1879 – 1966) - ein Leben als Anthropologe und Völkerkundler* (Leipzig: Evang. Verl.-Anst., 2002); and Maria Teschler-Nicola, "Aspekte der Erbbiologie und die Entwicklung des rassenkundlichen Gutachtens in Österreich bis 1938," in Gabriel and Neugebauer, eds., *Vorreiter der Vernichtung?*, 99–138, especially 106.

[22] Geyer, "Probleme der Familienanthropologie," 325; and Albert Harrasser, "Ergebnisse der anthropologisch-erbbiologischen Vaterschaftsprobe in der österreichischen Justiz," *Mitteilungen der Anthropologischen Gesellschaft Wien* 65 (1935), 204–232, especially 207-208.

[23] Ernst Klee, *Deutsche Medizin im Dritten Reich. Karrieren vor und nach 1945* (Frankfurt-am-Main: S. Fischer, 2001), 60–61.

[24] See Weninger, "25 Jahre Anthropologisches Institut," 202; Josef Weninger, "Menschliche Erblehre und Anthropologie (Zur Methode der Erbforschung)," *Wiener klinische Wochenschrift* 26 (1936), 1–17, 12.

[25] Geyer, "Probleme der Familienanthropologie," 326.

[26] Albert Harrasser to lawyer Franz Müller, 11 March 1934, Universität Wien, Department für Anthropologie, folder forensic correspondence.

[27] See Eickstedt, *Rassenkunde und Rassengeschichte der Menschheit*, 594; he considered anthropological paternity assessments "a typical case of successful application of initially purely cognitive results for practical purposes".

[28] Elvira Weisenburger, "Der "Rassepabst," in Michael Kißener and Joachim Scholtyseck, eds., *Die Führer der Provinz. NS-Biographien aus Baden Württemberg* (Konstanz: Universitätsverlag, 1997), 161–199.

[29] Gustav Kraitschek, *Rassenkunde* (Vienna: Burgverlag, 1923), 94–105; and Hans F. K. Günther, *Rassenkunde des deutschen Volkes*, vol. 1 (Munich: J. F. Lehmann, 1922).

[30] Eugen Fischer, *Anthropologische Erforschung der deutschen Bevölkerung* (Wittenberg: Herrosé and Ziemsen, n.d.), 1–9.

[31] Fischer, *Anthropologische Erforschung der deutschen Bevölkerung*, 2.

[32] Fischer, *Anthropologische Erforschung der deutschen Bevölkerung*, 3.

[33] Wilhelm Klenck and Walter Scheidt, *Niedersächsische Bauern*, part I (Jena: Gustav Fischer, 1929); Karl Saller, *Die Keuperfranken* (Jena: Gustav

Fischer, 1930); H. A. Ried, *Miesbacher Landbevölkerung* (Jena: Gustav Fischer, 1930); Karl Saller, *Die Fehmaraner* (Jena: Gustav Fischer, 1930); Walter Scheidt, *Physiognomische Studien an niedersächsischen und oberschwäbischen Landbevölkerungen* (Jena: Gustav Fischer, 1931); Walter Scheidt, *Alemannische Bauern in reichenauischen Herrschaftsgebieten am Bodensee* (Jena: Gustav Fischer, 1929); Karl Saller, *Süderdithmarsische Geestbevölkerung* (Jena: Gustav Fischer, 1931); Friedrich Keiter, *Schwansen und die Schlei* (Jena: Gustav Fischer, 1931); Herbert Göllner, *Volks- und Rassenkunde der Bevölkerung von Friedersdorf* (Jena: Gustav Fischer, 1932); Walter Scheidt, *Niedersächsische Bauern*, part II (Jena: Gustav Fischer, 1932); Rudolf Grau, *Die Questenberger* (Jena: Gustav Fischer, 1934); and Friedrich Keiter, *Rußlanddeutsche Bauern und ihre Stammesgenossen in Deutschland* (Jena: Gustav Fischer, 1934).

[34] In the 1930s, surveys of "German races" and the "Germans" were also conducted by Viktor Lebzelter, head of the Department of Anthropology at the Vienna Natural History Museum, including surveys of Germans in the southern Bohemian Forest and in Austria's Burgenland region, and of Transylvanian Saxons. The formulation of the issue at hand implied a differentiation between "Germans" and other, "inferior" populations; see also Eberhard Geyer, "Der Stand der rassenkundlichen Untersuchungen in der Ostmark," in Michael Hesch and Günther Spannaus, eds., *Kultur und Rasse* (Munich: J. F. Lehmanns Verlag, 1939), 80–87.

[35] Fischer, "Anthropologische Erforschung der deutschen Bevölkerung," 8.

[36] Fischer, "Anthropologische Erforschung der deutschen Bevölkerung," 4.

[37] Josef Weninger to *Deutsche Forschungsgemeinschaft* (hereafter DFG), 28 December 1936; *Bundesarchiv Koblenz* (hereafter BK), R 73/15621, 5.

[38] Josef Weninger to DFG, 28 December 1936; BK, R 73/15621, 5.

[39] Eberhard Geyer, "Anthropologie und Nationalitätenforschung," *Nation und Staat* 7 (1934), 323–327.

[40] Geyer, "Anthropologie und Nationalitätenforschung," 325.

[41] Josef Weninger to DFG, 29 December 1938; BK, R 73/15621; Friedrich Stumpfl carried out psychological investigations of families in Marienfeld.

[42] Eberhard Geyer, "Deutsche Vorposten in Rumänien," *Zeitgeschichte* 6 (1934), 161–168, especially 162.

[43] Robert Routil, "Familienanthropologische Untersuchungen in dem ostschwäbischen Dorfe Marienfeld im rumänischen Banat. I. Biometrische Studien," *Untersuchungen zur Rassenkunde und menschlichen Erblehre* 1 (1942), 1–82. See also Helga-Maria Pacher, "Anthropologischer Vergleich zweier mitteleuropäischer Bevölkerungsgruppen (St. Jakob im Rosental, Kärnten und Marienfeld, rumänisches Banat)," *Mitteilungen der Anthropologischen Kommission der Österreichischen Akademie der Wissenschaften*, 1, 1 (1952), 1–55.

[44] Eberhard Geyer, "Vorläufiger Bericht über die familienanthropologische Untersuchung des ostschwäbischen Dorfes Marienfeld im rumänischen Banat," *Verhandlungen der Gesellschaft für Physische Anthropologie* 7 (1935), 5–11, 6; and Weninger, "25 Jahre Anthropologisches Institut," 200.

[45] Geyer, "Vorläufiger Bericht," 10.

[46] Geyer, "Vorläufiger Bericht," 10.

47 See Maria Teschler-Nicola and Margit Berner, "Die Anthropologische Abteilung des Naturhistorischen Museums in der NS-Zeit; Berichte und Dokumentation von Forschungs- und Sammlungsaktivitäten 1938–1945," in Akademischer Senat der Universität Wien, ed., *Senatsprojekt der Universität Wien. Untersuchungen zur Anatomischen Wissenschaft in Wien 1938–1945* (Vienna: Akademischer Senat der Universität Wien, 1998), 333–358.

48 Josef Weninger and Hella Pöch, "Leitlinien zur Beobachtung der somatischen Merkmale des Kopfes und Gesichtes am Menschen," *Mitteilungen der Anthropologischen Gesellschaft Wien* 54 (1924), 232–270.

49 Josef Weninger, Typescript "Die Wiener Schule," 1938 (n. d.), Universität Wien, Department für Anthropologie, folder correspondence 1938/39, 1–3; see also Viktor Lebzelter, "Wozu und zu welchem Ende messen wir noch?" *Anthropologischer Anzeiger* 11 (1934), 1–2.

50 Weninger, "Die Wiener Schule," 3.

51 Emil Breitinger, "In Memoriam: Josef Weninger, 1886–1959," *Anthropologischer Anzeiger* 23, 2–3 (1959), 236–238; and Heinrich Hayek, "Josef Weninger," *Almanach der Österreichische Akademie der Wissenschaften* 109 (1960), 427–436.

52 Weninger, "25 Jahre Anthropologisches Institut," 199.

53 Weninger, "Menschliche Erblehre und Anthropologie," 2–3.

54 Weninger, "Menschliche Erblehre und Anthropologie," 10.

55 Weninger, "Menschliche Erblehre und Anthropologie," 16.

56 Weninger, "Menschliche Erblehre und Anthropologie," 16.

57 Josef Weninger to DFG, 28 March 1936; BK, R 73/15621, 2.

58 Josef Weninger to DFG, 28 March 1936; BK, R 73/15621, 1.

59 Josef Weninger to DFG, 28 March 1936, BK, R 73/15621, 7.

60 Maria Teschler-Nicola, "The Diagnostic Eye—On the History of Genetic and Racial Assessment in Pre-1938 Austria," *Collegium Antropologicum* 28 (2004), 7–29, 21.

61 Eugen Fischer to DFG, 3 June 1936; BK, R 73/15621.

62 Karl Thums to DFG, 30 June 1936; BK, R 73/15621, 1.

63 Karl Thums to DFG, 30 June 1936; BK, R 73/15621, 2.

64 *Reichs- und Preußisches Ministerium des Inneren* to *Deutsche Forschungsgemeinschaft*, 12 August 1936, BK, R 73/15621.

65 DFG to Josef Weninger, 9 September 1936; BK, R 73/15621; see also Josef Weninger to DFG, 28 December 1936; BK, R 73/15621.

66 Friedrich Stumpfl was based at the *Kaiser-Wilhelm-Instituts für Genealogie und Demographie der Deutschen Forschungsanstalt für Psychiatrie* in Munich; Josef Weninger to DFG, 28 December 1936; BK, R 73/15621.

67 Matthias M. Weber, *Ernst Rüdin. Eine kritische Biographie* (Berlin, Heidelberg: Springer Verlag, 1993); see also Peter Weingart, Jürgen Kroll and Kurt Bayertz, eds., *Rasse, Blut und Gene*, 2nd ed. (Frankfurt-am-Main: Shurkamp, 1996).

68 Ernst Rüdin to DFG, 12 March 1937; BK, R 73/15621.

69 DFG to Josef Weninger, 9 November 1937, BK, R 73/15621; Josef Weninger to DFG, 13 November 1937; BK, R 73/15621.

70 Rudolf Amon, experts' report, 31 January 1938; BK, R 73/15621; DFG to Rudolf Amon, 4 December 1937; BK, R 73/15621.

71 Könner's rather incoherent application for funds to prepare "hand and foot" analyses was rejected; for his part, Geyer started to pursue other interests of his own, requesting the Deutsche Forschungsgemeinschaft to subsidize a social anthropological project for collecting data on homeless Viennese families, and families with illegitimate children, to perform "racial and eugenic studies". The fact that his racial hygiene project aimed to "pay particular attention to the infiltration of foreign-race elements (the Jews in particular)" made him a "Nazi ideological hardliner" among the Austrian anthropological community, as is obvious also from his work as advisor and expert for the Reich *Sippenamt;* see Eberhard Geyer to DFG, 21 February 1938; BK, R 73/11229; Eberhard Geyer to DFG, 25 August 1938; R 73/11229; Dora Maria Könner to DFG, 7 March 1938; BK, R 73/12272.

72 DFG to Norbert Gürke (South-East Institute), 2 September 1938; BK, R 73/11229.

73 Josef Weninger to DFG, 28 December 1936; BK, R 73/ 15621, 1.

74 Josef Weninger to DFG, 29 December 1936; BK, R 73/15621, 2 and 3.

75 Josef Weninger to DFG, 29 December 1936; BK, R 73/15621, 3 and 4.

76 Margarete Weninger, "Familienuntersuchungen über den Hautleistenverlauf am Thenar und im ersten Interdigitalballen der Palma," *Mitteilungen der Anthropologischen Gesellschaft Wien* 65 (1935), 182–193, 190 and 192.

77 Routil, "Familienanthropologische Untersuchungen," 2.

78 Routil, "Familienanthropologische Untersuchungen," 28.

79 Routil, "Familienanthropologische Untersuchungen," 65.

80 Weninger, "25 Jahre Anthropologisches Institut," 202.

81 Weninger, "25 Jahre Anthropologisches Institut," 202 and 204.

82 Pöch, "Anthropologie der ukranischen Wolhynier," 289 and 290.

83 Eberhard Geyer, "Wissenschaft am Scheideweg," *Archiv für Rassen und Gesellschaftsbiologie* 37, 1 (1944), 1–6.

84 Renate Fricke-Finkelnburg, *Nationalsßozialismus und Schule. Amtliche Erlasse und Richtlinien 1933–1945* (Leverkusen: Leske and Budrich Verlag, 1998), 214.

85 Verena Pawlowsky, "Quelle aus vielen Stücken: Die Korrespondenz der Anthropologischen Abteilung des Wiener Naturhistorischen Museums bis 1938," in Gabriel and Neugebauer, eds., *Vorreiter der Vernichtung?*, 139–165.

86 Geyer, "Wissenschaft am Scheideweg," 3.

87 Maria Teschler-Nicola, "The Diagnostic Eye," 26; Maria Teschler-Nicola, "Der diagnostische Blick – Zur Geschichte der erbbiologischen und rassenkundlichen Gutachtertätigkeit in Österreich vor 1938," *Zeitgeschichte* 30 (2003), 137–149; Michael Grüttner, "Wissenschaft," in Wolfgang H. Benz, Hermann Graml and Hermann Weiß, eds., *Enzyklopädie des Nationalsozialismus* (Munich: Deutscher Taschenbuchverlag, 1997), 135–153, 152.

88 Geyer, "Deutsche Vorposten in Rumänien," 168.

89 Geyer, "Anthropologie und Nationalitätenforschung," 323.

[90] Geyer, "Anthropologie und Nationalitätenforschung," 325.

[91] Geyer, "Anthropologie und Nationalitätenforschung," 327; see also Geyer, "Probleme der Familienforschung," 319 and 321.

[92] Rudolf Amon to DFG, 16 April 1938; BK, R 73/15621.

[93] Routil, "Familienanthropologische Untersuchungen," 65.

[94] The complete material collected at Marienfeld is stored at the Department für Anthropologie, Universität Wien. I wish to thank Professors Horst Seidler and Harald Wilfing for permitting the use of correspondence and objects, and Michaela Zwölfer for her careful work of translation.

Of "Yugoslav Barbarians" and Croatian Gentlemen Scholars: Nationalist Ideology and Racial Anthropology in Interwar Yugoslavia

Rory Yeomans

In 1943, a Croatian translation of Ivo Pilar's 1918 polemic about the dangers of Serbian domination in the Balkans, *The South Slav Question*, was published to great acclaim. The Croatian Minister of Education, Mile Starčević (1904–1953), a former student nationalist, wrote in an article to mark its publication that Pilar's book had been the "bible" for his generation of Croatian nationalist youth at the University of Zagreb. With its theory of Serbian racial inferiority and the religious perils of Eastern Orthodoxy, Pilar's book had inspired them in their struggle against Belgrade in the 1920s.[1] In his introduction, Ferdo Puček, the translator of Pilar's opus, drew attention to the parallels between Pilar's racial ideas and those of the fascist Ustasha movement of which he, like Starčević, was an intellectual supporter. *The South Slav Question* showed, Puček continued, that the Serbs who lived in Croatia and Bosnia were "alien elements, Cincars, Greeks, Romanian and above all Balkan-Aromanian (Vlach) elements" that had fallen under the influence of the Eastern Orthodox Church and continually demonstrated their hostility to Croatia. This was in direct opposition to the "Western" outlook of the Croats, with their "Nordic Slavic-Gothic-Iranian" racial origins.[2]

In interwar Yugoslavia, as in the rest of Europe, questions of race and nationality dominated the political agenda. In the new Kingdom of the Serbs, Croats and Slovenes—renamed the Kingdom of Yugoslavia in 1929—the main ideological division was between separatist nationalists and Yugoslav integrationists. Separatists, like Croatian nationalists, believed that Serbs, Croats, and Slovenes were three distinct nations whose individuality and prosperity—indeed survival—could only be guaranteed if they existed as separate and independent nation-states; by contrast, supporters of Yugoslavism argued that the appellations "Serb," "Croat," and "Slovene" were merely tribal names that constituted an embryonic "Yugoslav race." Similar to nineteenth-century nationalists in Germany and Italy, they believed that by synthesizing and integrating the best national characteristics and qualities of the three tribes a new nation would be brought into being. Aside from this fundamental difference, Croatian nationalists and romantic Yugoslavs also differed in their approach to race and racial ideology. While both Yugoslav and Croatian racial biologists were influenced by political ideas emanating from elsewhere in Europe—most obviously those connected to the rise of fascism in Italy and National Socialism in Germany—it was writers and academics committed to the creation of a Yugoslav nation who embraced the technological possibilities of eugenics, both as a reflection of their progressive views and as a means of ending the tribal differences long impeding the creation of a "Yugoslav consciousness." In their conception of race and nation, Croatian nationalists largely rejected modern science and remained firmly entrenched in the anthropological and scholarly tradition of the late nineteenth century. Moreover, while the racial ideas of idealist Yugoslavists were officially supported by the state, the proponents of Croatian racial theories were, by and large, isolated individuals; they saw themselves as persecuted pioneers of Croatian racial utopias.

Yugoslav racial theories were characterized by the aim to create a new race that embodied the best qualities of the different South Slav races; therefore, even when their rhetoric became openly eugenicist, Yugoslav racial theorists were rarely threatening or aggressive. By contrast, Croatian racial theories did not allow for such a synthesis. Instead, Croatian race ideologists, envisioning the purification of the Croatian nation, were obsessed by perceived threats to their ethnic culture and "living space" by the demographic invasion of racially inferior "foreigners." Croatian racial theory, though far less influenced by the ideas of racial biology and eugenics, proved to be ultimately far more destructive than that of its neighbors.

Culture, Cars and Clean Bodies: Modernity and Eugenics in Yugoslavia

Yugoslav ideologues construed their nation as a thoroughly modern one. For them, history began in 1918. According to the poet Tin Ujević (1891–1955), the capital, Belgrade, was the representative of a "new, dynamic, explosive and frequently bombastic world."[3] Foreign visitors were equally impressed with Belgrade and its atmosphere of modernity and progress. The English writer David Footman, for example, visiting Belgrade in 1934, commented on its "modern and austere" apartments and impressive skyline, "like that of a young American city, with the beauty and vigour of youth."[4] The French diplomat Henri Pozzi was famously less impressed. In a still-incendiary study, he wrote of Belgrade as a city of "gilded lasciviousness," like a "nouveau riche who cannot stop dancing, yet spits ugly words at his poor relations that cluster around him."[5]

Even allowing for Pozzi's scathing attack on Yugoslavia's modern, urban, and cosmopolitan values, it was indeed true that the new Yugoslav state was extremely keen to embrace what it considered to be the enlightened and progressive practices that could be observed in Western Europe and America. Early Yugoslav ideologues aimed to build a modern, secular, and unitarist state which would wipe away the tribal divisions that separated Serbs, Croats and Slovenes. To do this, they felt that they needed to address the cultural and religious factors which had for so long impeded their unity. They set about this task with revolutionary zeal, setting out proposals for radical utopias and social experimentation. Writers, theologians and ideologues called for, among other things, an end to celibacy for Catholic priests; the establishment of a new national Yugoslav church which would synthesize the best attributes of Catholicism and Orthodoxy; the full emancipation of women; and a society from which all aspects of immorality, corruption, and injustice would be eradicated in order to make society healthy.[6]

Ideologues also wanted to build a new society in Yugoslavia by bringing modernity and technological advance to everyday life. In film, theatre, visual art, public transport and architecture, the conservative and cautious impulse found itself in noisy competition with the experimental and avant-garde expressions of a new generation of artistic and cultural visionaries. Ujević's portrait of the Yugoslav capital, a city of "speed, expansion and electricity," reflected the opinion of certain section of the Yugoslav intelligentsia that the new state should be founded on modern urban principles. In the poet's panegyric—entitled "Futurist Belgrade"—the brash and daring new world that Yugoslavia promised was encapsulated in the automobile, an essential part of the life of Belgrade and a celebration, as Ujević sardonically put it, of the "victory of petrol and money." Energy, rapidity, dynamism: these were the qualities lionized by the Italian Futurist, F. T. Marinetti (1876–1944). In the imagined Italy of the Futurists, the defining technological image was the aeroplane, hurtling the intrepid aviator at tremendous speed through time and space. To this, Ujević opposed the idea of the car—that roaring urban usurper of the horse and cart. Half joking though Ujević's article might have been, it nonetheless accurately described the mindset of those intellectuals and writers who belived that the power of mass technology and a futurist understanding of life—as much as enlightened civic attitudes—could help create a new Yugoslav utopia.

Other ideas were more obviously associated with eugenics. Some ideologues called for the reform of marital laws. One unnamed writer from Split, for example, complained that the clerical laws of the Habsburg era—which prevented some professors such as female teachers or civil sevants from marrying—had led to a large number of unmarried women who had been driven to prostitution, further resulting in venereal disease, hysteria, and the bearing of illegitimate children prone to criminality in later life. This had been compounded by the great loss of life among Yugoslav soldiers during the First World War. As a result, the Yugoslav race had been weakened, for men were "the motor of the national state engine."[7] For Andrija Štampar (1888–1958), the head of the newly created Yugoslav Health Service and a leading advocate of social medicine, a program of comprehensive sex education was crucial for addressing the escalation of venereal diseases, and for creating a healthier state. He argued that the sexual education of the young should begin before they became sexually active—perhaps as early as the age of eight—if such a policy was to have any chance of success.[8]

The Yugoslavia created in 1918 was an overwhelmingly rural state and Yugoslav ideologues saw one of the most pressing tasks to be the elevation of both the peasantry as well as the urban working classes. In 1919, Milan Pribićević envisaged the building of a "modern, great, cultured and social Yugoslavia" and "a progressive peasantry with clean respectable homes and villages, well fed and highly literate."[9]

Likewise, in 1929, the writer Bogumil Vošnjak argued that in order to create a strong Yugoslavia, the government needed not only to renew the economy, but also to improve the lives and conditions of workers and peasants. For there to be a healthy peasantry, the author believed that what he termed "social hygiene," as well as eugenics, needed to be introduced. For Vošnjak, as for many other radical Yugoslavs, the modernization of the village, as the centre of life in the new Yugoslavia, was essential to the development of the new state. This required not merely electricity and modern technology, but also the cultivation of an appreciation for culture by peasants through the endeavors of intellectuals whom, he contended, were uniquely positioned to unify the life of the town and the village.[10]

Additionally, there were proposals from a range of scientists within official institutions, whose opinions carried rather more influence. Some of the more important institutions in the Kingdom of Yugoslavia to address the question of health and hygiene were the Ministry of Public Health and the Central Institute for Hygiene in Belgrade, which produced its own journal, *Glasnik centralnog higijenskog zavoda* (The Journal of the Central Institute for Hygiene), edited by the director, Stevan Ivanić. Like Vošnjak, Ivanić believed that the need to introduce basic standards of hygiene and cleanliness to village houses was a pressing concern.[11] That the health and hygiene levels of villages in interwar Yugoslavia were depressingly poor is indisputable. As late as 1936, when the Croatian Peasant Party politician Rudolf Bićanić (1905–1968) toured the villages throughout Croatia and Bosnia, he found the living conditions of peasants to be deplorable.[12] Health and hygiene experts, like Vladimir S. Stanojević (1886–1978), agreed. In a textbook written for army hygiene classes in 1927, he stated that: "The health situation in our country and our nation is not good. In both the village and the town, our people are not educated in hygiene and, because of this, do not pay attention to personal hygiene."[13] Stanojević found that the situation was especially serious in the villages: "Many of our settlements are abandoned and neglected and, in the majority of cases, they do not have the most basic hygienic needs." Most peasants slept in one bed, and the houses they inhabited were infested with rats and mice, damp, dark, cramped, and flooded with fleas and bedbugs. Life for most city dwellers was no better, he cautioned, and, overall, the nation was still plagued by diseases almost unknown in the rest of Europe—including tuberculosis, dysentery, typhoid and malaria.[14]

Nonetheless, a leading Yugoslav racial biologist and president of the Association of Yugoslav Physicians, Svetislav Stefanović (1874–1944), argued in 1936 that, in terms of racial hygiene, the concerns of the village had been ignored, and that life in the villages remained characterized by a lack of medical aid and insufficient numbers of doctors and nurses.[15] Despite the best efforts of the government to address the concerns of the countryside, relatively little had been achieved. Stefanović called for comprehensive health education in the villages, better provision of doctors, midwives and nurses, as well as the material and moral protection of the family, in order for it to remain the cultural foundation of village life. This would be facilitated through the creation of a government department specializing in the health problems of the village within a newly established Ministry of Social Politics and National Health—reaffirming the belief that mother and child were the "future and the strength of the nation." Such a policy could have a positive social effect throughout the state, reducing divisions arising from the negative perceptions existing between city and countryside: "In this way the village would come into contact with the city and realize that the city was not just some monster that swallows the peasant and seizes his offspring and the hard fruits of his labour. Rather than the city being the carrier and hearth of some instinctual culture that estranges and degenerates the children of the peasant nation, it would be shown that the city also protects them and teaches them not only a better and more beautiful but also a more noble life."[16]

However, Stefanović also argued that such strengthening of social life would be racially profitable: citing the examples of the Soviet Union and Fascist Italy—where the policy of elevating the life of villages had been implemented—he pointed out that the newly created state institutions had indeed helped to defend both mother and child, while simultaneously promoting healthy breeding.

As was the case with Stefanović, many of Stanojević's solutions for the problems he described were eminently rational, reflecting the practices then fashionable elsewhere in Europe and the United States. He argued, for example, that workers required social security and a safe, clean environment in which to work and live. He also advocated the creation of trade societies where workers could purchase hygienic food and clothing. Furthermore, Stanojević proposed the establishment of preventative health clinics and quarantine stations; the mass distribution of health posters on trains, boats, and in public meeting places throughout towns and villages; the construction of cheap public baths; and the free availability of clean drinking water.[17]

However, Stanojević's concern with improving social conditions in Yugoslavia meant that he was also susceptible to eugenic arguments. In 1920, under the auspices of the Ministry for National Health, Stanojević wrote *Eugenika: higijena čovečeg začeća i problem nasledja* (Eugenics: The Hygiene of Human Conception and the Problem of Heredity), an official guide to the principles of eugenics for the uninitiated Yugoslav reader. Comparing eugenics to the Book of Revelation, in his introduction Stanojević asserted that the book would be especially useful for those embarking on marriage, and for those who wanted to find out about "the destiny of their home or the happiness of their children."[18] Racing through history, the author detailed how civilizations throughout the millennia had practiced eugenics: from the Spartans, who had practiced infanticide by throwing sick babies off the edge of cliffs to protect the purity and vitality of their race, to the Ancient Greek philosopher Plato, who had called for the killing of the weak and physically inferior in society; from the racial policies of the English "national master race" able to conquer and rule half the globe, to the eugenic movement in Germany and the United States, which, in some states in the latter, had led to the introduction of compulsory sterilization and the banning of marriage between criminals, epileptics, alcoholics, those with learning difficulties, the mentally ill and the disabled. The "American race," Stanojević wrote, had proved itself a "progressive and practical" nation. In introducing eugenic measures, he contended, the American people were ensuring for themselves "eternal youth, nobility, casting from themselves all that is damaging, and accepting all that is healthy and strong." In so doing, they were cultivating a "new ideal race" which would rule the entire world.[19]

In contrast to the attitudes of other advocates of racial biology and eugenics, Stanojević did not believe that the purity of the race made it any stronger. On the contrary, he argued that the more mixed a race was, the more virile and powerful it became: "Hygienic human breeding (...) means the rational utilization and control of existing racial characteristics and raw biological qualities on the one hand, and the improvement of all external conditions for breeding on the other."[20] The American race was powerful and destined for world domination precisely because it constituted a synthesis of a number of different racial groups—Anglo-Saxon Protestants, American Indians, and

Spaniards. In this way, Stanojević grafted modern Yugoslav concerns onto his understanding of American eugenics.

Stanojević further compared eugenics to the breeding of livestock, insisting that the rules of breeding applying to animals similarly applied to humans. Like the anonymous author from Split, he considered the reform of marital laws to be an urgent matter, complaining that marriages were not based on eugenic principles and thus could not guarantee a healthy or select family: "In the modern mating of human couples, there are no natural or eugenic conditions or incentives—everything is artificial, in everything non-eugenic concerns dominate (...) The modern marriage serves the Church, the State or tradition more than individuals or their offspring. Current matrimony is not just a slave, but the killer of all offspring. As well as the urgent need for the reform of marriage on eugenic principles, there is also the need for the introduction of widespread and deep propaganda for eugenic marriages."[21]

Stanojević called for the return of motherhood to its "classic Spartan role," and for women to "sacrifice and consecrate themselves" to the role of mother and housewife. Yet at the same time he demanded an end to military conscription and the idealization of military values that it represented, because it tore the young away from their familial homes to die in "dirty and unhealthy barracks," particularly from syphilis and other venereal diseases. Instead, a new society should be built on the principles of "Spartanism, Sokol and sport." Instead of militarism, militancy in a "modern form, geared towards the most contemporary needs of the state and society" was required. Stanojević also demonstrated the same social concern he had expressed elsewhere by reiterating his concept of aligning industry with racial hygiene in order to increase productivity. His study concluded: "The hygienic refinement and improvement of descendants—this is the future religion for the individual and the family as well as for the whole of cultured humanity."[22]

Writing in 1929, Bogumil Vošnjak compared the divisions of the different tribes comprising Yugoslavia to the nation-building problems experienced by Italy during the nineteenth century. Much like the Lombardians, Sicilians and Neapolitans, the Yugoslavs would overcome their differences and become one nation.[23] Likewise, the Serbian geographer Jovan Cvijić (1865–1927) believed that Serbs, Croats and Slovenes possessed different national characters. Cvijić considered that while the Serbs had a talent for "intuition and fantasy which, however, is not always disciplined," the Croats were gifted at "science, literature and art." The Slovenes, meanwhile, were rationalists, possessing an "unusually developed characteristic of ethnic endurance and toughness."[24] Yet the predominant psychic characteristics of all the three tribes would together contribute to the creation of a new Yugoslav civilization. Similarly, the Yugoslav critic Milan Marjanović (1879–1955), in the 1913 study *Narod koji nastaje* (The Nation that is Coming), noted the heroic and vengeful qualities of the Serbs as opposed to the intellectual and forgiving character of the Croats. In contrast to the contemplative Croat, the Serb was envisioned as a man of action.[25]

Although both Cvijić and Marjanović were from an older, pre-Yugoslav generation, they shared with their younger colleagues the belief that Yugoslavism meant the synthesis of three "South Slav tribes." A new generation drew on the ideas of Cvijić and Marjanović, adding the principles of eugenics to it, arguing that science could be used to create a new "Yugoslav race." Mijo Radošević, for example, hailed the "Yugoslav man" as not only the embodiment of an honourable soul, but also a "united ethnobiological type" with "an incredible talent, life force and militancy." The creation of this "new racial, cultural and political type" had been aided, according to the author, by the process of encouraging members of the three "Yugoslav tribes" to establish communities in regions of the Yugoslav Kingdom in which another group dominated. It was hoped that such a policy of ethnic mixing, which Radošević referred to as "internal colonization," would result in both a greater degree of inter-marriage and the creation of a generation of Yugoslavs imbued with qualities drawn from the superior elements of the three tribes. This, in turn, would ultimately lead to the eradication of "all tribal chauvinism and imperialism."[26]

The idea of using eugenics as a tool to create a Yugoslav master race was made most explicit in a 1924 article by J. Zubović, a contributor to the leading pro-Yugoslav, Croatian journal of the interwar period, *Nova Evropa* (New Europe). Announcing the imminent perfection of a new "Yugoslav man," he evoked the image of a Yugoslav superman around whom the nation, state, economy, political parties, families and culture would be built. The "Yugoslav man" was not simply new but a "man of better physical quality, stronger, more militant and healthier, more economically productive; (...) and, above all, great spirit, with beautiful motives and instincts, better habits, forceful will and more active, exuding higher intelligence and education."[27] How was this to be achieved? Zubović asserted that Yugoslavs first had to account for the peaceful assimilation of non-Yugoslav races with whom they lived side by side. Following this, "progressive" Yugoslavs should set to work on the creation of a "superior Yugoslav being." He proposed that marital and familial law should be altered to allow for mixed marriages, and that instead of religious considerations, civil marriages should be governed by a policy of national eugenics and social considerations:

> There can be no complete unity without blood unity, without the mixing of the various Yugoslav tribes (...). According to the laws of contemporary eugenics, progress is achieved by the process of mixing different but closely related tribes to produce a physically superior type. For us, it is of particular importance that the village rejuvenates the urban population; and that the urban middle-class mixes with workers, valley dwellers mix with mountain dwellers, people from the less educated regions with the educated. (...) This mixing will be profitable to us. It is apparent, for example, that the bony, stocky and militant Dinaric type will strengthen the average Yugoslav person just as the strongly evolved Slovenian women will. In the same way, the great industriousness of the Dalmatian or the Likan or the agrarian culture of the Slovenian, Croat or Vojvodinan will elevate the level of Yugoslav diligence and agriculture generally.[28]

Not all eugenic suggestions regarding the improvement of the race were so benign; some stressed not so much the betterment of the race as the prevention of breeding among biologically "degenerate" sections of society. At the Seventeenth Congress of the Association of Yugoslav Doctors in 1935, for example, Svetislav Stefanović reaffirmed his opposition to abortion, declaring that in "racial-hygienic and racial-biological terms abortion destroys the health of the mother, endangers her life and destroys the life of her descendants and, in the most extreme cases, leads to the degeneration of the race and accompanies or determines all other symptoms of collapse or decay of a nation or race."[29] Moreover, he pointed out that restrictions on the number of offspring had a "degenerative effect" on race. These included far higher rates of suicide, as well as higher rates of other "degenerative phenomena"—including incest, marriage between close relatives, and a greater number of pathologies and genetic defects which were subsequently passed on to following generations.

Stefanović proposed that, in order to avoid social degeneration,

Yugoslavia should look to Mussolini's Italy and Stalin's Soviet Union where the state intervened—with striking results in both cases—to protect both child and mother, while reinforcing the principle of natality. Although Stefanović was opposed to abortion, he did believe that it could be justified in exceptional circumstances; for example, he advocated that, for the sake of "mental-hygienic defense," human life could be artificially terminated through abortion and sterilization. Stefanović considered it paradoxical that many doctors who advocated abortions and birth control for social reasons simultaneously opposed abortions for reasons of "racial and spiritual hygiene that would reduce and prevent births among the mentally defective." Stefanović cited Germany's example, where the exponential increase in numbers of alcoholics, the mentally disabled, the mentally ill, and the genetically deaf and blind between 1870 and 1935 meant that the German government now spent the equivalent of twenty million dinars caring for them, thereby imposing an impossible burden on the budget: "If we take into account the fact that the physically defective breed more prolifically than those who are physically superior, it is obvious from the perspective of racial and spiritual hygiene that we should bear this fact in mind, even if we do not agree with the idea of forced sterilization especially if we consider that the combined figures of a few thousand or a few hundred thousand [forced sterilizations] are as nothing in comparison to the millions of violent and artificial abortions which are carried out year after year."[30]

The experience of Germany showed, Stefanović believed, that there was a danger of the racially inferior and the degenerate becoming socially dominant. To avert this possibility in the future, the only possible justification for abortion should be on the grounds of "medical, racial and spiritual hygiene." In such circumstances, the health of the mother was paramount, and the operation was to be carried out under the most stringent hospital procedures in order to demonstrate that the state was not indifferent to the health of woman. As it was not clear exactly when Yugoslavia would be in a position to enact such legislation the medical profession should be proactive, taking the initiative in creating a Directorate for the Defense of the Mother and the Child, in addition to forming sub-committees in the centre of each region of the state for the co-ordination of this initiative, never deviating from the thought that the creation and the defense of the child and motherhood is "one of the most important questions, not only for a government and for a politician, but also for the entire nation."[31]

In a similar vein, Stevan Ivanić called for a policy of racial hygiene to encourage simultaneously the selection of the "racially strong types" and the segregation of the "racially and genetically degenerate." Like Stefanović, he cited the example of Germany and wrote approvingly of the racial laws that the new Nazi regime had recently introduced. As he saw it, racial hygiene had three tasks: to create superior humans; to advance means with which to defend the healthy population; and "to root out from the healthy community the genetically inferior (the insane, epileptics, the deaf, the congenitally blind, congenital criminals, alcoholics, tramps and so on)." In the future, such people should be sterilized and thus prevented from breeding; at the same time, they should be segregated from the healthy community to prevent too much mixing.[32]

"We, the Yugoslav Barbarians!" The Rhetoric of Anti-Civilization and the Dinaric Superman

The well-known interwar Yugoslav ethnographer Vladimir Dvornikovíć (1888–1950) commented that the Yugoslavs as a race were "one of the most naturally gifted peoples of Europe," leading "all other peoples in brain size."[33] Moreover, as a synthesis of the three Yugoslav tribes, Dvornikovíć found that the Yugoslav man possessed "dynamism, rhythm, strong temperament, expressiveness and the constructive ability of fantasy."[34] Dvornikovíć was one of many ethnologists and anthropologists at the time who believed that the Yugoslavs constituted a race. In his massive study of 1939, the extensive sociological study of the peoples of Yugoslavia, *Karakterologija Jugoslovena* (The Characteriology of the Yugoslav), the author combined poetry, folklore, ballads, geography, as well as the most modern eugenicist and racial thinking—including, inevitably, much writing on race and nation that had proved popular with the Nazis—to produce a prototype of the ideal Yugoslav man and woman. The purest expression of the "Yugoslav race" was to be found, he argued, in the rocky Dinaric region, inhabited by the "Dinaric race." The idea of a Dinaric race was not new, having been championed by Jovan Cvijić at least as far back as the turn of the century; indeed, much as the book was influenced by the precepts of racial biology, in many ways it could also be seen as a recapitulation of many of Cvijić's ideas and theories. According to Cvijić in one of his last articles of 1930, the Dinaric people were "young, full-blood-ed and keenly alive to natural phenomena." He also believed that they were full of "kindness, good feeling, a sense of justice and a readiness to sacrifice themselves both as a nation and as individuals." The most characteristic feature of the Dinaric region was the presence of "forceful, violent and fiery men in whom the most unrestrained qualities of the race find their highest form of development. They are impulsive and act without any consideration." Sometimes sentimental, among them existed, nonetheless, men who "think nothing of sacrificing their lives for moral ideas or for the benefit of the race."[35]

Dvornikovič focused extensively on the Dinaric peoples, lauding them as a prototype for the future Yugoslav person. For Dvornikovič, the manliness and virility of the Dinaric man was unsurpassed not only in the South Slav region but throughout Europe: "The Dinaric type is the prototype of the male warrior, perhaps the most outstanding amongst all the white races: his ideas embody this type (…). This Illyrian man must be raw, strong and martial. The violence, which is constantly remarked upon when one talks about the Dinarics, emerges in the Illyrian in an even more elemental form (…). A. Geljan writes that the look of the Illyrian is so terrible and fascinating that it could 'kill a man'."[36]

At the same time, however, this did not imply that the "Dinaric man" was a primitive brute. On the contrary, the epic poems of the South Slavs lauded the Hajduks, the feared brigands and highwaymen in the Balkans who had terrorized and robbed travelers. Their spirit was captured by the "Dinarics" as the "idol and only hope of an enslaved nation," demonstrating the psychic connection they enjoyed with the people and with the land.

Dvornikovič argued that many of those who wrote about the "Dinarics" were anthropologists who had failed to enter their world; yet without such direct experience of their "patriarchal morals and ethical ideals" they could never hope to understand the "Dinarics." Those who spoke only of their plundering and thievery had not properly comprehended the soul of the "Dinarics" any more than "the superficial foreign tourists for whom the people are no more than thieves. Some of our writers are 'western', alien to these people: it is as if they had never experienced his world." Dvornikovič further pointed out that, despite the patriarchal and heroic social milieu from which the "Dinaric man" emerged, "Dinaric women" were far from submissive and displayed the Amazonian qualities that one might expect in the female companion of the "Dinaric warrior." The "Dinaric woman" had masculine ten-

dencies and a "masculine aura."[37] In any case, the author believed
that the brutal living style of this warrior prevented any form of "altruistic sentimentality and the feelings of consideration towards others."
Although a disposition to strong feelings was expressed in many
"symbols and forms of national life, in certain traditions and superstitions, national poems, proverbs and sayings," the style of "Dinaric"
life, especially in the southern regions, restricted the range of these
feelings to "a hard and rudimentary form." The "Dinaric race" thus
remained fundamentally warlike and pagan, "a warrior of the Balkan,
not Slav-Christian soul."[38]

Other ethnologists embraced the idea of the "Dinaric man" as the
prototype for a "Yugoslav superman," including the ethnologist and
government physician Branimir Maleš. For him, the "Dinaric man"
was far superior to his European counterparts. In fact, Maleš characterized the "Dinaric man" in a similar manner to Dvorniković. He argued
that the "Dinaric man" was an independent and unique racial type, related neither to "Alpine" nor to "Nordic" racial types. In 1935, Maleš
declared: "All his characteristics are exclusively Dinaric, harmonious-
ly joined and constituting one biological essence."[39] For this ethnologist, the key to the racial uniqueness of the "Dinaric man" was to be
found in his body shape and skull formation. The skull shapes and
bodies of the Alpine and Nordic races were allegedly completely different to those of the "Dinaric race," as was their temporal and frontal
lobes. Unlike the round faces and short stature of the Alpine race,
Maleš explained, the long face of the "Dinaric" person was in complete harmony with his "long body and all other body parts." In addition, he rejected the contention of some anthropologists and writers
that the "Dinaric race" was either a genus of the central "Armenian-
Alpine race" or a combination of the "Armenian" and "Nordic" races.
It was erroneous, he continued, to group together all those with brachycephalic skulls and dark complexions, and worse still to group the
"Dinaric race" with the "Alpine race (...)" and with American Indians
and Asiatic Mongols, part of the great yellow racial group."[40]

Given the dark hair and long bodies of the majority of the "Dinarics," Maleš also argued that there was a variant of "Dinarics" with
blond hair (*Blond Dinarics*). This also set them apart from the "Alpine
race," among whom blond hair was almost unknown. In Yugoslav
regions, he wrote, it was common to find people with red or blond hair
and blue eyes. Despite this "all their other features, both morphologi-

cal and physiological, are purely Dinaric." Maleš's fieldwork in Montenegro had shown that this phenomenon was actually quite common. Although he could not say with any certainty whether the "blond Dinarics" were a special species or a variant of the prototype "Dinaric" racial type, there was no doubt in his mind that they were related. This was proved, he insisted, by the fact that many blond "Dinaric" children became darker as they grew older. It remained to be seen whether both dark-haired "Dinarics" and blond "Dinarics" were related to the Nordic group. However, it was beyond doubt to Maleš that the "Dinarics" were closer to the "Nordic race" than any other "European race."[41]

In some respects, the popularity of the theory of a "Dinaric race" among a certain strata of intellectuals and academics reflected the desire—common throughout Europe, especially in the interwar period—to give notions about national identity a scientific basis and therefore a grounding in "fact." If academic and scientific enquiry could prove the existence of a Yugoslav race and, moreover, one that had existed long before the establishment of the Kingdom of Yugoslavia, then who could oppose a union of the Serbs, Croats and Slovenes? It also reflected the belief, prevalent among a largely urbanized nationalist élite throughout Europe, that the "authentic" culture of the nation was to be found in villages, among the peasants, rather than in cities. Indeed, as Svetislav Stefanović noted, "while the city demonstrated great interest in the folklore and clothes [of the village], it did not seem so interested in its life and health, its births and deaths, its homes and families."[42] On the other hand, it was also symptomatic of the general faith in the capability of science and technology to advance social progress and address national and social problems. For example, in a 1933 study assessing the health of adolescent "Dinaric" girls in villages and towns in Belgrade and its surrounding villages, Maleš used scientific means to establish which girls should be excluded from the survey on the basis of their "non-Dinaric anthropological characteristics." This included examining the shape of their faces and heads, inspecting coloring and complexion, as well as measuring their height.[43]

Dvorniković and Maleš were joined in their investigations into the "Dinaric race" by other anthropologists and scientists, who spent much of the 1920s and 1930s analyzing the racial characteristics, as well as the culture, music, clothes, language, folklore and religion, of various ethnic groups in Yugoslavia, especially those communities living in the frontier regions of the new state and those just outside its borders. Many of these studies amounted to more than just the accumulation of anthropological knowledge, and had a clear political agenda. Through such studies, writers aimed not only to provide a scientific basis for the Yugoslav race, but also to legitimate Yugoslavia's claim to territories currently under dispute.[44]

Despite the faith in science and technology shared by many Yugoslav racial anthropologists—a faith exemplified by the theory of "Dinaric" racial origins—this does not mean that they accepted all, or even most, of the values of the modern society from which many eugenic principles had originated. On the contrary, at the same time as they appropriated many of the racial ideas of modern European society, Yugoslav racial anthropologists simultaneously rejected many of its other supposed values. In particular, they opposed what they perceived to be the soulless nature of the "West"—embodied in its urban capitalist system—with the heroism and humanity of the eastern Slavs. Dvornikovič, for one, not only eulogized the East and envisioned a messianic calling of the Slavs as an alternative to the excessive rationalism of the West, but also held that the Slavs could save the West from degeneration and decay. In Dvornikovič's case, the embracing of "Dinaric" racial theory reflected his belief that the "Dinaric man" was a Balkan superman, virile and energetic, who could racially revive a torpid and exhausted Europe. An important element of this belief structure was a rejection of the supposedly civilized values of the West in favor of what was assumed as distinctly Balkan, particularly its alleged savagery, wild instincts and aggressiveness. This was a view shared by a sizeable intellectual constituency in Yugoslavia. Such hostility towards the cultural superiority of Europe was encapsulated in a memorable verse from the poem *Na Kale-Mejdanu* (At Kalemegdan) by the Slovenian poet Anton Aškerc (1856–1912): "Thus we protected you, Europe/ from the blows of wild hordes/ Ah, thus we spent our youth, / we—the Yugoslav barbarians!"[45]

In the period immediately following the creation of the Yugoslav state, various artistic groups utilized the image of the Balkan barbarian evoked by Aškerc in deliberate opposition to those Yugoslav ideologues seeking to imitate the practices and fashions of the West. Thus the Zenithist movement of Boško Tokin (1893–1954) and Ljubomir Mičić (1882–1942) declared in its 1922 manifesto that the Zenithist artist was the "new type of constructed barbarian" who honored the Balkan race: "We admire their grown-up barbarianism because absolute bar-barians are geniuses." For Mičić, this modern Balkan savagery was also to be praised for its "unsentimental vitality—its pure thought—unfalsified aims and open heart."[46] In his 1922 book *Kola za spasa-vanje: zenističke barberogenija u 30 činova* (The Circle for Salvation: Zenithist Barbarism in Thirty Acts), Mičić announced that the aim of the movement was to balkanize Europe: "We are in awe of the awak-ened barbarism. (...) Zenithism is the most rebellious act of the young barbaric race!"[47] Not surprisingly, their views on Croatian nationalist culture, with its scorn for Balkan and "Eastern" values and its aspira-tions to be part of the "civilized" West, were not so complimentary. According to Mičić, in its desire to emulate the values of the West, Croatian culture was "the illegitimate child of an unnatural marriage between a trained monkey and a parrot whose real name and address is Most Esteemed Sir, Office of the Imitation of Culture, Zagreb." How could this possibly compare with the glories of Balkanism, a synthesis, he claimed, of "young wild Slavism and the ripe fruits of Hellenism"?[48]

Mičić stated baldly that the Zenithists were opposed to the culture of the West: "Will we continue to remain slaves defending Lloyd George, Briand, Foch and D'Annunzio?" he asked. "No! Out with the Latins! It is a time of heroism! (...) we can by ourselves be pioneers and part of the creation of a culture for all humanity that carries in itself the spirit of the oriental man from the Urals and the Balkans—born in the cradle which we call Russia."[49] In a similar fashion, his Zenithist colleague Boško Tokin wrote of the "animalism and dynamism" of the new Balkan race and the "aestheticization of dynamism" which had a racial component, creating something that was both Balkan and uniquely Yugoslav.[50] Meanwhile, Rade Drainac (1899–1943) made similar pronouncements in his Hypnist manifesto: "It is time that to the Balkans came good spiritually. We have had enough of licking the boots of Catholicism, the Pope in Rome and the Gallic waves of Paris."[51]

Not all rejections of Western European culture and ideals were so irreverent. Both the writer on aesthetics Vladimir Vujić, and the play-wright Vladimir Velmar-Janković (1895–1976), vehemently rejected the idea of the Western conception of man in favor of the Balkan man. For Vujić, the Europe of 1929 did not represent culture and civilized values, but the brutal incarnation of violence. By contrast, the East, personified especially by Serbia, was imbued with the spirit of love, justice and humanity.[52] Vujić questioned the alleged progress of the

West and argued that in order to create an autonomous Yugoslav culture, it was not necessary to "direct our eyes to the West and seek out a model there which we will be compelled to blindly imitate." Indeed, he spoke disdainfully of the progress of the West, mentioning the "loss of faith and religious duty," and the "straying from moral principles, the development of a decadent professional class, sex as the foundation of life and the interpretation of life, hypocrisy, lies, the cruelty and perfect hypocrisy of everyday life."[53] Vujić also criticized the popular perception of the East among visitors and travelers from the West. For the rational West, the East was a fabled land of "darkness, ignorance, imprisonment, slavery, indolence and filthiness." He resented the yearning of those who traveled in the Balkans in search of spirituality and enlightenment, a sentiment born of a stereotypical perception of the East. The East was not backward or primitive, or a land of "vagueness, legends, despotism" and superstition. Nor was it, as some Westerners would have it, a "joyous empire of poetry (...) dreams and beauty, hashish and opium, peace and connection."[54] Vujić declared with satisfaction: "No, we are not Europe and it is good that we are not. We are neither Europe nor the West by our spiritual understanding of the world, by our spiritual style, by our view of the world and life."[55]

Much of Vujić's attack was directed against what he perceived to be the nihilistic and decadent societies of the West, evidenced most notably in its large anonymous cities. He wrote that "mechanical rationalism" signaled the end of civilization, since if spirituality ceased, so did civilization. For him, Western culture was irreversibly moribund. By contrast, Yugoslav culture—as a new spiritual phenomenon—would help to regenerate Europe. Yugoslav culture would draw on its epic national poetry to create an independent culture which, with its "racial genius," would save Europe from nihilism, decadence, materialism and moral equivalence. The Yugoslav cultural conception would not be "economic, bourgeois, European or Parisian." Rather, "the ideology of this culture must emanate and emerge from the eternal knowledge of our racial soul."[56]

For the dramatist Vladimir Velmar-Janković, writing in 1938, the new Balkan man was embodied in the male revolutionary of the Serbian uprising of 1804—the "Kalemegdan" man. He was certainly not a European, nor did he aspire to be. On the contrary, he was an Eastern male and "a man of a virile Belgrade persuasion," epitomized by the Serbian peasant. Owing to long centuries of harsh Turkish rule,

Velmar-Janković insisted, the Serb peasant had been protected from the Renaissance and Humanism. Instead, the development of the Serbs was shaped by different influences—including familial (*zadruga*) and tribal society. After 1918, the Serbs had been exposed to increasingly modern and materialistic influences, which had conflicted with the heroic-patriarchal view of life; and for this, he blamed intellectuals who had fallen under the spell of European culture in its entirety.[57]

Despite relating the manner in which the Yugoslav village had been allowed to fester and sink into backwardness and ignorance, Svetislav Stefanović perceived villages to be "the foundation and greatest reservoir of all culture," and he echoed Dvorniković in conceptualizing the "Dinaric race" as representative of a new race of supermen. Stefanović argued that in the contemporary world, where, as a result of the First World War, the solidarity of the white race had broken down, the Dinaric race would inevitably take the place of the Nordic race as the dominant European racial type. In Stefanović's thinking, the racial superiority of the "Dinaric man" and his harsh environment also imbued him with certain psychological qualities, as well as a philosophical outlook on life far removed from the Western European world. Unlike the "Nordic racial type," the "Dinaric man" would not be a capitalist owner. He would be a heroic being, "not a hero of commerce, hard, cruel, brutal, egoistic and inhuman, but a hero of satisfactory kindness, of a tender heart and soul, who not only uses his intellect to rule but also heroically bears all the slings and arrows that the ruling type, blessed with a hedonistic understanding of life, does not know." In short, this "Dinaric man" would be *"humanitas heroica."*[58]

It is perhaps unsurprising that much of the rhetoric, if not the ideology and practice, of those at the forefront of racial hygiene and race theory in interwar Yugoslavia appears to have been close to the rhetoric of Nazi racial biology and eugenics. Not only was eugenics viewed in much of interwar Europe as the science of choice for any state which desired to be seen as progressive and modern, and a science, moreover, in which Nazi Germany was undoubtedly a pioneer, but the totalitarian regimes in Italy, Germany and, to a lesser extent, the Soviet Union, were perceived to be the epitome of modernity in many other spheres of political life and social organization. While some anthropologists and scientists, like Dvorniković, appropriated some of the racial ideas of Nazism and at the same time explicitly condemned the discriminatory uses to which the Nazis put their science, many others had, by the late 1930s, moved to the far right of the political spectrum. Stevan Ivanić and Vladimir Velmar-Janković, for example, became prominent members of the semi-fascist pan-Yugoslav party, the Zbor Yugoslav National Movement, led by Dimitrije Ljotić, when it was founded in 1934. When the Germans occupied Serbia in 1941, among those who served in the various administrations of Colonel Milan Nedić's collaborationist government were Vladimir Velmar-Janković and Stevan Ivanić; the latter became minister for Health and Social Policy. For his part, Svetislav Stefanović, as chairman of the Association of Serbian Writers and the PEN Club, railed in the collaborationist press against the destructive work of Jews and their Communist supporters. In 1944, following the liberation of Serbia from Nazi rule, he was executed by the Communist resistance.[59]

Vlachs, Cincars and Other Deviations from the Racial Norm: Croatian Gentlemen Scholars Envisage Race and Nation

In contrast to the technological and scientific pretensions of Yugoslav racial ideology, rooted in a belief in the Eastern Slavic messianic tradition, Croatian racial concepts were rooted in nineteenth-century ideas of nationalist exclusivity more common to the West. At the heart of Croatian racial ideology was the nature of the relationship with the Serbs living in Croatia and Bosnia, and what extreme Croatian nationalists often perceived to be a struggle for survival against them. Unlike Yugoslav racial ideologues, who often worked at state research institutes and whose work was informed by twentieth-century notions about progress and technology, Croatian ethnologists and anthropologists were often solitary scholars whose ideas—at least until the 1930s—remained outside the political and cultural mainstream. Intellectually speaking, Croatian nationalists existed in the rarefied world of nineteenth-century gentlemen scholars, informed by the largely obsolete values of the Habsburg era. To understand their mentality and ideas, it is necessary to explore the intellectual milieu in which their values were formed.

The father of Croatian nationalism and the founder of the nationalist Croatian Party of Rights, Ante Starčević (1823–1896), was the author of a number of influential books on the national question. Whereas

Starčević, had written that the Muslims of Bosnia were the "purest" of Croats, he sometimes refused to acknowledge the existence of the Serbs as a race, and even contended that many historical Serb figures were, in fact, Croats. On the other hand, he argued that the Serbs were a degenerate and inferior race, the ancestors of nomadic Vlach shepherds that had settled in Croatia and Bosnia in the fourteenth century along with the arrival of the Ottoman army, acting as water carriers and slaves.[60] For Starčević, drastic measures should be taken against all Serbs. In order to save Croatia from annihilation and destruction, they would have to be "exterminated from the nation."[61]

The idea that the Serbs in Croatia and Bosnia were "racially inferior Vlachs" and nomads gained common currency not only among nationalist intellectuals and writers, but also at the popular street level in cities and towns, including the Croatian capital, Zagreb. Racist slogans became a defining characteristic of the violent anti-Serb riots by the followers of Josip Frank's Pure Party of Rights, a virulently nationalist and pro-Habsburg political faction at the turn of the twentieth century. Indeed, such rhetoric was a standard feature of the Party's newspapers.[62] Even the leader of the Croatian Peasant Party, Stjepan Radić (1871–1928), argued that Bosnia was ethnically and racially Croatian territory. In the 1908 study *Živo hrvatsko pravo na Bosnu i Hercegovinu* (The Clear Right of Croatia to Bosnia and Herzegovina), he argued that Croatia had a historical and ethnographical claim to Bosnia and Herzegovina. Bosnian Muslims, he concluded, were Croatians according to the laws of "ethnographical science;" they were "our people, by their blood and by language [and] their disposition and integrity."[63] Bosnia was today "an organic part of the rest of Croatia and it will soon, God willing, be an established fact."[64]

In 1904, anthropologist, archaeologist, and director of the Bosnian Agricultural Museum in Sarajevo, Ćiro Truhelka (1865–1942), published a notorious text which reflected the distillation of these prejudices, *Hrvatska Bosna: Mi i oni tamo* (Croatian Bosnia: Us and Them over There). His study provided an intellectual justification for the expulsion of the Serbs from Bosnia, as advocated by extreme Croatian nationalists. Truhelka emphasized the importance of Bosnia as a cultural borderland between East and West, as well as between "blond-haired Slavs and dark-skinned Vlachs, [and] between culturally passive and active tribes, Croats and Serbs, progress and stagnation, life and death."[65] He also developed a pseudo-scientific racial definition of the supposed Vlach origins of the Serbs. Truhelka alleged that Vlachs were, like the Jews, recognizable at a hundred paces: "whether he is from Romania, Šumadija, or the Banat, from the Lika, or Srijem, or Bosnia or Herzegovina, dress him in the clothes of the Shah of Persia, and as soon as every intelligent child sees him [they] will exclaim: 'There's a Vlach!'" He noted that in Bosnia Muslims and Croats were blond and blue-eyed, while the Orthodox Serbs were dark-skinned and brown-eyed. The latter were also pigeon-chested; this contrasted with the wide chests of the Balkan Muslims and Catholics. Moreover, the shape of their skulls—dolichocephalic as opposed to the brachycephalic skulls of Muslims and Catholics—implied that they were representatives of an older, culturally less worthy race than were the Croats. Anthropologically, Truhelka explained, Catholics and Muslims were identical in genetic makeup, while Orthodox inhabitants represented "a black-skinned, overwhelmingly dark, physically degenerate type."[66]

Biology and race were also linked to character. If one ventured to the villages, one could observe that the percentage of "fair, blue-eyed and physically developed tribes" increased at an alarming rate, whereas in cities, where capitalism was rife, the number of black-haired, black-eyed and physically weak tribes was much greater: "This is the best proof that the dark blood of the Orthodox of Bosnia spreads from the towns and cities, from 'plutocrat' circles to the village," he claimed.[67] As a corollary, Truhelka wrote that the love of profit was a key feature of the personality of the Vlach. He described how the Vlach made money at the expense of others, buying property and becoming a landlord, acquisitive businessman, banker or broker, lending money "at one hundred per cent interest on unsown corn, not undertaking any work that is not motivated by at least fifty per cent pure interest."[68] Many affluent citizens had fallen victim to the avaricious impulses of the Vlachs, and many a Beg or Aga fallen on hard times had found his land seized by the "dirty fingernails" of the medieval Morloch, who would boast in the "cafés or taverns in the towns of Bosnia" how the time would come when the present Aga would work for him.[69]

Truhelka claimed that his study was not a polemical work. Yet, at the same time, he believed that it should be made clear to patriotic Croats that the Vlachs were socially dangerous. On arriving in the Balkans, they had come across the homes of Croats and poured their "dark blood" into the veins of one section of the native population. Like some animals, the Vlachs were also incapable of evolving and were cultural parasites, unable to contribute anything meaningful to society. Despite their cunning nature they were "sterile, stereotypical and anthropologically inflexible, and susceptible to endemic tuberculosis and sterility, possessing weak physical and psychological constitutions."[70]

With the assassination of the heir to the Habsburg throne, Archduke Franz Ferdinand, in Sarajevo by a Serb member of Young Bosnia in 1914 and the subsequent declaration of war against Serbia, there was a return to the extreme anti-Serb rhetoric that had characterized much of Croatian nationalist discourse at the turn of the century. This was accompanied by mass riots against Serbian civilians, businesses and properties that resembled, according to a Sarajevo newspaper, "the aftermath of the Russian pogroms."[71] During the next few months, nationalist newspapers in Croatia returned to the aggressive rhetoric used earlier against the Serb community in 1902, with one newspaper stating that Croatia had declared "a war of life and death against the Serbs and their permanent exile from Bosnia and Herzegovina."[72] In Bosnia itself, a campaign of terror was being directed against the Serb population. Large numbers of citizens were arrested and incarcerated in concentration camps where many later died, while others (including priests, teachers and merchants), suspected of being nationalist sympathizers, were hanged and their bodies left on the gallows as a warning to others. Thereafter, troops formed from local Muslim and Croat populations, the *Schutzkorps*, *Freikorps* and Frankist legions, indulged in acts of killing, violence and sadism against Bosnian Serb civilians.[73]

Although it is unlikely that the young soldiers guilty of committing atrocities against the Bosnian Serb community had read Truhelka's inflammatory arguments, his study, nonetheless, directly influenced other writers. In 1917, Ivo Pilar (1874–1933), a lawyer from Tuzla, published his study *Die Südslawische Frage* (The South Slav Question), which appropriated many of Truhelka's messages. Pilar argued that the South Slav problem was in fact a Serb problem—the Balkan and Byzantine temperament of the Serbs had a negative impact upon the development of the region. Like Truhelka, Pilar argued that Bosnia was ethnically and politically Croat and that the Serbs of Bosnia were not Serbs at all, but Vlachs brought to the region by the Turkish Ottomans. As a result, Croatia was fighting a "race war, a national, social and economic war."[74] For him, the Serbs' nomadic Vlach origins also bestowed upon them certain distinctive personality traits. The Serbs,

he wrote, had a covetous and pilfering nature, and they were untrust-
worthy businessmen in much the same way that the Jews had been,
constantly lusting after the goods and property of others. It was also
a common feature of Vlachs to swindle others. This dishonesty was
reflected most notably in the Serbs' claim to Bosnia: their envious and
greedy urges had led them to try and deny its status as a constituent
and legitimate part of Austro-Hungarian territory, and to grab it for
themselves. Conversely, the nature of Orthodoxy, with its aims of
regional domination, had also led to the ghettoization of the Serbs who
could not bear to live with other people of different faiths. This was
evident among the Serb population in Bosnia who had set themselves
apart from the Catholics and the Muslims.

Pilar further contended that Vlachs only had the power and talent to
destroy things and this was why, in a period when "destructive, anti-
social and evil" instincts held sway, the Serbs had been recruited especially for the Ottoman army. Similarly, he believed that, as eternal
wanderers, the Serbs presented a mortal danger to the Balkans because
they were the leading instigators of plots, conspiracies and revolutions:
"How often was state conflict and regicide carried out in a peripatetic,
impatient, way!" he exclaimed. Worse still, their "migratory spirit"
meant that many of them had now settled in Zagreb, where they were
sure to outbreed the Croats. In short, there were no limits to Serb ambitions; in order to avoid Serb domination, Bosnia should be annexed to
Croatia as soon as possible.[75]

After the creation of the Yugoslav state in 1918, writers like Pilar
and Truhelka faded into relative obscurity and, in Pilar's case, interest
in his work only persisted among nationalist students. Yet Pilar's views
about the ensuing battle between the civilized West and the savage,
primitive East endured. Throughout the existence of interwar Yugoslavia, writers argued passionately about this subject. Some writers,
such as Miloš Djurić, believed that Yugoslavia should endeavor to
bring together the prime attributes associated with the West and the
East;[76] in contrast, Milan Radeka wrote that Yugoslavs should simply
rise above the paradigms of East and West and create a new, Yugoslav
culture "as the only possible exit from the opposing ideologies of East
and West in whose shackles we still languish."[77] However, some Croatian writers actively embraced the East. The novelist Dinko Šimunović
(1873–1933) lauded the East for its primitiveness and "legends, poems
and mysticism."[78] Others, such as the literary critic Albert Haler

(1883–1945), were undecided, but remained convinced that there were no social or cultural distinctions between the East and the West.[79]

More commonly, however, interwar Croatian nationalists persisted in adhering to Pilar's nineteenth-century view that Croatia's inclusion in a Yugoslav state represented not merely its subjugation to a Greater Serbia, but victory for the values of the Balkan East over those of the European West. The most well known protagonist of the Croatian clash of culture theory was the nationalist intellectual and former professor of Anthropology at the University of Zagreb, Milan Šufflay (1879–1931). In his writings, Šufflay, was obsessed by the ideas of "blood," "soil," and the "white race," which he believed was under threat by the "yellow and Asiatic hordes." Much like Pilar, he was influenced by the writings of Oswald Spengler (1880–1936)—particularly his theory of the decline and degeneration of the West—which Pilar adapted to the situation in Yugoslavia. The idea that Serbs and Croats were too different ever to live in the same state, let alone produce some kind of ethnic synthesis, was a characteristic motif throughout his writings. For him, Croats belonged to the civilized West and the Serbs to the East. In one fairly typical essay of 1922, he wrote that Croatian national feeling should listen to the voice of its blood, thus rejecting its Balkan alliance with the Serbs: "The Croat name, the blood of Croatdom, does not mean simply a nation! Here the blood of Croatdom means civilization. Croatdom is a synonym for all that is beautiful and good that the European West created! (…) But in Dušan's empire, in which it now finds itself, it sees something that is worse than death, it sees the Balkanization of the Croat nation (...) If Dušan's empire were to become a federation, it would be a purely Balkan creation. In it Croatia would lose the very thing which the Croatian Party of Rights considers its best attribute and which Radić also considers its best attribute—it would lose its instinct for Western culture and for humanity."[80]

Šufflay also believed that Croat nationalism was far superior to the nationalism of other nations. He argued, for example, that "Croatian nationalism represents something far more important than the nationalism of some borderless nation and is of far more worth to humanity than integral Yugoslavism." While philosophers, with some justification, would maintain that nationalism was a negative force because it resulted in "the division of humanity and the halting of progress" and thus set humans against one other, so long as there remained "a fatal gulf between the medieval East and the West"—and therefore, by extension, between Europe and Asia—Croatian nationalism was to be applauded. Unlike other nationalisms, the Croat type was not simply a form of regional patriotism, but also distinguished itself as one of the strongest bulwarks of a Western civilization under threat from the East. It meant "not only love towards the mother earth and the Croat homeland," but also "loyal service to the white West."[81]

In 1931, members of the youth group Young Yugoslavia assassinated Šufflay. His murder occasioned international condemnation and outrage from, among others, Albert Einstein, and Heinrich Mann of the German League for the Rights of Man. Shortly afterwards, students from the Croatian University Club Association produced an English-language pamphlet, *How the Croatian Savant, Professor of University Dr. Milan Šufflay, was Murdered by the Serbian Royal Dictatorship*, in which they appealed to European public opinion to recognize that the murder of Šufflay was the result of endeavors to build an absolutist South Slav state. In such a state, the "European culture" of Croatia and the Croatians was to be replaced by Serbian suzerainty, in which "orthodox, byzantine, oriental-asiatic and oldturc (sic) political and social traditions" had been subsumed into a system "not only contrary, but also odious to European culture." The pamphlet explained that Šufflay needed to be murdered by Slav nationalists, both as a personality who had had a great impact on the direction of Croatian national affairs, and as an intellectual who had "understood better than anyone else the abyss which separates European culture from the Balkan Byzantine region."[82]

By the late 1930s, the relationship between Serbs and Croats had deteriorated even further. For many Croatian nationalist scholars and writers, Truhelka's ideas about the Vlach origin of Bosnian Serbs once again gained widespread popularity, as nationalists became ever more explicit in their desires for an independent Greater Croatia that would include Bosnia. The idea that Serbs were the descendants of the Vlachs was a traditional Croatian nationalist belief, extending back generations. However, this does not mean that Yugoslav scientists and anthropologists did not recognize the existence of the Vlachs. For writers such as Vladimir Ćorović (1885–1941), the nomadic sheep-herding Vlachs—called "Black Vlachs," "Karovlachs" or "Morlachs"—were absorbed into the Slav society in which they settled. While there were still traces of the Vlachs in Yugoslav place names, over time the term "Vlach" came to mean anyone involved in shepherding.[83] Others were not as certain, including Dušan Popović, an academic at the University of Belgrade. The publication of his controversial 1937 study about the Vlachs of central Serbia appeared to confirm the negative personality and racial characteristics ascribed to the Vlachs, while also suggesting that many of the current Serb rulers and influential personalities in the arts, military and politics were descendants of Vlachs and Cincars.[84] Although Popović made it clear that the Vlachs had many good qualities too, his book nevertheless outraged some groups. Jovan Tomić of the Serbian Royal Academy of Sciences and Arts accused Popović of exaggeration. Popović replied with the publication of a second volume, containing new findings and a long list of Vlach families with illustrious members in Serbia. "They told me I am exaggerating 'Cincarism'. Here is my answer," he declared testily.[85]

One of the student signatories of the pamphlet protesting against the murder of Šufflay, Mladen Lorković (1909–1945), had been a student of law at the time; later he became a prominent member of the fascist Croatian party, the Ustasha movement. In 1939, on the eve of the creation of a semi-independent Croatian Banovina, Lorković published a geo-political and ethnic study of Croatia under the title *Narod i zemlja Hrvata* (The Nation and Lands of the Croats). The book reinforced many existing ideas concerning racial biology in Croatia. He wrote, for example, that Serbs in Croatia and Bosnia had descended from Vlachs and Cincars brought to the region by the Turkish army and converted to the Serbian Orthodox Church. However, he also alleged that many other Serbs in Croatia and Bosnia were actually Croats, who had been forcibly converted to Orthodoxy in Herzegovina after Catholic priests had fled.[86] Lorković's study was not alone in alleging that a large majority of Catholics had converted to Orthodoxy with the arrival of the Ottomans in the sixteenth century. Two years earlier, Krunoslav Draganović, a Catholic priest, had published a study in which he asserted that Serbs had outnumbered Croats in Bosnia through their conversion to Orthodoxy. Draganović further explained that Orthodox priests in Bosnia, accompanied by "Vlach shepherds and frontier Morlochs," had taken "our living space." The settlers, "Slav-Romanian-Albanian hybrids," were of a "violent Dinaric type," and were "dark nomadic Slav elements of a very alien blood."[87] There were also accusations from other clerical writers that the Yugoslav government was deliberately trying to destroy the Catholic Croats demographically through a variety of means, such as encouraging Catholic girls to marry Orthodox Serb soldiers garrisoned in Croatia and Bosnia; building Orthodox churches in the centre of Catholic areas; under-funding Catholic cultural and educational initiatives; banning Catholic youth organizations; and preventing the opening of new Catholic schools.[88] Other nationalist writers accused the Yugoslav government of using other methods, including military service and promoting the concept of the modern family, all to prevent breeding and reduce the birth rate of the native Croat population.[89]

As a corollary to this argument, Lorković also argued that Bosnian Muslims were the purest of Croatians, and that Bosnia was a Croat land "by virtue of its ethnicity, people, its language, state regions and traditions." For Lorković, the fact that, in their history, the Muslims and Catholics of Bosnia had fought each other was "the greatest tragedy of the Croatian past that had serious consequences which we are still coming to terms with, and while the Islamic and Christian Croats shed blood, part of their land was settled by non-Croat elements." At the same time, the bravery with which Croats had fought for both Islam and Christianity showed that they were a strong race: "A people of weak blood, of polluted racial stock, of small lands and poor numbers would not have been able to demonstrate evidence of their life force and greatness in the way in which the Croats of the two faiths [Muslim and Catholic] did as they battled on opposite sides of the world barricades."[90] Lorković also suggested that the Croats might be of Iranian descent mixed with Slav blood. He based this idea on the etymological origin of the word *Hrvat* (Croat). The author claimed that it was based on the Iranian word "Hu-urvatha," meaning "friend." The Croat love of horses, which they shared with the Iranians rather than the Slavs, supposedly proved the Iranian origin of the Croats. The Bogumils—a heretical Christian sect in medieval Bosnia and Herzegovina whose faith, Bogumilism, was said to have been practised by many Bosnians who later converted to Islam—were also, the author wrote, an Iranian sect. There also existed, he continued, many words of Iranian origin in the Croat language.[91]

However, Lorković's main concern was not the historical greatness of the Croats but rather their current demographic weakness—at least in relation to the alien Serbs—in regions that were historically Croat. Due to a combination of factors, including the emigration of Croats and the exponential birth rate among Serbs, especially in Bosnia and

Dalmatia, the very existence of the Croatian nation was perceived to be threatened.[92] Lorković pointed out that there were more Croats living in America than in Dalmatia and Istria, and that 24 per cent of Croats lived outside Croatia. Moreover, 2.5 million non-Croats lived in the heart of the Croat lands, with slightly less than 2 million Croats living elsewhere. This meant that lands the Croats had considered to belong to them for centuries had now been surrendered to other nationalities, resulting in the "collapse of their biological strength." To remedy this parlous situation, Lorković recommended colonizing Croatia's borders in the East, since this would help to create a culturally and ethnically united nation: Croats, he wrote, could not protect the purity and strength of their nation if they were to give up a quarter of their national living space or, as Lorković put it, their "national organism."[93]

Reviews of Lorković's book suggested that the themes discussed therein had struck a chord among nationalists. The travel writer and cultural commentator Dragutin Gjurić lauded the book for rejecting Pan-Slav romanticism and instead developing the concept that the Croats were an original Iranian-Slav race with an independent history and the right to an independent state. Gjurić was especially keen to highlight how Lorković's book had discussed the problem of Croatian cultural and ethnic unity in the face of pressure from "foreign influences." Although the strength of Croat national consciousness had enabled them to withstand five centuries of Turkish rule, as a result of their precarious position on the border between East and West, under the rule of the Ottomans the Vlach element had become entrenched in the region. In supporting Lorković, Gjurić pointed out that it was necessary to "return the Croat soil to the Croat people."[94]

The contention that the Serbs of Bosnia and Croatia were a racially alien and unstable ethnic element, one endangering the cause of Croatdom, was an idea rooted in nineteenth-century Croatian nationalist ideology. This does not mean to say that Croatian nationalists were immune to the influence of contemporary political ideologies like fascism and, above all, National Socialism, in the articulation of their ideology. For example, extreme nationalists such as Stjepan Buć (1888–1975), editor of *Nezavisnost* (Independence), and the clerical journalist Kerubin Šegvić (1867–1945)—much to the amusement of certain sections of the state press—wrote books declaring that Croats were not Slavs but were instead people of Gothic and Nordic origin, destined to rule the world.[95]

With the fortieth anniversary of Starčević's death in 1936, the separatist intelligentsia in Croatia took the opportunity to reassess his work. In a speech to university students in 1936, Buć reinvented Starčević as a National Socialist. Analogous to Hitler's attitude towards the Jews, the father of Croatian nationalism had recognized that the Serbs were not only of "foreign blood" and "sick racial mongrels," but that their degenerate behavior was an expression of the "voice of their blood." In order for Croatia to be liberated from these foreigners, thereby saving the nation from annihilation and extermination, the Croats would first need to develop "blood and race pride"; and second, they would need to build a future Croatian race on the basis of the finest "biological elements" of the nation.[96] A year later, Filip Lukas, a prominent member of Croatia's most prestigious cultural organization *Matica Hrvatska* (The Croatian Queen Bee), declared that the nation was a blood community rooted in spiritual and biological unity. He proclaimed Starčević to be a warrior against the degeneracy, disease and decadence of urban life as exemplified by the Serbs, in addition to championing the village as the source of the physical renewal of the nation. An opponent of the theory of the Slav origin of the Croats and an advocate of an ethnically pure Croatia, Lukas asserted that Starčević comprehended the need to "compress from the biology of our race the conditions of our most necessary existence."[97]

Nevertheless, the hostility towards the Serbs in Croatia and Bosnia implicit in Croat racial theories during the 1920s and 1930s became more aggressive and threatening following the establishment of the Croatian Banovina in September 1939, when members of the Serb community, led by leading personalities of the Serbian Orthodox Church, began to demand regional autonomy. The idea provoked outrage among extreme Croat nationalists, especially because large parts of Bosnia had not been included in the Banovina. Thus when Serbs from the Vukovar region demanded to be allowed to join "Dunavska Banovina," the leading nationalist newspaper in Croatia, *Hrvatski narod* (The Croatian Nation), fumed that this was out of the question as, ethnologically speaking, the area had always belonged to the Croats. The newspaper pointed ominously to many historical examples of the "resolution of the question of national minorities," including the expulsion of the Greeks from Turkey.[98] Other articles from the same journal concentrated on the foreignness of the Serbs, either as colonial settlers after the end of the Great War, or as Gypsy lackeys of the invading Turks.

Despite what Serbs called themselves now and what erudite professors of history insisted that they were, the truth was, argued another article, they were what they had always been: "Turkish mercenaries," "Morlochs," a "Turkish horde" and an "émigré minority" who had been settled in order to "terrorize" the Catholic population. In the coming Croat state, these Vlachs would be afforded all the rights accorded to national minorities, "but this national minority would not call itself 'Serb' but 'Vlach' which in truth it is."[99]

The journalist Luka Grbić pointed out that—despite the treacherous nature of the Vlach-Serbs who had always sided with Croatia's enemies—this group had gained positions in the Banovina government of Vladko Maček where they, yet again, deliberately worked to the detriment of the Croatian nation. Since Croatia was to be an ethnically homogeneous state, the Serbs, like all other national minorities, should be happy with the rights accorded to them: "Whoever is not content with this, let him emigrate from the Croatian lands because the Croats did not invite them to come to their lands but, on the contrary, they arrived without their knowledge and against their wishes. They need to be grateful to the Croats who suffered them and gave them bread, and not dare to initiate anti-Croatian propaganda intended for the exercise of power. The Croat nation will not allow foreigners to rule over them in their own house."[100]

During the late 1930s, Croatian nationalists intensified their efforts to persuade public opinion that Bosnian Muslims were in fact Croats, and that Bosnia should be included in any future Croat state. They took their lead from racial anthropologists like Truhelka who, unlike Pilar, had continued to publish his scientific findings in specialist journals. Truhelka was persistent in his assertion that Bosnian Catholics and Muslims were racially distinct from the Greek Eastern Orthodox ethnic group, by virtue of the latter's dark hair.[101] Nationalist students at the University of Zagreb continued to disseminate the idea that Muslims were in fact Croats. On 21 April 1939, as Vladko Maček (1879–1964), leader of the Croatian Peasant Party, and Dragiša Cvetković (1893–1969), the Yugoslav prime minister, met to negotiate the creation of an autonomous Croatia, radical student clubs in Zagreb, led by Grga Ereš (1912–1947), Jusuf Okić and Muhamed Hadžijahić (1918–1986), convened a conference to declare that Bosnia was a Croatian land and should be included in any future Croatian state. A resolution declared that otherwise a section of the inhabitants of Bosnia would fall under the power of "immigrant Serb elements." In an effort to avoid pervasive Serbian influence, the resolution further stated that it would not allow Bosnia to be separated in the event of any agreement between Belgrade and Zagreb. These radical students insisted that a united Bosnia within a greater Croatia was essential to the liberation of the Croatian people; without it, a Croat state was doomed to failure. For the students, the racial kinship of Catholics and Muslims was also important. They argued that Catholics and Muslims—the most ethnically pure segment of Croatdom—were blood brothers, comprising a clear majority over the "immigrant Vlach-Greek Eastern or Serb elements."[102]

Other Croatian nationalist youth groups expounded similar initiatives; their aim was to convince Muslim youths that they were in fact Croats. In 1938, a group of students calling themselves the "Croatian Youth of Herzeg Bosnia" began publishing a journal, *Sbornik hrvatske omladine Herceg-Bosne* (The Anthology of the Croatian Youth of Herzeg-Bosnia), dedicated to Croatia's rightful claim to Bosnia and Herzegovina. In one article, the editor, Munir Šahinović-Ekremov (1900–1945), wrote that if Bosnian Muslims were not Croats, then "Croat politics has no serious prospect of lasting establishment in this land."[103] However, he also wrote that Croat politicians were making great efforts to gain the allegiance of Muslims. Important circles of Croat society placed great stress on the Muslims as "a special and particularly important part of the Croatian nation." Despite the work of nationalist intellectuals, journalists and politicians, Šahinović-Ekremov insisted that they needed to do more to reach out to the Muslims in order to underline the racial kinship existing between Croats and Muslims. Despite Croatia's historical claims to Bosnia, Muslims were also racially Croat; therefore, the claim to Bosnia was not merely historical but based on "racial-biological knowledge."[104]

Writing in the pages of *Plava Revija* (The Blue Review), Halid Čausević, a young Muslim intellectual and member of the sub-committee of *Matica Hrvatska*, affirmed the view that the Serbs in Bosnia were "Cincars, Vlachs, Serbs, Black Vlachs and others" who had occupied Croat living space.[105] Yet in the same journal, H. Bošnjak complained that outside of a few metropolitan centers such as Tuzla, Sarajevo and Mostar, few Muslims believed that they were Croats and were largely unconscious of their Croat identity; for this, he blamed the attitude of many Catholics towards Muslims. Čausević argued that a new, conscious young Muslim public worker, freed from the modern vices of selfishness and corruption, was needed to guide the majority of young Muslims towards a greater Croat consciousness. Entering into public life as nationally conscious Croats, Muslim youths would in time realize that they were indistinguishable from Croats "in racial characteristics, speech, culture, feelings and writings."[106]

By the late 1930s, such ideas were not simply the currency of a nationalist intelligentsia or groups of radical students, but were shared more widely across Croatian society. For example, like many on the right of the Croatian Peasant Party, Maček believed that most Serbs in Croatia and Bosnia were Vlachs; Muslims, by contrast, were ethnically pure Croats. Speaking on the occasion of the visit of a delegation of Muslim peasants to Zagreb in 1937, Maček affirmed that Bosnia was part of the old Croat Kingdom and that Muslims were Croatian blood brothers. He urged Muslims to be loyal to Croatdom, proclaiming that: "they carried in their souls Croatian national consciousness, imbibed with their mothers' milk."[107]

By 1939 Croatia was an autonomous state within Yugoslavia, where such notions had become an integral part of political discourse. In the increasingly separatist atmosphere of the Croatian Banovina, the activities of the extreme nationalist intelligentsia and its supporting political parties were initially tolerated, as was the promotion of fascist and National Socialist ideas; yet at the same time, the government of the Banovina favored harsh and repressive measures against Communist and left-wing groups, as well as minority groups within the nascent state. By 1940, however, government officials were reporting with evident concern that, not only were Ustasha sympathizers and nationalist followers infiltrating economic, social, and cultural organizations in Croatia, but radical nationalist parties and organizations were becoming increasingly popular with the Croatian masses, especially among the peasantry and working classes—an assertion confirmed in the local elections that summer. In such a politically unstable climate, the government in the Banovina arrested leading nationalists and Ustasha activists within Croatia, who were deemed dangerous contenders for the leadership of Croatia by prominent members of the Peasant Party. Simultaneously, however, the Peasant Party attempted to co-opt the nationalist message and dilute its appeal. In a nationalist semi-independent state in which Serbs were being treated as aggressive foreigners by many of the nationalist newspapers in Croatia as well as, increasingly, being denied their elementary rights, it is small wonder that when the fascist Ustasha movement came to power in April 1941, they found a receptive audience for their ideas about race and nation.

Conclusions

Croatian and Yugoslav theories of race were profoundly different, and in many respects inimical to one another. What the racial philosophy of integral Yugoslavism wished to create, Croatian nationalism sought to destroy, with the latter conceptualizing the Yugoslav "nation" as an artificial, pseudo-scientific Trojan horse for Greater Serbia. In opposition to the creation of a new nation, Croatian nationalists sought the return of the medieval Croatian kingdom, denoting a return to a period before "Vlachs" and "Gypsies" polluted Croatian living space. Rather than relying on the modern concepts of racial science as propagated by state health and hygiene institutes, Croatian nationalist ethnography relied on the pioneering work and research of respected individuals like the professional anthropologist Truhelka, and the amateur ethnographer Pilar. For all the rhetorical adoption (and in some cases ideas) of the National Socialist biologically determined understanding of nation and race, these men were arguably influenced far more by traditional nationalist stereotypes and arcane folklore than by the certainties of science.

In other respects, Croatian and Yugoslav racial theories resembled each other more closely. For example, the Croatian nationalist embrace of Bosnian Muslims and the belief that they were ethnic and racial "blood brothers" to some extent replicated the official Yugoslav slogan of brotherhood, harmony, and the mixing of the races. Ultimately, however, the two ideologies significantly differed. While Yugoslav racial ideology far more openly embraced eugenicist and biological concepts of race, its ideas also reflected the utopian and ambitious vision of academics, scientists and intellectuals, who believed that science and modernity could be used to create a synthetic Yugoslav nation from which all traces of tribal hatred and internecine struggle could finally be eradicated and replaced by a new Yugoslav person.

By contrast, Croatian racial theory reflected the mindset of those who felt threatened, marginalized and powerless. Largely eschewing the advances of science, Croatian nationalist anthropologists, academics and amateur scientists together believed that Croatia was endangered by alien and racially inferior foreigners, whose very presence would result in the annihilation of the nation from within. Only an ethnically and racially pure Croatia could preserve its national identity. The writings of Truhelka, Šufflay and Pilar were influenced by the conviction that Serbs and Croats could not live together due to their disparate cultural practices and opposing racial characteristics. Needless to say, the idea that Croats emerged from a racially superior and more civilized culture did indeed dominate their writing. Ultimately, their ideas were far more dangerous than those of Yugoslavism, since they were predicated on the demonization—and to some extent the dehumanization—of a specific ethnic community within Croatia, and Yugoslavia, generally. Šufflay, Truhelka and others did not create the fascist Ustasha movement. After all, their basic ideas concerning race and the nation were part of a wider nationalist consensus by 1941. However, their philosophies, as Mile Starčević pointed out, directly inspired the racial notions of Croatian fascism, and, more importantly, their writings created an intellectual atmosphere in which genocide could be legitimated. One question thus persists: Could it be that an eccentric group of nineteenth-century gentlemen scholars inspired a movement of Croatian barbarians?

Endnotes:

[1] Mile Starčević, "Krik izmučene hrvatske duše: uz hrvatski prievod Süd-landova (Pilarova) 'Južnoslavenskog pitanja'," *Hrvatska revija* 17, 9 (1944), 457–471.

[2] Ferdo Puček, "Predgovor prevodioca," in Ferdo Puček, ed., *Južnoslavenske pitanje: prikaz cjelokupnog pitanje* (Zagreb: Matica Hrvatska, 1944), xi.

[3] Tin Ujević, "Futuristički Beograd," *Novosti* 2, 9 (1922), 1.

[4] David Footman, *Balkan Holiday* (London: William Heinemann, 1935), 195.

[5] Henri Pozzi, *Black Hand over Europe Consisting of War is Coming Again* (London: The Francis Mott Company, 1935), 53–54.

[6] L., "Jugoslovenska žena," *Nova Evropa* 4, 1 (1922), 1–3; Jedan svečenik [i.e. a priest], "Za slobode mišljenja u katol. Crkvi," *Jugoslovenska njiva* 3, 29 (1919), 348–349; Jedan Svečenik, "Apostazije katolickih svečenika," *Jugoslovenska njiva* 3, 22 (1919), 461–462; and Edo Marković, "Moralno ozdravljenje našega društva," *Jugoslovenska njiva* 3, 19 (1919), 296–298.

[7] M. A., "Za reformu braćnog prava," *Jugoslovenska njiva* 3, 23 (1919), 363–65.

[8] Andrija Štampar, "Seksualna pedagogika kao sredstvo u borbi protiv spolnih bolesti," *Jugoslovenska njiva* 3, 14 (1919), 222–223. See also *Serving the Cause of Public Health: Selected Papers of Andrija Štampar*, ed. by M. D. Grmek (Zagreb: Izdavački zavod Jugoslavenske akademije znanosti i umjetnosti, 1966).

[9] Milan Pribićević, "Naš idejal Jugoslavije," *Književni jug* 3, 1 (1919), 1–2.

[10] Bogumil Vošnjak, *Pobeda Jugoslavije* (Belgrade: Sveslovenska knjižara, 1929), 56.

[11] Stevan Z. Ivanić, "Kuća na selu," *Glasnik centralnog higijenskog zavoda* 17, 4 (1934), 145–154.

[12] See Rudolf Bićanić, *Kako žive narod? Život u pasivnim krajevima* (Zagreb: Tipografija, 1936).

[13] Vladimir S. Stanojević, *Higijena za sredne i stručne škole* (Belgrade: Skerlić, 1927), 166–167.

[14] Stanojević, *Higijena za sredne i stručne škole*, 166–176.

[15] Svetislav Stefanović, "Za zdravstveno podizanje sela," *Jugoslovensko lekarsko društvo* (1934), 3.

[16] Stefanović, "Za zdravstveno podizanje sela," 10.

[17] Stanojević, *Higijena za sredne i stručne škole*, 166–176.

[18] Vladimir S. Stanojević, *Eugenika; higijena čovečeg začeča i problem nasledja* (Belgrade: Ministarstva narodnog zdravlja, 1920), viii.

[19] Stanojević, *Eugenika*, 149–151.

[20] Stanojević, *Eugenika*, 123–124.

[21] Stanojević, *Eugenika*, 154–155.

[22] Stanojević, *Eugenika*, 155–156.

[23] Vošnjak, *Pobeda Jugoslavije*, 103–104.

24 Jovan Cvijić, "Osnovi jugoslovenske civilizacije," *Nova Evropa* 7, 1 (1922), 212–218.

25 Milan Marjanović, *Narod koji nastaje* (Rijeka: Trbojević, 1913), 3–58.

26 Mijo Radošević, "Etnobiološki problem," in Mijo Radošević, ed., *Osnovi savremene Jugoslavije: političke ideje, stranke i ljudi u XIX i XX veku* (Zagreb: Zadružena štamparija, 1935), 613–614.

27 J. Zubović, "Jugoslovenski čovjek," *Nova Evropa* 10, 6 (1924), 151, 153.

28 Zubović, "Jugoslovenski čovjek," 152.

29 Svetislav Stefanović, "Rasna higijena, abortus i zaštita matere," *Evgenika* 1, 5 (1935), 1.

30 Stefanović, "Rasna higijena, abortus i zaštita matere," 7.

31 Stefanović, "Rasna higijena, abortus i zaštita matere," 7.

32 Cited in Nenad Petrović, "Arijevci sa Balkana," *Duga* (18 May 1996), 95.

33 Vladimir Dvorniković, *Karakterologija Jugoslovena* (Belgrade: Kosmos, 1939), 608.

34 Dvorniković, *Karakterologija Jugoslovena*, 514.

35 Jovan Cvijić, "Studies in Yugoslav Psychology," *Slavonic and East European Review* 9, 26 (1930), 378–384.

36 Dvorniković, *Karakterologija Jugoslovena*, 208, 295–296.

37 Dvorniković, *Karakterologija Jugoslovena*, 797.

38 Dvorniković, *Karakterologija Jugoslovena*, 768.

39 Branimir Maleš, "Nekoliko napomena o dinarskoj rasi," *Socijalno-medicinski pregled* 7, 2 (1936), 1–7.

40 Maleš, "Nekoliko napomena," 3.

41 Branimir Maleš, "O Dinardirima svetle kompleksija," *Glasnik centralnog higijenskog zavoda* 17 (1934), 136–144.

42 Stefanović, "Za zdravstveno podizanje sela," 3.

43 Branimir Maleš, "Menarha sela i varoši – utičaj socijalnih prilika," *Glasnik centralnog higijenskogzavoda* 8 (1933), 25–28.

44 See, for example, Milenko S. Filipović, "Visočka cigani," *Etnobiološka biblioteka* 16 (1932), 1–20; Milenko S. Filipović, *Golo brdo: belješke o naseljima, poreklu i stanovništva narodnim životom i običajima* (Skoplje: Južna Srbija, 1940); and Milenko S. Filipović, "Severna velješka sela," *Srpski ethnografski zbornik*, 51, 1 (1935), 489–573; Dušan Popović, *O cincarima: prilozi pitanju postanka našeg gradjanskog društva* (Belgrade: Drag. Gregorić, 1937).

45 Anton Aškerc, "Na Kale-Mejdanu," in Mirko Deanović and Ante Petrović, eds., *Antologija savremene jugoslovenske lirike* (Split: Knjiznica Vinko Jurić, 1922), 70.

46 Ljubomir Mičić, "Kategorici imperative zenističke pesničke škole," *Zenit* 13, 2 (1922), 1.

47 Ljubomir Mičić, *Kola za spasavanje: zenističke barberogenija u 30 cinova* (Belgrade and Zagreb: Zenit, 1922), 3, 5.

48 Ljubomir Mičić, "Papija i monopol: hrvatska kultura," *Zenit* 3, 24 (1923), 1–2.

49 Ljubomir Mičić, "Delo Zenitizma," *Zenit* 1, 8 (1921), 2.

50 Boško Tokin, "Izložba jugoslavenskih umetnika u Parizu," *Plamen* 1, 13

(1919), 25; and Boško Tokin, "Sedam posleratnih godina naše književnosti," *Ljetopis Matice Srpske* 38, 3 (1928), 380.

51 Rade Drainac, "Program Hipniza," *Hipnos* 1 (1922), 2–3.

52 Vladimir Vujić, "Vidovdanska razmišljanja o kulturi: naša tragičnost," *Narodna odbrana* 24, 7 (1929), 452.

53 Vladimir Vujić, "O ljudima i zapada i o nama," in Vladimir Vujić, *Sputena i oslobodjenja misao: ogledi* (Belgrade: Geca Kon, 1931), 162–163.

54 Vujić, "O ljudima i zapada i nama," 167.

55 Vujić, "O ljudima i zapada i nama," 166.

56 As cited in Petrović, "Izgubljena mladost Evrope," *Duga* (2 June 1996), 96.

57 Vladimir Velmar-Janković, *Pogled s Kalmegdana: ogled o beogradskom čoveku* (Belgrade: Gregorić, 1938).

58 As cited in Petrović, "Arijevci sa Balkana," 96.

59 Regarding the racial and social ideology and the political development of the Zbor Movement, see, for example, Jovan Byford, "The Willing Bystanders: Dimitrije Ljotić, Shield Collaboration and the Destruction of Serbia's Jews," unpublished paper.

60 See, for example, Ante Starčević, *Pasmina Slavosrbska po Hervatskoj* (Zagreb: Lav Hartmann i družba, 1876), esp. 27–31, 33, 38; and Ante Starčević, "Ime Serb," in Ante Starčević, *Djela Dr Ante Starčević*, vol. 2 (Zagreb: Antun Scholz, 1894), 72–73, 165.

61 Ante Starčević, *Nekoliko uspomene* (Zagreb: Narodna tiskara, 1870), 342.

62 See, for example, (anonymous), "Protovláski pokretu Zagrebu," *Hrvatsko pravo* (1 September 1902), 3.

63 Stjepan Radić, *Živo pravo na Bosnu i Hercegovinu* (Zagreb: Pučka seljačka stranka, 1908), 29–59.

64 Radić, *Živo hrvatsko pravo na Bosnu i Hercegovinu*, 29–59.

65 Ćiro Truhelka, *Hrvatska Bosna: (Mi i oni tamo)* (Sarajevo: Hrvatski dnevnik, 1907), 2–6.

66 Truhelka, *Hrvatska Bosna*, 12–13.

67 Truhelka, *Hrvatska Bosna*, 16.

68 Truhelka, *Hrvatska Bosna*, 20–21.

69 Truhelka, *Hrvatska Bosna*, 52–53.

70 Truhelka, *Hrvatska Bosna*, 27–29.

71 Vladimir Ćorović, *Crna knjiga: patnje Srba Bosne i Hercegovine za vreme svetskog rata, 1914–1918* (Sarajevo: Dj. Djurdjević, 1920), 27–51.

72 Ćorović *Crna knjiga*, 27–51.

73 Ćorović *Crna knjiga*, 52–82; Pero Slijepčević, "Bosna i Hercegovina u svjetskom ratu," in Pero Slijepčević, ed., *Napor Bosne i Hercegovine za oslobodjenja i ujedijenje* (Belgrade: Narodna odbrana, 1929), 219–277.

74 L.V. Südland, *Die südslawische Frage und das Weltkrieg: Übersichtliche Darstellung des gesamt Problems* (Vienna: Manzschen k.u.k Hof Verlags und universitäts Buchhandlung in Wien, 1918), 148–156, 289–296.

75 Südland, *Die südslawische Frage*, 357.

76 Miloš, Djurić, *Slovenskim vidicima: prilozi filozofiji slovenske kulture* (Belgrade: Državna stamparija, 1928), 65.

[77] Milan Radeka, "Istok i zapad u nama," *Javnost* 2, 32 (1936), 700–703.

[78] Bogdan Radica, "Kod Dinka Šimunovića," *Hrvatska revija* 2, 6 (1929), 438–441.

[79] Albert Haler, "Istok u zapad," *Nova Evropa* 20, 5 (1929), 129–132.

[80] Milan Šufflay, *Hrvatska u svijetlu svjetske historije i politike* (Zagreb: Merkantile, 1928), 28–29.

[81] Šufflay, *Hrvatska u svijetlu svjetske historije i politike*, 37–41.

[82] Mladen Lorković, Branimir Jelić, Vilko Pečnikar, Josip Milković, Ante Valenta and Ivan Košutić, "Appeal of the Croatian Academicians to the World of Civilization," in Mladen Lorković, Branimir Jelić, Vilko Pečnikar, Josip Milković, Ante Valenta and Ivan Košutić, eds., *How the Croatian Savant, Professor of University Dr. Milan Šufflay, was Murdered by the Serbian Royal Dictatorship* (Zagreb: Croatian University Academic Association, 1931), 4–5.

[83] Vladimir Ćorović, "Vlasi i mavrovlasi," in Stanoje Stanojević, ed., *Narodna enciklopedija srpsko-hrvatsko-slovenačka*, vol. 4 (Zagreb: Bibliografski zavod D.D., 1929), 1113–1134.

[84] Popović, *O Cincarima*, 12–15, 26, 30–35, 270–283, 300–307.

[85] Popović, *O Cincarima*, 313–480.

[86] Mladen Lorković, *Narod i zemlja Hrvata* (Zagreb: Matica Hrvatska, 1939), 67–73.

[87] Krunoslav Draganović, *Massenübertritte von katoliken zur "orthodoxie" in kroatischen Sprachgebiet zür Zeit der Turkenherrschaft* (Rome: Pontifiscom institutom orientalism studorium, 1937), 7–8, 69.

[88] Ivo Guberina, "La formazione cattolica della Croazia" reprinted in anonymous, *Croazia sacra* (Rome: Tipografija Agostiana, 1943), 23–31.

[89] M.S. [Mile Starčević?], "Srpski apetit," *Nezavisna Hrvatska Država* (24 December 1938), 1.

[90] Lorković, *Narod i zemlja Hrvata*, 37–47.

[91] Lorković, *Narod i zemlja Hrvata*, 3–37.

[92] Lorković, *Narod i zemlja Hrvata*, 159–165.

[93] Lorković, *Narod i zemlja Hrvata*, 106–110, 219–232.

[94] Dragutin Gjurić, "Životni problemi hrvatske nacije," *Hrvatska revija* 13, 3 (1940), 153–156.

[95] See, for example, Kerubin Šegvić, *Die gotische Abstammung der Kroaten* (Berlin: n. p., 1936); Stjepan Buć, *Naši službeni povjesničari i pitanje podrijetla Hrvata (jedno predavanje iz god. 1940)* (Zagreb: Hrvatski državni tiskarski zavod, 1941); and (Anonymous), "K. Šegvić kao 'gotski' Hrvat," *Javnost* 5, 2 (1936), 36, 787.

[96] Stjepan Buć, *Temeljne misli nauke Dra. Ante Starčevića: predavanje održano dne 15 veljače 1936 hrvatskoj sveučilišnoj omladini* (Zagreb: Danica, 1936), 25–31.

[97] Filip Lukas, *Dr Ante Starčević: govor prof. Filip Lukas održano na komemoraciji prigodom 41-godišnjice starčevićeve smrti u dvoranu hrvatskog glazbenog zavoda, dne 28 veljače 1937* (Jastrebarsko: Ivan Lesnik, 1937), 6–7, 17–19, 24–27.

98 (Anonymous), "Zahtjevi nekih dosljenika vukovarskog kotara," *Hrvatski narod* (17 November 1939), 1.

99 M. O., "Vlasi i ne Srbi," *Nezavisna Hrvatska Država* (1 June 1940), 1

100 Luka Grbić, "Još o Srbo-Cincaro-Vlasima," *Nezavisna Hrvatska Država* (4 November 1939), 2.

101 Ćiro Truhelka, "O podrietlu bosanskih Muslimana," *Hrvatska smotra* 7, 7–10 (1934), 249–257.

102 Grga Ereš et al., "Ne damo Bosnu!" *Nezavisna Hrvatska Država* (3 June 1939), 1.

103 Munir Šahinović-Ekremov, "Bosansko-hercegovački muslimani kao komponeta hrvatske državnosti," in Emil Lasić, ed., *Sbornik hrvatske omladine Herceg-Bosne* (Sarajevo: Kranjćević, 1938), 49.

104 Šahinović-Ekremov, "Muslimani u prošlosti i budućnosti hrvatstva," in Lasić, ed, *Sbornik hrvatske omladine Herceg-Bosne*, 27–40.

105 Halid Čausević, "Herceg-Bosna i mi," *Plava revija* 1, 2 (1940), 42–43.

106 H. Bošnjak, "Nacionalna izgradnja Muslimana i muslimanka omladina," *Plava revija* 1, 1 (1940), 27–30.

107 Vladko Maček, "Govor predsjednika Vladka Mačeka hrvatskim seljacima muslimanske vjera prilikom njihovog posjeta Zagrebu, kao glavnom gradu svih Hrvata," in Vladko Maček, *Bit hrvatskoja seljáckoja pokreta* (Zagreb: Zadruga hrvatskihjk seljáckih kulturnich radnika, 1937), 23.

Anthropological Discourse and Eugenics in Interwar Greece

Sevasti Trubeta

During the 1920s, the decade in which anthropology was developed and institutionalized in Greece, the discipline was—as was the case in the rest of Europe—inextricably linked with politics. Moreover, anthropology was connected to the eugenic movement, as well as to population and racial studies. From its inception, Greek anthropology reflected both national ideas and those notions common in the rest of Europe, particularly in France and Germany, which had a significant impact on the evolution of anthropology in Greece.[1]

This chapter will analyze and discuss the context in which anthropology and eugenics emerged and developed in interwar Greece, in addition to considering the most important events in the establishment of anthropology as a discipline, its institutionalization, and its leading proponents. The Greek Anthropological Association (*Ελληνική Ανθρωπολογική Εταιρεία*), and its contribution to the dissemination of anthropological discourse, will be the central focus of this chapter. This association was a unique institution in which science and politics intermingled, and within which concepts of the nation were discussed in relation to physical anthropology and race. Moreover, close attention will be devoted to those members of the Greek Anthropological Association who participated in debates on eugenics.

In Greek, "phili" (*φυλή*), which is usually translated as "race," represents the merger of racial and national ideas. As a term, "phili" was associated with the concept of the nation, but it also had naturalistic, biological, and racial connotations.[2] In the interwar period, the racial implications of "phili" were stressed. On the one hand, this reflected the intellectual trends relating to race prevalent in Europe at the time; however, on the other hand it resulted from attempts to refute theories about Greek racial inferiority and racial impurity. In this sense, anthropology was used to assert the link between modern Greeks and their supposed Hellenic ancestors.

The Emergence of Greek Anthropology

Typical of the emergence of physical anthropology elsewhere in Europe, the interplay of archaeology and medicine proved of primary importance for the development of anthropology in Greece. In short, physicians institutionalized the discipline. However, archaeologists may be considered as scientific precursors to anthropologists because they provided the foundation for anthropological research. The interplay between archaeology and medicine is clearly reflected in the activities of individuals who played a central role in the establishment of anthropology as a scientific discipline in Greece, such as Clon Stephanos (1834–1915) and his successor, Ioannis Koumaris (1879–1970).[3]

After studying medicine in Athens, Stephanos completed his education in Paris with the study *La Grèce au Point de Vue Naturel, Ethnologique, Anthropologique, Démographique et Médical* (Greece from a Natural, Ethnological, Anthropological, Demographic, and Medical Point of View).[4] For the rest of his life Stephanos devoted himself to anthropological and archaeological investigations, including directing excavations and collecting skeletal remains from various sites in Greece. In 1886, he established a Museum of Anthropology at the University of Athens, which he directed until his death in 1915. In the same year, a chair in anthropology was created in the faculty of medicine at the University of Athens in Stephanos's honour, and in recognition of his commitment to scientific research. Stephanos died, however, before he was able to take up the chair. Ioannis Koumaris was promoted to a professorship in his place, although he did not occupy the chair "for political reasons," as he noted in his autobiography.[5] Ten years later, in 1925, another independent chair of physical anthropology was established at the University of Athens.

Koumaris studied medicine, anatomy and surgery in Athens; later he continued his medical education in Berlin (1906–1908) and Paris (1908). He gradually became familiar with physical anthropology, prehistoric anthropology and ethnology in Paris, Brussels, Vienna, Munich, Berlin and Rome.[6] The work of Clon Stephanos had provided the rudiments necessary for establishing and institutionalizing physical anthropology in Greece, while Ioannis Koumaris shaped its development well into the 1970s. Indeed, Koumaris held many positions: he was director of the Museum of Anthropology (1915–1950); he held the first Chair of Physical Anthropology at the University of Athens

(1925–1949); and he was co-founder and director (1922–1928) of the journal *Ιατρική* (Medicine), the publication of the Association of Medical Scientists (*Ιατρική Εταιρεία*). One of his most important achievements was the founding of the Greek Anthropological Association in 1924, of which he became life president.

Koumaris aimed to model Greek anthropology on the latest scientific developments in the rest of Europe. By way of acknowledging the varied character of the discipline, he advocated dividing anthropology into sub-disciplines. His primary motivation was the establishment of different museums for physical anthropology, ethnology and palaeontology, using the material held in the collections of the Museum of Anthropology in Athens as a basis for each sub-discipline. Koumaris also campaigned to rename the existing Museum of Ethnology (*Εθνο–λογικο Μουσείο*) the National and Historical Museum (*Εθνικό και Ιστορικό Μουσείο*) in an attempt to stress the distinct character of these scientific disciplines. The importance he attributed to this transformation is attested by Koumaris's constant collaboration with state institutions, as well as by publicized efforts in both the scientific and popular press.

The Chair of Anthropology at the University of Athens

In his inaugural lecture as professor of anthropology (22 May 1925), Koumaris outlined the purpose of the chair and of anthropology in general.[7] He also expressed his heartfelt respect for his predecessor Stephanos, not merely as the founder of the discipline in Greece, but also as its most important anthropologist. With the exception of Stephanos's work, Koumaris contended, there had been no anthropological studies of the "Hellenic phili."[8] Stephanos's greatest achievement, according to Koumaris, was his accumulation of the basic materials necessary for the establishment of Greek anthropology, and of the Museum of Anthropology, which soon became a leading institution in the field.[9]

Koumaris consistently referred to anthropology as a science meant to serve the Greek nation. Furthermore, he considered the Chair in Physical Anthropology to be a prerequisite for the establishment and development of this discipline in Greece, and, thereby, for the education of a new generation of researchers. In addressing the overlap between archaeology, human medicine and anthropology, Koumaris delineated the latter's tasks and distinguishing characteristics. Crucial to the development of anthropology was its interaction with medicine, despite fundamental differences between the two. Although medicine offered important cognitive and practical tools for anthropology, the latter prevailed over the former by investigating man not as an individual but as a collective, as part of a larger group.[10] In spite of this fundamental difference, both disciplines complemented each other because anatomy formed the basis of physical anthropology, thus providing necessary skills for anthropological exploration.

The new generation of anthropologists Koumaris hoped to train under his direction would be able, after having acquired the necessary skills, "to select somatic information on contemporary Greeks, but also on mores and traditions that have ceased to exist in the course of the civilizing process."[11] The tasks of these new scientists would include fieldwork not only about the "dead," but also about "living compatriots"—those that were "either members of isolated groups or populations living on the borders with other peoples, our current friends or enemies, or perhaps one-time relatives or strangers of our phili. They have to look for elements of ancient civilization and to preserve the remains of our remote ancestors."[12] Koumaris argued that in training anthropologists in this fashion, the Chair of Physical Anthropology would ultimately contribute not only to Greek, but also to international, science.

The Foundation of the Greek Anthropological Association

The founding of the Greek Anthropological Association on 1 June 1924 brought the natural sciences and the humanities under one institutional roof, one later developed into a forum for integrated science and the discussion of European trends. Compared to the rest of Europe, the founding of an anthropological society in Greece occurred relatively late.[13] The Greek Anthropological Association was thus heralded as an institution both filling the existing gap between Greek and European science, and contributing to the modernization of Greek science. Commenting upon this relatively late development, Koumaris claimed that the Greek association, with its forty-eight founding members, was a high-quality institution in comparison with older European associations, which were notably smaller.[14]

Indicative of the auspicious beginnings of the Greek Anthropological Association is the fact that the prime minister of Greece, Alexandros Papanastasiou (1876–1936), presided over its inaugural session. Papanastasiou, co-founder of the association, liberal politician and one of the leading proponents of modernism, was also an influential intellectual and founder of the Sociological Association (*Κοινωνιολογική Εταιρεία*), which, in 1916, became the Association of Political and Social Studies (*Εταιρεία των Κοινωνικών και Πολιτικών Σπουδών*). In his opening speech Papanastasiou remarked that the founding of the Anthropological Association was important not only in scientific but also in national terms: one of its main tasks was to "ascertain the various philetic elements which constitute the modern Greek nation."[15] The association was to conduct systematic anthropological studies so as to provide scientific evidence for proving the philetic continuity of the Greeks "who, though they had always been mixing with other peoples, had never lost their continuity."[16] The Anthropological Association was thus conceived as an institution that would fulfil a public desideratum. Papanastasiou hoped that the Association of Political and Social Studies and the newly founded Anthropological Association would co-operate and complement each other's work.[17]

Anthropology and Medicine

To be sure, anthropology was conceived in broad terms in the 1920s. It was regarded as a universal science, concerned with the "exploration of human nature in its entirety."[18] This state of affairs was reflected in Koumaris's intention, as expressed at the founding session of the Greek Anthropological Association, to bring a wide spectrum of disciplines under the umbrella of anthropology, including "physical" and "mental" anthropology, "human palaeontology," "zoological anthropology," "ethnogeny," "comparative ethnography," "ethnology," "prehistoric anthropology," "linguistics," "religious studies," "ethics," "social psychology," "criminal anthropology," "eugenics," and "racial hygiene." Such an extensive and elaborate conception of anthropology annoyed Papanastasiou, who responded to Koumaris's inaugural speech by suggesting that the Greek Anthropological Association should instead concentrate on its "own field."[19]

Papanastasiou's remark is symptomatic of the then prevailing disagreement about the scope of anthropology, in Greece as well as in
Europe. Koumaris explained his standpoint one year later when he took
up the Chair of Physical Anthropology at the University of Athens.[20]
He began his inaugural speech by outlining the varied definitions of
anthropology and its scope, and suggested that anthropology was situated in the "broader field of biology," thereby constituting a sub-discipline which he called the "biology of the human species." Attributing a
universal character to anthropology, Koumaris defined it as "humanity's
natural history in every time and every place," one able to embrace
those disciplines that did not deal with the individual but were concerned with "human groups (*Hominiden*), their past and future, as well
as collective human acts."[21] Koumaris may well have been influenced
by the definition of anthropology as the "natural history of hominids in
their spatial and temporal expansion" offered by the German anthropologist Rudolf Martin (1864–1925).[22] However, Koumaris criticized
Martin for defining anthropology too narrowly since, as the former
argued, this discipline should not be regarded as limited to the "nature
or morphology of man" but should go beyond these limits in order to
investigate the "psyche, that means the expressions of the intellect."[23]
Koumaris further distinguished between *psychological anthropology*
on the one hand, including ethnology, psycho-sociology, folklore,
religious studies, linguistics and criminal anthropology; and *physical
anthropology* on the other, which encompassed the disciplines of "zoological anthropology, anatomical anthropology and others."[24]

Anthropology, for Koumaris, was a "disinterested science" aiming
at the proliferation of theoretical knowledge and its practical implementation in racial hygiene and eugenics. Eugenics was a matter of
"practical targets" and, as such, it strove for "the correction of the
human species by means of selection and after freeing the world from
superstitions and the consequences of unreasonable traditions."[25]
Koumaris's understanding of race and eugenics followed the anthropological discourse developed in the aftermath of the First World War, in
which anthropology was understood as a discipline allied to population
science (*Bevölkerungswissenschaft*).[26] Koumaris helped popularize the
discipline in so far as he believed that anthropology should be extended beyond the limited academic sphere to broader society. In order to
implement this idea, he disseminated information about anthropology
to the general public by various means, including the press.[27]

The Greek Anthropological Association: Membership and Internal Debates

The Greek Anthropological Association attracted a wide range of intellectuals and other prominent figures. Founding members included the diplomat, politician and intellectual, Phillipos Dragoumis (1890–1980); the left-wing intellectual and pedagogue, Dimitris Glinos (1882–1943); and the Minister of Education, Ioannis Lymperopoulos. Those on the honorary membership list included August Rutot (Brussels); Georges Papillault (Paris); André Forster (Strasbourg); Carl M. Fürst (Lund), and Prince Peter of Greece and Denmark. In 1928, the chairman of the Austrian Speleological Society for the Investigation of Caverns, Adalbert Markovits (Markovich) (1897–1941), joined the association. On Koumaris's recommendation, Ilse Schwidetzky (1907–1997) also became a member, officially joining at the session of 7 April 1937 while visiting the Museum of Anthropology in Athens.

With the exception of Markovits, who participated actively in association sessions by giving several lectures,[28] the members mentioned above only appeared in the membership list by name and did not contribute to debates within the association. Although the membership of the association was composed predominantly of physicians and natural scientists, social scientists and scholars from the humanities also joined the association. Its prominent members included Byzantine specialist Konstantinos Amantos (1870–1960)—recorded in the membership list, however, as an "ethnologist"[29]; Phaidon Koukoules (1881–1956),[30] the linguist and editor of the Historical Dictionary of Greek Language, Nikolaos P. Andriotis (1906–1976)[31]; and professor of criminology at the University of Athens, Konstantinos Gardikas (1896–1984). Archaeological and other relevant scientific issues were often discussed in the sessions, as illustrated by the presentations given by Georgios Mylonas (1898–1987)—who later became professor of archaeology in the US—and G. Pournaropoulos (1908–1992), a physician concerned with the relationship between archaeology and anthropology.

Many of the physicians involved in the Greek Anthropological Association were affiliated with state and educational institutions and exerted a strong influence on politics. Some of the most influential of these physicians included Emmanuel Lampadarios (1882–1943),[32] professor of pediatrics and, after 1911, head of the Office for the School of Hygiene (founded in 1908) at the Ministry of Education; Konstantinos

Savvas (1861–1929), the first professor of hygiene and microbiology at the University of Athens; and the anatomist Giorgos Sklavounos (1869–1954). Notwithstanding these other luminaries, until he died in 1970 Koumaris remained the central figure in the Greek Anthropological Association. His death marked the end of an era for anthropology in Greece, and, above all, for the Greek Anthropological Association.[33]

Eugenics and Race

Debates within the Greek Anthropological Association reflected, to a great extent, the state of affairs in anthropological discourse in the rest of Europe, since most of the Association's members were affiliated with scientific institutes abroad. Matters related to race, eugenics and population policies discussed under the heading "anthropology" thus became central to Greek scientific discourse at the time.

The matter of eugenics had been raised and debated by Stavros Zurukzoglu (1896–1966) at the first session of the Greek Anthropological Association in May 1925.[34] Zurukzoglu was born in Smyrna in 1896 and studied medicine in Berlin, Berne, Geneva and Munich. In 1927, he submitted his doctoral dissertation on bacteriology and hygiene, and qualified as a university lecturer in Berne. His research focused on eugenics, population studies, alcoholism and criminal biology.[35] In 1938, the Swiss government appointed him director of the Department for Statistics at the Swiss Ministry of the Interior, praising his contribution to the enactment of legislation banning the consumption of alcohol.[36]

The lecture on eugenics delivered by Zurukzoglu in 1925 was extracted from his book *Biologische Probleme der Rassenhygiene und die Kulturvölker* (Biological Problems of Racial Hygiene and the Civilized Peoples).[37] The lecture was an extensive introduction to eugenics, and in it Zurukzoglu addressed the origins and development of the discipline in addition to surveying disputes between evolutionists and eugenicists. Clearly Zurukzoglu sided with the latter group. Emphasizing the emancipation of eugenics from its original connection with Darwinism and evolutionism, he advocated treating it as an independent science. Zurukzoglu lamented the lack of interest in eugenics in Greece by the state and indigenous scholars, accusing both of failing to realize the necessity of embracing and implementing eugenics.

Zurukzoglu also outlined the existing relationship between social hygiene and eugenics, arguing for the superiority of the latter on account of its conception of people in terms of collectives rather than as individuals. He further identified the fundamental difference between hygiene and eugenics: while the former protected and promoted every individual, the latter sought to anticipate the proliferation of "hereditarily negative elements that counteracts the protection they enjoy from individual hygiene."[38] Zurukzoglu believed that eugenics, in practice, involved the promotion of physically and mentally capable people in a "teleological attempt to maintain the nation, humanity and civilization."[39] However, given the uneasy relationship between hygiene and eugenics, any first step in applying eugenic measures should be undertaken, Zurukzoglu argued, through a combined implementation of individual and social hygiene.

Zurukzoglu then discussed theories of race and, in particular, the attempts by European theorists of race to discredit the "Greek phili." Using Rudolf Martin's ideas, Zurukzoglu tried to refute Arthur de Gobineau's assertion that the "dolichocephalic blond race" was superior, by arguing that every advanced race has developed a civilization, and therefore currently primitive races would be able to do so in the future.[40] Besides, Zurukzoglu argued, the "blond race" was well represented in Greece and equaled its counterparts in southern Germany, central France, and Italy. A possible temporary deficit in "superior intellects" did not indicate degeneration (or "genocide," as he termed it), because the nation was able to generate "superior intellects" if given optimal conditions for doing so.[41] Such circumstances were held to apply in the case of the Greek nation, and Zurukzoglu's argument was based on the presupposition that there existed a "philetic core"—a view widespread among Greek intellectuals in the interwar period. According to this theory, a culturally or intellectually superior race (phili) is able to assimilate alien "philetic elements," while retaining a stable "philetic core."[42]

A popular version of Zurukzoglu's lecture appeared in the Greek journal *Υγεία* (Health).[43] Unlike the lecture, the article focused on eugenic measures rather than racial issues, and included the additional chapter "Practical Applications" (*Πρακτικόν Μέρος*).[44] This chapter detailed proposals for applying eugenic measures in Greece in order to combat epidemics, endemics, and infant mortality. However, in order to achieve these aims, eugenics needed to be supplemented by "individual hygiene." Zurukzoglu suggested the internment of "damaging individuals" in special institutions where "they could be dealt with in a productive way."[45] He was also in favor of the sterilization of "hereditarily degenerate individuals" in principle, provided that they or their relatives agreed to the measure. Yet Zurukzoglu was against forced sterilization and regarded such a measure as unrealistic for "scientific as well as social reasons." With respect to marriage between "degenerate individuals," he suggested the "voluntary abstinence from childbearing" rather than the prohibition of marriage. The reduction of both hereditary and environmental degeneration, and the promotion of social hygiene, required raising awareness of the importance of hygiene among working people, which would have an immediate effect in tackling venereal disease, tuberculosis, malaria, and alcoholism—even though the latter "had not yet reached Greece."[46]

Zurukzoglu's lecture delivered to the Greek Anthropological Association was a detailed introduction to eugenics, but he was not the first to deal with this topic in Greece. At the beginning of the 1920s, another physician, Moisis Moisidis, wrote extensively on eugenics in a series of publications,[47] most notably the 1922 *Ευγονική και Γάμος* (Eugenics and Marriage),[48] and his 1925 *Ευγονική και Παιδοκομία παρά τοις Αρχαίοις Ελλήσιν* (Eugenics and Childbearing among the Ancient Hellenes).[49] The latter was based on an award-winning study acknowledged a year earlier by the Medical Society (*Ιατρική Εταιρεία*). Nevertheless, in his review of the book Zurukzoglu criticized Moisidis for blurring the distinction between hygiene and eugenics.[5] He also refuted Moisidis's claim that the origins of eugenics could be traced back to Hellenic Antiquity and, particularly, to Sparta, where eugenics was not an idea but a practical measure. In this respect, Zurukzoglu's position was revisionist and exceptional: most Greeks involved in this debate claimed that the idea of eugenics was first formulated and implemented in Ancient Greece.

One of those who accepted this perspective was Koumaris, as demonstrated in several of his publications.[51] Moreover, Koumaris stressed the need to reintroduce eugenics in modern Greece. In the article "A National Issue: On Eugenics," published in 1931 in the newspaper *Εστία* (Vesta), Koumaris raised questions about the practical implementation of eugenics and suggested the introduction of marriage certificates by the state. Young couples could be enticed to undergo eugenic testing by means of material benefits or, conversely, testing could be facilitated by the threat of punishment for those that did not volunteer.[52]

Eugenics was also addressed from a criminal-psychological perspective. For instance, at a session held on 22 December 1932 at the Greek Anthropological Association, the psychologist Georgios Sakellariou considered eugenics in relation to criminology and psychology.[53] After highlighting a wide variety of crimes, the role played by the "mental capability" of criminals, and the distinction between hereditary and environmentally determined criminality, Sakellariou suggested ways to combat crime. He advocated legal reform and the improvement of the Greek correctional system rather than the introduction of eugenic measures like sterilization.

After the 1933 Nazi seizure of power in Germany and the subsequent enactment of "race laws" and legislation concerning compulsory sterilization, the discussion about eugenics in Greece intensified. Debates in the press and scientific journals involved physicians, natural scientists, intellectuals and the clergy. Generally, the Greek debate followed existing developments in Western Europe; however, it was largely characterized by skepticism regarding the implementation of eugenic measures due to Greece's "backwardness" in terms of social care, the welfare state, and scientific institutions. The central argument in this debate was that the necessary infrastructure existing in other European countries was only just evolving in Greece.[54]

Even though opponents of eugenics and sterilization were not absent,[55] most participants in the debate confined their criticism to compulsory sterilization and embraced the idea of eugenics and voluntary sterilization.[56] One of the most prominent advocates of compulsory sterilization was the physician Konstantinos Moutousis, professor of hygiene at the University of Athens.[57] In his inaugural lecture on 9 November 1933, he argued that the implementation of eugenics by the state would result in the improvement and perfection of its "human material," and contribute to the maintenance of humanity as a whole.[58] Positive effects of "eugenic hygiene" included "the reduction of costs of the ill or infirm," as well as "epileptics," the "mental a-normal," "psychopaths," "criminals," and other "disabled individuals."[59] Like many of his contemporaries, Moutousis believed that eugenics originated in Hellenic Antiquity and that this was evidenced in the works of Plato and Aristotle. In its practical implementation—which ensured the creation of a modern state (*Πολιτεία*)—eugenics was to be guided by a combination of ancient Hellenic ideals, Christian principles, and current achievements of the science.[60] Employing Christian principles in this fashion was not, however, a persuasive argument as far as the church was concerned.[61]

Publications and statements from the judicial sphere were also rampant. Konstantinos Gardikas (1896–1984), for instance, was professor of criminology at the University of Athens and an active member of the Greek Anthropological Association.[62] In 1934 he published the article "Eugenik im alten Griechenland" (Eugenics in Ancient Greece) in the *Monatschrift für Kriminalpsychologie und Strafrechtsreform* (Criminal Psychology and Criminal Legal Reform Monthly).[63] In the article, Gardikas argued that eugenics originated in Greece, tracing it from the "heroic age" to Aristotle without, however, explicitly referring to the contemporary debate about eugenics or the issue of sterilization.

A further discussion about the implications of eugenics appeared in 1934, in *Καθημερινή* (Daily Newspaper), under the title "The Problem of the Greek Correctional System and Hitler's Sterilization Law."[64] The article rejected sterilization on the grounds that, although sterilization measures might be feasible in a country with a large population and developed programs of social hygiene, it was not possible in Greece.[65] The author of that article also considered that the modernization of the correctional system and the education of the personnel employed in the correctional institutions, were fundamental preconditions for the reception and implementation of eugenic measures—especially the sterilization of criminals in Greece.[66]

It was in the context of this intensified debate on eugenics that Moisis Moisidis's book *Eugenic Sterilization: Principles–Methods–Application* (*Ευγονική Αποστείρωσις. Αρχαί–Μεθόδοι–Εφαρμογή*), appeared in 1934—after the passing of the sterilization law in Germany—providing an account of the international debate about sterilization.[67] Moisidis also provided a comprehensive overview of the discussion about eugenics in Greece, its opponents and advocates, and the standpoints of each. His attitude towards sterilization was ambiguous, for although he accepted the core principles of eugenics he was nevertheless unable to side with one of the parties involved in the debate without "running the risk of confronting the others."[68] Despite his initial skepticism, Moisidis sided with the "moderate eugenicists,"[69] advocating voluntary and remedial sterilization with the assent of the persons involved, on the condition that it would contribute to combating degeneration and protecting the "phili" (race).[70]

The immediate connection between eugenics and racial issues was one of Koumaris's central concerns. For example, he opposed racial mixing.[71] Typical examples of Koumaris's convictions about race and eugenics are statements made in debates at the Greek Anthropological Association (most notably in the session on 14 May 1939) and an article he wrote in 1939 for the National Socialist journal *Ziel und Weg*, under the title "Rasse und Gesundheit" (Race and Health).[72]

Given his acceptance of German racial science on the one hand, and his self-identification as a Greek nationalist on the other, Koumaris found himself in a difficult position of conflicting loyalties. His conviction that "racial purity" was an ethical value, an ideal rather than an achievable goal, may have served as a compromise solution to his divided loyalties.[73] Thus, even though Koumaris accepted the central tenets of contemporary German racial discourse, he strove to refute assertions about the inferiority of the "Greek race" put forward by contemporary German racial theorists. In doing so, he was critical of theories about the superiority of the "Nordic race," replacing the notion of "racial purity" with his own idea of the "fluid stability" ($\rho\acute{\varepsilon}o\nu\sigma\alpha$ $\sigma\tau\alpha\theta\varepsilon\rho\acute{o}\tau\eta\tau\alpha$) of race; that is, its capacity to assimilate other races while remaining stable at its "core," despite external modifications. Koumaris remained convinced that the application of preliminary eugenic measures in Greece, such as racial cleansing of hereditary illness, was necessary. Moreover, in order to legitimize these measures it was necessary to indoctrinate the populace along the lines of the German model. Yet, this was a preliminary "half measure," to be followed by more intense measures in the future.

In "Το πρόβλημα της φυλής" (The Problem of Race), a lecture delivered to the Greek Anthropological Association in May 1939, Koumaris also addressed the contemporary debates on the "Jewish race" and the persecution of Jews. Although he expressed his understanding of efforts to avoid racial mixture with Jews, he was nevertheless surprised not by the rejection of Jews as such—according to him this a common practice in any historical period—but rather by the "intensity" of contemporary measures against the Jews. Koumaris critically compared modern persecution of Jews with "medieval religious fanaticism," and claimed that the latter should under no circumstances be replaced by "racial fanaticism."[74]

Conclusions

This investigation of the development of Greek anthropology and eugenics in the interwar period reveals that both scientific disciplines were dominated by national concerns. The national community was supposedly bound together by primordial and civil ties. The implications of this hypothesis became all the more obvious through attempts to harness anthropology with means by which to modernize Greek science and, more importantly, the Greek state and society. Consistent with developments elsewhere in Europe, Greek scholars involved in anthropological discourse endeavored to establish a body of ideas that would assist political discourse, popularize science, and raise popular awareness of "social and racial hygiene." The ultimate goal was the maintenance and gradual advancement of the nation, or race (phili). The merger of primordial and civil concepts in this definition of the Greek nation was facilitated by eugenics and its connection to the "cult of antiquity," according to which the origins of eugenics were to be found in ancient Hellenic culture.

As far as conclusions are possible on the basis of current research, in the Greek case racial politics and eugenics were never fully developed or implemented and, therefore, remained restricted to intellectual and scholarly spheres. This was due to several factors, including, most notably, a non-existent welfare state, and the divide between scientists and society at large. Ioannis Koumaris's aim to popularize anthropology and eugenics as sciences remained unfulfilled.

Endnotes:

[1] See Kostas Krimpas, *Θραύσματα Κατόπτρον* (Athens: Themelio, 1993).

[2] The meaning of the term "phili" (*Φυλή*) has been transformed over time, from the time of its origin in Hellenic Antiquity, through to the time of the Ottoman Empire, when it was adapted to the twentieth century. It was useful because it could be employed to describe large communities, potentially including aspects central to the definition of an "imagined community." In church texts of the eighteenth century "phili" was used in reference to all Orthodox Christians in the Ottoman Empire. Later, the term was used to describe distinct religious and linguistic communities within the Empire and was enriched by national connotations. Indicative of this transformation is the usage of the term "philetismos" (*Φυλετισμός*) by the patriarch of Constantinople in reference to the Bulgarian Schism. See Paraskevas Matalas, *Έθνος και ορθοδοξία. Οι περιπέτειες μιας σχέσης. Από το Ελλαδικό στο Βουλγαρικό σχίσμα.* (Herakleion: Panepistimiakes Ekdoseis Kritis, 2002). By the end of the eighteenth century, the term had acquired additional racial connotations. This was due to a number of reasons, including the increased collaboration between Greek and European scholars and the rising significance of race in European scientific discourse. On the intellectual popularity of the term *phili* in the first half of the twentieth century, see Dimitris Tziovas, *Οι μεταμορφώσεις του Εθνισμού και το ιδεολόγημα της ελληνικότητας στο μεσοπόλεμο* (Athens: Odysseas, 1989); and Mario Vitti, *Η γενιά τον Τριάντα* (Athens: Hermes, 2000).

[3] For a short overview on the emergence and development of physical anthropology in Greece, see Theodoros Pitsios, "Εκλαϊκευτικό και ερευνητικό περιεχόμενο της φυσικής ανθρωπολογίας. (Ανθρώπινη Βιολογία)," in *Ανθρωπολογία (Δελτίο της Ανθρωπολογικής Εταιρείας των Φίλων του Ανθρωπολογικής Μουσείου)* 1 (1993), 33–41.

[4] Clon Stephanos, *La Grèce au Point de Vue Naturel, Ethnologique, Anthropologique, Démographique et Médical* (Paris: G. Masson, 1884).

[5] Ioannis Koumaris, *50ετηρίς*, part I (Athens: n.p., 1951), 12.

[6] Koumaris, *50ετηρίς*, 11–12.

[7] The text of this inaugural lecture was published in the journal of the Greek Medical Association, *Ιατρική*. See Ioannis Koumaris, "Ανθρωπολογία ως πανεπιστημιακόν μάθημα," *Ιατρική – Ελληνική Ιατρική Επιθεώρησις* 3, 3 (1925), 71–77; and *Ιατρική – Ελληνική Ιατρική Επιθεώρησις* 3, 4 (1925), 99–105.

[8] Koumaris, "Η Ανθρωπολογία ως πανεπιστημιακόν μάθημα," 103.

[9] Koumaris, "Η Ανθρωπολογία ως πανεπιστημιακόν μάθημα," 101–102.

[10] Koumaris, "Η Ανθρωπολογία ως πανεπιστημιακόν μάθημα," 99.

[11] Koumaris, "Η Ανθρωπολογία ως πανεπιστημιακόν μάθημα," 104.

[12] Koumaris, "Η Ανθρωπολογία ως πανεπιστημιακόν μάθημα," 104.

[13] For an account of the founding of anthropological associations in different countries, see Wilhelm E. Mühlmann, *Geschichte der Anthropologie* (Wiesbaden: Sammlung Aula, 1986), 96.

[14] Koumaris, "Η Ανθρωπολογία ως πανεπιστημιακόν μάθημα," 104.

[15] See the opening lecture by Alexandros Papanastasiou in the minutes of the founding session of 1 June 1924. *Ελληνική Ανθρωπολογική Εταιρεία, Founding Session* (1924), 5–6.

[16] *Ελληνική Ανθρωπολογική Εταιρεία, Founding Session* (1924), 6.

[17] Papanastasiou suggested that the Association of Political and Social Studies assign its conference hall to the Greek Anthropological Association. However this never occurred: the minutes of the meetings reveal that the Anthropological Association held most of its sessions either in the hall of the Archaeological Society (an honorary member of the Anthropological Association since 1924) or at the medical faculty of the University of Athens.

[18] Mühlmann, *Geschichte der Anthropologie*, 162. See also Andreas Lüddecke, *Rassen, Schädel und Gelehrte. Zur politischen Funktionalität der anthropologischen Forschung und Lehre in der Tradition Egon von Eickstedts* (Franfurt am Main: Peter Lang, 2000), 25–49.

[19] *Ελληνική Ανθρωπολογική Εταιρεία, Founding Session*, (1924), 15.

[20] Koumaris, "Η Ανθρωπολογία ως πανεπιστημιακόν μάθημα," 22.

[21] Koumaris, "Η Ανθρωπολογία ως πανεπιστημιακόν μάθημα," 22.

[22] Rudolf Martin, *Lehrbuch der Anthropologie in systematischer Darstellung* (Jena: Verlag von Gustav Fischer, 1914), 1.

[23] Koumaris, "Η Ανθρωπολογία ως πανεπιστημιακόν μάθημα," 76.

[24] Koumaris, "Η Ανθρωπολογία ως πανεπιστημιακόν μάθημα," 76.

[25] Koumaris, "Η Ανθρωπολογία ως πανεπιστημιακόν μάθημα," 100.

[26] Lüddecke, *Rassen, Schädel und Gelehrte*, 38.

[27] See a series of instalments of articles by Koumaris in the newspaper *Βραδυνή*, such as Ioannis Koumaris, "'Ανθρωπολογία και Θρησκεία," *Βραδυνή* (1932), 1, 3; Ioannis Koumaris, "Μεντελισμός και Ανθρωπότης," *Βραδυνή*, part 1 (10 March 1937), 1; part II (11 March 1937), 3; part III (12 March 1937), 1, 7; part IV (13 March 1937), 1, 3; and part V (16 March 1937), 1, 7. After 1940, Koumaris published numerous articles in the journal *Ήλιος*.

[28] Markovits's lectures were translated into Greek as "Από τα σκότη των σπηλαίων," minutes of the meeting, 24 April 1928, *Ελληνική Ανθρωπολογική Εταιρεία Πρακτικά Συνεδρίων του Έτους*, (1928), 45–61; "Περί των μέχρι σήμερα ερευνών επί της λιθικής περιόδου της Ελλάδος," minutes of the meeting, 3 December 1929, *Ελληνική Ανθρωπολογική Εταιρεία Πρακτικά Συνεδρίων του Έτους* (1929), 114–134; "Περί της βιομηχανίας του πυρίτου κατά τα τελευταία 500 έτη," minutes of the meeting, 29 November 1930, *Ελληνική Ανθρωπολογική Εταιρεία Πρακτικά Συνεδρίων του Έτους* (1929), 44–50; "Μικρόλιθοι μετ' ακάνθης," minutes of the meeting, 30 December 1931, *Ελληνική Ανθρωπολογική Εταιρεία Πρακτικά Συνεδρίων του Έτους*, (1931), 115–124.

[29] Konstantinos Amantos delivered a speech at the session held on 22 February 1926, "Σλάβοι και Σλαβόφωνοι εις τας ελληνικάς χώρας." See minutes of the meeting, *Ελληνική Ανθρωπολογική Εταιρεία Πρακτικά Συνεδρίων του Έτους* (1926), 10–31.

30 Phaidon Koukoules, "Από την μεσαιωνικήν διαπόμπευσιν," minutes of the meeting, 12 April 1927, *Ελληνική Ανθρωπολογική Εταιρεία, Πρακτικά Συνεδρίων του Έτους* (1927), 59–68.

31 Nikolaos Andriotis, "Προϊστορικοί τάφοι εν Σαμοθράκη," minutes of the meeting, 12 March 1929, *Ελληνική Ανθρωπολογική Εταιρεία, Πρακτικά Συνεδρίων του Έτους* (1929), 54–63.

32 Emmanuel Lampadarios, "Η σωματική εξέλιξις του Έλληνος μαθητού (ανθρωπολογική αυξησιολογία)," minutes of the meeting, 21 February 1928, *Ελληνική Ανθρωπολογική Εταιρεία, Πρακτικά Συνεδρίων του Έτους* (1929), 19–37.

33 After Koumaris died in 1970, the Greek Anthropological Association disintegrated, although it was soon revived. Several of its members, however, joined Aris Poulianos in founding another association in 1970, the Anthropological Society of Greece (*Ανθρωπολογική Εταιρεία Ελλάδος*).

34 Stavros Zurukzoglu, "Περί, ευγονίας," minutes of the meeting, 11 May 1925, *Ελληνική Ανθρωπολογική Εταιρεία, Πρακτικά Συνεδρίων του Έτους* (1925), 14–40. Parts of this lecture were published in three instalments: *Ιατρική* 3, 2 (June 1925), 66–69; *Ιατρική* 3, 3 (June 1925), 93–94; *Ιατρική* 3, 4 (June 1925), 122–128.

35 See Isidor Fischer, ed., *Biographisches Lexikon der hervorragenden Ärzte der letzten fünfzig Jahre* vol. 2 (Berlin: Urban & Schwarzenberg, 1933), 1732.

36 According to the Greek encyclopaedia, *Νεώτερον Εγκυκλοπαιδικόν Λεξικόν Ήλιος*, vol. 17 (Athens: Helios, n.d), 933, Stavros Zurukzoglu was the editor of a series on alcoholism and its social consequences. See also Stephanos Zurukzoglu, ed., *Die Alkoholfrage in der Schweiz* (Basel: Benno Schwabe Verlag, 1935), 8. In the introduction it is mentioned that Zurukzoglu helped in the drawing up of the law banning the consumption of alcohol.

37 See Stavros Zurukzoglu, *Biologische Probleme der Rassenhygiene und der Kulturvölker* (Munich: Verlag von J. F. Bergmann, 1925). A review of Tsakalotos's book appeared in the medical journal *Ιατρική* 3, 9 (1926), 267–268.

38 Zurukzoglu, "Περί ευγονίας," 16.

39 Zurukzoglu, "Περί ευγονίας," 16.

40 In the Greek language there is no distinction between culture and civilization.

41 Zurukzoglu, "Περί ευγονίας," 37.

42 This argument was applied not only by Greeks but also by philhellenists in order to refute the theory that there was no racial continuity between ancient Hellenes and contemporary Greeks. See H. Rübel, "Rassenkräfte in der hellenischen Geschichte," *Volk und Rasse* 17, 10 (1942), 169–175; Roland Hampe, "Griechenland, das Land der Gegensätze," *Volk und Rasse* 16, 7–8 (1941), 117–121.

43 This article, "Η Ευγονία," was published in three instalments in the journal: *Υγεία* 11 (1 June 1925), 221–124; 12 (15 June 1925), 246–148; 13 (29 June 1925), 265–267.

44 *Υγεία* 13, 265–267.

⁴⁵ *Υγεία* 13, 266.

⁴⁶ The psychiatrist Symeon Vlavianos opposed this view some years later. Alcohol was, for Vlavianos, a central issue that had to be dealt with, as argued in the newspaper *Ιατρική Εφημερίς*, of which he was the publisher. See Simeon Vlavianos, "*Ανταλκοολικός Αγών,*" *Ιατρική Εφημερίς* (1931), 2; and Simeon Vlavianos, "Κατά του Αλκοολισμού, Ο Αλκοολισμός εν Ελλάδι" *Ιατρική Εφημερίς*. (1934), 1–4. However, Vlavianos was a passive member of the Greek Anthropological Association. On his contribution to the development of psychiatry in Greece, see Panagiota Kazolea-Tavoulari, *Η ιστορία της Ψυχολογίας στην Ελλάδα (1880–1987)* (Athens: Ellinika Grammata, 2002), 93.

⁴⁷ Moisis Moisidis was co-editor of *Υγεία*, a journal of "practical medicine" that was supported by the Greek Red Cross and that aimed to reach "all social classes." He joined the Greek Anthropological Society in 1928.

⁴⁸ Moisis Moisidis, *Ευγονική και γάμος* (Athens, n.p. 1922).

⁴⁹ Moisis Moisidis, *Ευγονική και Παιδοκομία παρά τοις αρχαίοις Έλλήσιν* (Athens: n.p., 1925).

⁵⁰ The review, by Stavros Zurukzoglu, appeared in *Ιατρική* 3, 2 (1925), 32–33

⁵¹ Ioannis Koumaris, "Ενα εθνικόν ζήτημα. Δια την ευγονίαν" (1931), 1; Ioannis Koumaris, "Rasse und Gesundheit," *Ziel und Weg* 9, 12 (1939), 386–388.

⁵² Ioannis Koumaris, "Ενα εθνικόν ζήτημα. Δια την ευγονίαν," *Εστία* (1931), 1.

⁵³ Georgios Sakellariou, "Περί του εγκλήματος εν Ελλάδι από ψυχολογικής," minutes of the meeting, 27 January 1932. *Ελληνική Ανθρωπολογική Εταιρεία, Πρακτικά Συνεδρίων του Έτους* (1932), 3–24. See also Georgios Sakellariou, *Το έγκλημα εν Ελλάδι: Μελέτη ψυχολογική και κοινωνική* (Athens: n.p. 1932).

⁵⁴ See Antonis Liakos, *Εργαδία και πολιτική στην Ελλάδα του Μεσοπολέμου. Το Διεθνές Γραφείο Εργασίας η ανάδυση των κοινωνικών θεσμών* (Athens: Idryma Erevnas kai Paideias tis Emporikis Trapezas tis Ellados, 1993).

⁵⁵ An example of an important scholar who rejected eugenics was Dimorthenis Eleftheriadis (1885–1964). Eleftheriadis was professor of social biology at the Panteion Institute for Political Science between 1937 and 1943. The Panteion Institute was founded in 1930 as an institution for training civil servants.

⁵⁶ See Simeon Vlavianos, "Ο νέος νόμος του Χίτλερ δια την στείρωσιν των μεταδοτικών νόσων κληρονομικώς," *Ιατρική Εφημερίς* (1933), 4.

⁵⁷ Moutousis joined the Greek Anthropological Association in 1927, but he was not an active member.

⁵⁸ His speech was published a couple of days later in the medical review *Κλινική* under the title "Η Υγιεινή και η ευγονία εις τας συγχρόνους Κοινωνίας," *Κλινική* 47 (1933), 867.

59 Moutousis, "Η Υγιεινή και η ευγονία εις τας συγχρόνους Κοινωνίας," 867.

60 Moutousis, "Η Υγιεινή και η ευγονία εις τας συγχρόνους Κοινωνίας," 873.

61 See the reaction published in the journal *Εκκλησία* (Church): (Anonymous), "Η αναγκαστική στείρωσις," *Εκκλησία* 47 (1933), 371.

62 Konstantinos Gardikas was involved in state administration, particularly in the Ministry of the Interior, and oversaw the organization of the police. He joined the Greek Anthropological Association in 1930. By 1936 he had delivered four lectures: "Η επίδρασις της ώρας του έτους και του κλίματος επί της εν Ελλάδι εγκληματικότητος," minutes of the meeting, 28 December 1933. *Επηνική Ανθρωπολογική Εταιρεία, Πρακτικά Συνεδρίων τον Έτους* (1933), 28–33; "Η συχνότης των διαφόρων τύπων των δακτυλικών αποτυπωμάτων εν Ελλάδι," minutes of the meeting, 28 December 1933, *Ελληνική Ανθρωπολογική Εταιρεία, Πρακτικά Συνεδρίων τον Έτους* (1933), 33–35; "Επάγγελμα και εγκληματικότης," minutes of the meeting, 7 December 1934, *Ελληνική Ανθρωπολογική Εταιρεία, Πρακτικά Συνεδριών του Έτους* (1934), 36–47; "Η εγκληματικότης εν Ελλάδι κατά τόπους," minutes of the meeting, 23 May 1936 *Ελληνική Ανθρωπολογική Εταιρεία, ΠρακτικάΣυνεδρίων του Έτους* (1936), 21–41. Two of these lectures were published in a German journal of criminology. See Konstantinos Gardikas, "Beruf und Kriminalität in Griechenland," *Monatschrift für Kriminalpsychologie und Strafrechtsreform* 25, 10 and 12 (1934), 549–557; and Konstantinos Gardikas, "Über den Einfluß der Jahreszeiten auf die Kriminalität in Griechenland," *Monatschrift für Kriminalpsychologie und Strafrechtsreform* 25, 1 (1934), 24–29.

63 Konstantinos Gardikas, "Eugenik im alten Griechenland," *Monatschrift für Kriminalpsychologie und Strafrechtsreform* 25, 6 (1934), 340–346.

64 Giorgos Ap. Katopodis, "Το ελληνικόν σωφρονιστικόν πρόβλημα και ο χιτλερικός νόμος της στειρώσεως," *Καθημερινή* (1934), 1.

65 Katopodis, "Το ελληνικόν σωφρονιστικόν πρόβλημα και ο χιτλερικός νομός της στειρώσεως," 1.

66 This article reflected the contemporary state of affairs in the Greek judicial system as well as in the studies of criminology in which a series of reforms were announced. See Konstantinos Gardikas, "Ein griechischer Fortbildungskurs der Strafrechtswissenschaft," *Monatsschrift für Kriminalpsychologie und Strafreform* 25, 1, (1934), 47–48. Gardikas reported that, among other Greek and foreign scientists, Professor Aschaffenburg was invited to give a lecture on "Die Bedeutung der Psychopathen für die Gesellschaft und das Recht, Anlage und Umwelt, Sterilization." The precarious situation of prisoners in Greece and the reforms in progress were reported also by Dimitris Karanikas, "Berichte aus Griechenland," *Monatsschrift für Kriminalpsychologie und Strafreform* 25, 2–3 (1934), 153–154.

67 Moisis Moisidis, *Ευγονική Αποστείρωσις. Αρχαί - Μεθόδοι–Εφαρμογή* (Athens and Alexandria: Publishing House A. Kasigoni, 1934); and Moisis Moisidis, *Ο μαθουλσιανισμός άλλοτε και νύν* (Athens: n.p. 1932).

The latter book was honoured with an award by the Medical Association in 1931.

[68] Moisidis, *Ευγονική Αποστείρωσις*, 19.

[69] Moisidis, *Ευγονική Αποστείρωσις*, 19.

[70] Moisidis, *Ευγονική Αποστείρωσις*, 67, 68.

[71] As he put it: "Wir sind gegen die Geschlechtsverbindung zwischen andersrassigen Menschen." In Koumaris, "Rasse und Gesundheit," 386, 387. See also Koumaris, "Το πρόβλημα της φυλής," minutes of the meeting, 14 May 1939, *Ελληνική Ανθρωπολογική Εταιρεία, Πρακτικά Συνεδρίων του Έτους* (1939), 10–26, 17.

[72] Koumaris, "Rasse und Gesundheit." In his memoirs, Koumaris claimed that the journal *Ziel und Weg* reprinted the article without his permission. See Koumaris, *50ετηρίς*, 87.

[73] In his own words: "ob die Reinheit und völlige Isolierung der Rasse zu ihrer Verbesserung wirklich beiträgt u.a.m., das sind Fragen, die immer noch diskutiert werden, und auf die eine endgültige Antwort leider noch nicht möglich ist." See Koumaris, "Rasse und Gesundheit," 386.

[74] Koumaris, "Το πρόβλημα της φυλής," 24. The question of whether Jewish and Greek races are related was discussed at length at the session of October 1943 on the occasion of a special lecture on "The History of the Jewish Community of Tripoli in the Peloponnese" delivered by Nikos A. Bees. See Nikos Bees, "Εκ της Ιστορίας της Ιουδαϊκής Κοινότητος της Τριπόλεως της Πελοποννήσου," minutes of the meeting, 8 October 1943, *Ελληνική Ανθρωπολογική Εταιρεία, Πρακτικά Συνεδρίων του Έτους* (1943), 50–59, discussion 59–65. In this session, both Bees and Koumaris claimed that in Greece racial mixing between Jews and Greeks had never occurred. Georgios Sklavounos, however, insisted that such mixing of blood was possible and not condemnable because there are no "pure races" in biological or genetic terms.

Part II

Eugenics and Racial Hygiene in National Contexts

Eugenics, Social Genetics and Racial Hygiene: Plans for the Scientific Regulation of Human Heredity in the Czech Lands, 1900–1925

Michal Šimůnek

Contemporary interest in the history of eugenics is not only reflected in current discussions on the "new eugenics," "neo-eugenics" or "back-door-eugenics" but also at the political and ideological level.[1] Today it is clear that, when assessing the history of eugenics, it is necessary to build on the existence of multiple parallel eugenic movements as well as several modes of eugenic thinking, and that an important aspect of the current research focuses on a deeper analysis of their mutual inter-actions. In *Toward a Comparative History of Eugenics* (1990), Mark Adams identified six dimensions of the historical development of eugenics needing to be further explored. The first is "scientific"; the second "disciplinary"; the third "professional"; the fourth "institution-al"; the fifth "popular and pedagogical"; and the sixth "ideological and political."[2] With these criteria in mind, this chapter discusses the adop-tion of eugenics by Czech intellectuals at the turn of the twentieth cen-tury. I shall not, however, analyse the adoption of eugenics among German-speaking intellectuals in the Czech lands. This is a topic that still requires substantial research.[3]

Local Sources and External Influences (1900–1918)

Like intellectuals elsewhere in Western and East-Central Europe, the Czechs aspired to create a "healthier" society in the face of the crisis of individual and national "degeneration." This was the point at which the movement for the "reform of life" (*Lebensreform*) in Central Europe began.[4] Although it is possible to find numerous explications of "degeneration" before 1914, the problem was mostly understood as a consequence of certain long-term processes that had led to deviations from the optimal norm of mental and physical skills of individuals, nations and races. During that period it was, however, perceived that "degeneration" could take two forms: mental and physical. The consequences were treated as a "diagnosis," and specific "symptoms" were then attached to them. These "symptoms" were confirmed by extensive statistical research, which provided arguments for the necessity of collective "therapy," largely by means of governmental regulations. Pessimistic deliberations and debates concerning ongoing "degeneration"—its forms, manifestations, and consequences—characterized pre-1914 eugenics.

The first efforts to establish "Czech" eugenics as a body of knowledge occurred in Bohemia—then part of the Austro-Hungarian Empire—at the end of the nineteenth century.[5] In 1900, the future professor of neuropathology at the Faculty of Medicine of the Czech section of the Charles-Ferdinand University in Prague, Ladislav Haškovec (1866–1944), addressed the subject of practical proto-eugenic measures.[6] At that time he gave several popular lectures in which he drew attention to the importance of medical examinations before marriage. Later these measures became known as "eugenic marital revision" (*eugenická sňatková revise*). In 1901, Haškovec proposed his idea to local Bohemian authorities in the form of a bill making health certificates obligatory. His efforts were summarized in an essay published in 1902 under the title "Snahy veřejného zdravotnictví ve smlouvě manželské" (Public Health Efforts and the Question of the Marriage Contract).[7] In the same year, Haškovec requested that the *Société de neurologie de Paris* create an international committee composed of physicians, lawyers and sociologists in order to "fight human degeneration [and] pathological heredity."[8] After many discussions, the society refused to further consider his suggestion. But Haškovec did not give up, and he referred again to the significance of pro-eugenic arguments on the occasion of the International Congress of Neurology and Psychiatry, which took place in Lisbon in 1906.

Haškovec's views were not atypical of Czech physicians at the time, as exemplified by František Lašek (1898–1947) from eastern Bohemia. In 1910 and 1916, Lašek published two articles, "O dědičnosti a jejím významu pro úpadek a zachování lidstva" (On Heredity and Its Importance for the Decline and Preservation of Humankind), and "Zušlechtění lidstva (eugenika)" (Refining Humankind, Eugenics), respectively.[9]

Two years after Lašek's first study was published, Czech readers learned in greater detail about the English and American schools of eugenics and Mendelian principles of heredity. In 1910, another pioneer of Czech eugenics, the botanist and future first Czechoslovak professor of genetics, Artur Brožek (1882–1934), began his experiments on plant hybridization.[10] After several years of experimentation, Brožek published a comprehensive overview of American eugenic thinking, *Zušlechtění lidstva* (Refining Humanity).[11]

Brožek's case, and that of his university teachers Alois Mrázek (1868–1923) and Bohumil Němec (1873–1966), reveals the "instructory" role played by eugenic-prophylactic measures. The most important of these had been implemented by individual states in America since the end of the nineteenth century and had a great impact on European and Czech eugenic thinking.[12] Czech eugenicists assigned great importance to the "quality of population" and family (marital) issues. In 1912, influenced by American developments in health care, Brožek suggested: "Measures that the state or the society in general could apply everywhere, where the self-love of ill or degenerated individuals stood in the way of improving the health of the population; for if society has the right to punish its members by death, it certainly has the right to isolate those members who are the ailing part of its body for the period of their fertility, or to forbid marriages of individuals innately ill, such as the mad and idiots; or to introduce compulsory medical examinations of engaged couples by a doctor paid by the government, or to establish state genealogical registrar offices and other similar eugenic measures. In cases where these measures could be deemed insufficient, society would certainly have the right to apply artificial infertility, as has already been the case in some American states."[13]

Brožek was the author of one of the earliest and most pertinent assessments of eugenics prior to 1914. According to him, eugenics was "a discipline concerned with refining humanity and improving its health, based on the rules of heredity."[14] English and American eugenics became one of the most important models for Czech eugenicists. The first step on the way towards the practical application of eugenics in the Czech lands before the First World War occurred on 12 June 1913, when the eugenics office (or "central eugenics bureau") was "attached" to the lunatic asylum, the *Ernestinum*, in Prague-Hradčany.[15] This was due to the initiative of four leading Czech eugenicists: Brožek; the biologist and later physician Karel Herfort (1871–1940); Jindřich Matiegka

(1862–1941), the first Czech professor of physical anthropology at the Charles-Ferdinand University in Prague; and the professor of pedagogy at the Faculty of Arts and Philosophy at the same university in Prague, František Čáda (1855–1918). The office focused on the detailed genealogical and statistic processing of data from individual patients. As in the American case, Czech eugenicists borrowed from Mendelian genetics and Francis Galton.[16] The pioneers of *Ernestinum* prepared the so-called Family Record Questionnaire (*rodopisný dotazník*), which they distributed among relatives of the patients.[17] They were interested in documenting human mental defects, as well as disorders such as epilepsy, *dementia praecox*, and so on. Some of these authors published in a special eugenic supplement of Haškovec's *Revue: Neuropsycho-pathologie, lékařství sociální, dědičnost a eugenika, therapie* (Review: Neuropsychopathology, Social Medicine, Heredity and Eugenics, Therapy), which was transformed in the 1920s into *Československá psychiatrie* (Czechoslovak Psychiatry). Haškovec's supplement was re-titled "Heredity and Eugenics," and aimed to inform readers about eugenics and genetic research.

Inherited Health Disorders

Heredity and inheritance remained central features of all eugenic preoccupations in the Czech lands between 1900 and 1925. The field of systematic exploration of heredity and inheritance was broad, encompassing different types of methodological approaches as well as research strategies.[18] There were at least four scientific disciplines that contributed to the systematic research of heredity in the following areas: cytology, animal breeding, biometrics, and, after 1900, Mendelian genetics (or "Mendelism").[19] Interest in the last sub-discipline resulted in an increase in the systematic research on heredity.

The role and importance of inherited factors was at one point the main topic of concern for many scientists and was explored in various contexts. In 1909, the Danish biologist Wilhelm Johannsen (1857–1927) introduced the term *gene* in his widely read textbook *Elemente der exakten Erblichkeitslehre* (Elements of Exact Heredity Teaching), and made the fundamental distinction between *genotype* and *phenotype*.[20] During and after the First World War, so-called chromosome genetics, which analysed the "carriers of heredity" at the cell level, and population genetics, tended to work primarily with large population statistics. However, in the 1920s, after the American zoologist Thomas Hunt Morgan (1866–1945) formulated the theory of new chromosome heredity, experimental research moved to a higher level, looking at more complex aspects of heredity. Before the seeming disparities between Mendel's theory of heredity and Darwin's theory of evolution were overcome, the theory of acquired characteristics still attracted adherents.[21]

"Pathological heredity" (*patologická dědičnost*) was one of the issues deemed most pressing by physicians and others in the Czech lands. In 1912, for instance, Haškovec described the main problems regarding the latest discoveries in human heredity: "It is questionable whether medical science has collected enough reliable detail to be able to deduce certain laws of pathological heredity. If such laws exist there is another question: What lessons should humanity learn from them, and how could they be used by public health authorities for the well-being of humanity? Another question is whether it is possible to protect humanity from hereditary weaknesses, and what role should public health play in doing so. Today humanity rightly demands answers to these questions."[22]

Mendel did not discuss the applicability of his laws. But those following the Mendelian tradition could now search for evidence in pathological manifestations. While Haškovec was an early eugenicist interested in mental diseases and their causes, the second leading figure of the Czech eugenic movement, and the first professor of general biology at the Medical Faculty of the Charles-Ferdinand University in Prague, Vladislav Růžička (1872–1934), tackled pathological heredity at a more theoretical level.[23] In 1923, his list of "hereditary diseases" contained more than 190 items.[24]

Many contemporary Czech physicians and social reformers, including eugenicists, considered alcohol a "poison" that harmed human "germ plasm" and caused serious hereditary damage. During the 1890s, Czech anti-alcohol campaigners adopted new scientific arguments based on the presumption that alcohol caused hereditary transmissive degeneration. In 1908, the sociologist Břetislav Foustka (1862–1947), for example, wrote *Die Abstinenz als Kulturproblem mit besonderer Berücksichtigung der österreichischen Völkerstämme* (Abstinence as a Cultural Problem with Particular Consideration to Austrian Nationalities), in which he stated that "Everywhere, one calls for regeneration; and not only political, ethical, and religious types, but also biological regeneration."[25] Other arguments concerning the effects of alcohol on insanity, criminality and prostitution were added. Many anti-alcohol campaigners cited the "degenerative" effect of alcohol as an argument for giving up drinking or introducing legal prohibition.

In the Name of Heredity

At the turn of the twentieth century, hereditary factors became important components not only of everyday life but also in terms of health, social, and racial issues.[26] In 1921, Haškovec explained the Czech eugenic point of view to an American audience as follows:

> I remain convinced that through well-conducted and developed eugenic efforts we may strengthen and increase in power the life of our nation, both internally and externally. Eugenics is concerned with the health, not only of the individual, but also of the entire nation. A healthy nation, as healthy as possible from the psychical and moral point of view, and with well-developed altruistic sentiments, holds in check perverse and anti-social instincts and works for the fraternity of nations, for peace and true liberty (...) I am certain that eugenic progress among the civilized peoples of the world can alone prepare the way for a permanent society of all nations.[27]

But the connection between heredity and the nation was not new. Already in 1908, at the Fourth Congress of Czech Physicians and Researchers of Nature (*Sjezd českých lékařů a přírodozpytců*), Haškovec stated: "On the basis of the rules of heredity, especially psychological heredity, we claim special privilege for all nations in their striving for the preservation of national types and traits, national peculiarity and language; the respect and preservation of these characteristics is the duty of every nation."[28] This was an important statement, especially considering that, in Central Europe, the mutual cohabitation of several nationalities was marked by extreme nationalism, which in turn influenced political decisions, including population and family policy. The eugenic dimension of long-term generational planning now reduced every newborn individual to a cog in the "chain of generations of the nation"; according to Galton: "A person is therefore more important as a probable progenitor of many others more or less like him in constitution, than as a mere individual."[29] As such, heredity could be used as a justification for national unity.

Eugenics was thus closely connected to the issue of population, or the "population question" (*populační otázka*), and this question was itself part and parcel of a larger "national question." As in other European countries, the general trend of declining birthrates was observed in Bohemia, Moravia and Silesia at the beginning of the twentieth century. Czech eugenicists interpreted these trends in the context of "social degeneration." Moreover, these arguments constituted the core of the discussion about planed population policy, which was later intensified and politicized during the First World War.

Organisation and Action

The First World War permanently altered the activity of the Czech eugenic movement. The need to protect both the "quality" and the "quantity" of the Czech population—defined after 1918 in the hereditary sense—became more pressing than ever before. Comparing the situation in Bohemia and Moravia in 1918 to that which followed the battle of White Hill (*Bílá Hora*) in 1620, Czech eugenicists declared that the effects of the 1914–1918 war were far more dysgenic for Czech social elites. According to the physician and health administrator František Kulhavý (1863–1931), "there is a huge difference between these two disasters. The disaster of *Bílá hora* occurred when the nation was fully, primitively healthy. The disaster of today is influencing the nation, whose organism was being destroyed for a long time by diseases."[30] It was at that moment that Czech eugenicists decided to act.

On 2 May 1915, the Czech Eugenics Society (*Česká eugenická společnost*) was established in Prague under the auspices of the Czech Provincial Commission for the Protection of Children (*Česká zemská komise pro ochranu dítek*) and the Protection of Youth (*Ochrana mládeže*). This was done on Haškovec's initiative in co-operation with the Association for the Establishment of a People's Mental Sanatorium of the Kingdom of Bohemia (*Sdružení pro ustanovení lidového snatoria pro duševně choré v království Českém*). The programme and goals of the society were developed in an introductory lecture delivered to the opening session by Čáda.

Haškovec was elected the Society's first chairman; Vladislav Růžička (1870–1934) the first vice-chairman; Brožek the first secretary (there was just one elected); while one of Růžička's students, Jaroslav Kříženecký (1896–1964) took the first minutes of the meeting. Among its first members the Eugenics Society also included illustrious figures such as Edvard Beneš (1884–1948), the Czechoslovak president (1935–1938 and 1945–1948); Jiří Guth (1861–1943), the first president of the Czechoslovak Olympic Committee; Antonín Benjamin Svojsík (1876–1938), founder of the Czech Boy Scouts movement (*Junák*); and Břetislav Foustka (1862–1947), professor of sociology at the Faculty of Arts and Philosophy of the Czech part of the Charles-Ferdinand University in Prague.[31]

During the war, the Eugenics Society continued to promote eugenic ideas. The main tasks and aims of the society were defined as follows: a) the special study of biology; b) the dissemination of knowledge about the conditions of physical and psychological health among all classes of the population; c) the "fight" against hereditary diseases and infant mortality; d) the encouragement of care for both the woman in confinement and the newborn; and, finally, e) the "fight" against alcoholism and tuberculosis, as well as against venereal diseases and all other factors perceived to be destroying the roots of the nation.[32]

The Eugenics Society pursued these aims through public lectures and assorted publications, in addition to the circulation of eugenic propaganda materials in schools. It was decided that a special museum of hygiene would be established in Prague as a centre of hygienic study and instruction. Czech eugenicists preferred to remain in the background of these actions; however, they openly and enthusiastically advocated eugenic principles in the public domain. This campaign was necessary for the adoption of special eugenic legislation and governmental regulations, both enabling prophylactic measures in favour of the eugenic concern for the "protection of future generations."

The Eugenics Society appeared to be an independent organization, but in reality it was connected to a network of several national organizations and social organizations such as the Czech Heart (*České srdce*), established in 1915, and the Republican League for the Moral Revival of the Nation (*Republikánská liga pro mravní obrodu národa*), established in 1919.[33] At a scientific level, the Society closely co-operated with Růžička's Institute of General Biology (*Ústav všeobecné biologie*) at the medical faculty in Prague, and both formed the institutional foundation of the Czech eugenic movement.[34]

Eugenicists were active members in other political societies as well. In 1917, the first and last Austrian Ministry for Public Health (*Ministerium für Volksgesundheit/Ministerstvo veřejného zdravotnictví*) was established in Vienna under the direction of Jan Horbaczewski (1854–1942), professor of medical chemistry in Prague. In the same year, Czech eugenicists devised a resolution covering the issue of eugenics according to the principles adopted earlier by the Czech Eugenics Society; however, due to wartime circumstances, the proposal remained dormant.[35]

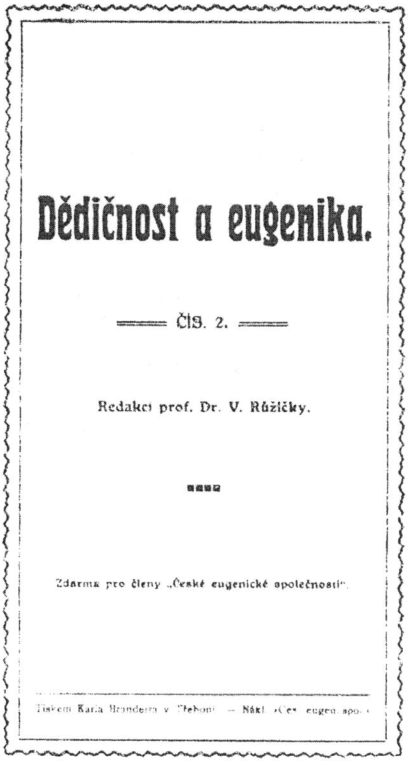

Dědičnost a eugenika.

Čís. 2.

Redakcí prof. Dr. V. Růžičky.

Zdarma pro členy „České eugenické společnosti“.

Tiskem Karla Brandejsa v Třeboni. — Nákl. „Čes. eugen. spol.“

Fig. 1. The cover of Vladislav Růžička's article on "Heredity and Eugenics" (1920)

On 28 October 1918, the Czechoslovak state came into existence and members of the Czech Eugenics Society submitted a similar proposal to the newly elected president of the republic, Tomáš Garrigue Masaryk (1850–1937), and to the Czechoslovak government in Prague. This resolution requested: a) the creation of a national institute of eugenic research; b) the adoption of special records for registering the health of the population; c) the foundation of central eugenic stations; d) the creation of institutes for studying the development of human psychology, as well as a museum of comparative genetics; e) the protection of infants; f) the reform of midwifery; g) the reorganization of the system of teaching modern hygiene, especially in terms of sex education; h) support for eugenic instruction in society by means of public discussions, theatrical and cinematographic performances, and, in particular, the establishment of a Museum of Hygiene as the centre-point of all instruction; and, finally, i) the compulsory issuing of a health certificate before marriage.[36] All of these would become key themes in the eugenic agenda during the 1920s and 1930s.

The Need for Eugenic Reform (1919–1925)

The end of the First World War rapidly changed the context within which eugenics in Bohemia, Moravia and Silesia, as well as in Slovakia, could develop. This was considered a time when many projects envisaged in the late nineteenth century could become reality. In the Czechoslovak case, especially in the period following the end of the war, eugenics was presented not as a demand of everyday life, but as an important scientific strategy.[37] The most important Czech theoretical work on eugenics written at the time was the 1923 *Biologické základy eugeniky* (The Biological Foundations of Eugenics) by Vladislav Růžička,[38] which became a standard text for all Czech eugenicists during the 1920s and early 1930s.

The doctrine of heredity was regarded by Růžička as an aspect of genetics conceived as a science not only of "internal factors" transmitted by heredity, but also of "external" factors such as social environment, education and so on. In this conception, heredity was identical with the biochemical entity of life, and was founded on the ability of metabolism to regenerate the specific structure of the living substance. According to Růžička, the main aim of eugenics was the improvement of the "social biological efficiency" (*sociálně biologická zdatnost*) of mankind, in addition to increasing the sense of "responsibility towards the community and future generations."[39] Eugenics thus aimed at regulating those factors controlling the health of a population and influencing reproduction and the development of the embryo. Růžička emphasized that a proportionate development of all social and biological virtues must become the principal demands of eugenics; from this standpoint he discussed the German concept of racial hygiene, the issue of sexual selection and decreasing birthrates.[40]

During this period of major socio-political transformations it was not surprising that eugenic thinking rose in importance on an ideological level. Strongly hereditarian arguments were repeatedly used against the German- and Hungarian-speaking aristocracy. The most well known arguments posited were the hereditary diseases of members of the ruling Habsburg dynasty,[41] including endogamy, exogamy, and inbreeding.[42] In the new Czechoslovak state, practical eugenics was presented as a conglomerate of several "democratic" and "progressive" social strategies: "Furthermore, we are, in Bohemia, always in touch with new, progressive ideas, with humanitarianism and with new social and scientific

Fig. 2. Contemporary illustration of the "typical hereditarian traits of the Habsburg family" (c. 1920)

ideas, so we were well prepared for new eugenic ideas, even before the war."[43] Shortly thereafter, eugenics became an important part of the new republican ideology of the "young," "healthy" and "plebeian" nation.[44]

Czech eugenicists strongly and repeatedly opposed German racial hygiene or, as they called it, "selective national eugenics."[45] They considered the German application of pure principles of selection as aristocratic and undemocratic: "Selective national eugenics perfectly corresponds with the foundations of German thinking. Therefore it is not surprising that the majority of German eugenicists such as Schalllmayer, Ploetz, Ammon, von Ehrenfels, Siemens, Lenz and so on, are very much in favour of it (...); it is clear that the entire nation agrees with this position."[46] Those "external conditions" were of greater importance for Czech eugenicists than the idea of pure selection of the "carriers" of heredity. Another argument against German racial hygiene was that it considered selection only—dealing with "complete racial traits" and not with the question of their origin—which was a limited task.[47]

As mentioned above, Czech eugenicists had been working on their own definition of eugenics since 1917. Between 1919 and 1925, "national eugenics" assumed the status of official doctrine, an idea largely based on the notion of a "national eugenics" developed by Galton and Pearson. In the Czech case, however, eugenics was concerned with the "preservation and promotion of the biological uniqueness of a nation and prevention of a decline of its biological organization."[48] Růžička defined "biological uniqueness" as "the harmony between hereditary constitution and influences of the external environment."[49] It is clear how important the non-Mendelian component of "biological uniqueness" was to Czech eugenicists, in contrast to the contemporary school of German racial hygiene.

Another similar concept, "eubiothics" (*eubiotika*), was developed by Stanislav Růžička (1876–1946).[50] In 1923 he defined the relationship between eugenics and eubiothics as follows: "Eubiothics (*Eubi-otik*) completes the refinement of mankind that is caused by eugenics (*Eugenik*) when new individuals are born. It is done through the construction of a particular type of life (*Lebenstypus*) that follows the natural physiological rules of the body (*Körper*) and soul (*Geist*). Together eugenics and eubiothics represent the rational scientific way of how to refine the living material of the nation."[51] It can thus be said that during the early 1920s Czech eugenicists more often than not professed "euthenics" rather than "eugenics."

Despite the diversity of definitions, heredity and inheritance were seen as phenomena worthy not only of systematic research but also of practical measures. In 1921, Haškovec described this eugenic "necessity" as follows: "Research on heredity is of enormous importance in all spheres of life, family, society, nation and state. Research on heredity and the prophylaxis of both inherited and inborn disorders concerns not only the physician, but also the educator, teacher, lawyer, sociologist and politician, and is today of great importance also for lawmakers."[52]

From Theory to Practice

After the end of the war, a eugenic approach is detectable in discussions about social policy. Three of these discussions regarding the necessity of demographic reform (concerning changes in marital regulations and the question of eugenic sterilization and abortion) were particularly important for Czech eugenicists.[53] All of these were considered, however, to be "negative eugenic" measures. On the other hand, "positive eugenic" measures were proposed in the field of family and childcare policy as well as population planning. After 1918, forms of eugenic thinking in Czechoslovakia became part of a new branch of eugenic "quantitative" research into the effects of the First World War.

Yet it was "national eugenics" that was repeatedly declared as representing the "most vital interest for the nation," and as needing to be applied to society through a new specialized institution, the Institute of National Eugenics (*Ústav pro národní eugeniku*). The work of such an institute was to be guided by certain principles formulated in 1917, when the idea of a eugenic institute was first advocated. After 1918, however, it was argued that the institute should consist of three special departments and one museum: a) the Department for the Research of Genetics in Man; b) the Department for the Research of the Ecology of Human Ontogenesis; c) the Department of National Psychology; and d) the Museum of Comparative Genetics.[54] An ambitious research programme was devised—to include biometry, hybridization, vital statistics, family pedigrees and so on—which focused on both the hereditary and biological constitution of the population within the territory of Czechoslovakia after 1918. "Special pedigrees," alongside other additional information of a statistical nature about the hereditary status of man, were collected, in addition to the preparation of special health certificates.

The institute was not established in 1919. However, at the Second International Congress of Eugenics, organized by the American Museum of Natural History in New York between 1 and 28 September 1921, it was agreed that the next eugenics congress was to be held in Prague. After their return home, leading Czech eugenicists, like Růžička, suggested that the creation of a special institute was necessary in order to demonstrate the efficiency of the Czechoslovak eugenic movement to the international scientific community.[55] The institute was thus established in Prague in 1923 and was called the Czechoslovak Institute of National Eugenics (*Československý ústav pro národní eugeniku*). Both Charles University in Prague and the Ministry for Public Health and Physical Education (*Ministerstvo veřejného zdravotnictví a tělesné výchovy*) participated in the establishment of the Institute. In contrast to earlier proposals, the Institute was neither large nor independent, functioning as an affiliate of Růžička's Institute of Biology at the Faculty of Medicine in Prague.[56] Also, after 1920, Czech eugenicists worked in the newly established Masaryk Work Academy (*Masarykova Akademie Práce*), where they organized a Eugenics Commission (*Eugenická komise*).[57]

New Allies and Opportunities

Although Czech eugenics was projected and developed as a "national" programme, the eugenicists perceived themselves as contributors to "eugenic universalism," in which the "betterment of mankind" was possible only through the "betterment" of smaller, national organized entities.[58] After 1918, Czech eugenicists became part of an international eugenic network and contributed to international debates and congresses. In 1919, for example, Haškovec participated in the Inter-Allied

Congress for Social Hygiene (*Interallié Congrès de Hygiénique sociale dans la Reconstruction des pays dévastés par la guerre*) on behalf of the Czechoslovak Ministry for Health and Physical Education. The congress centred upon the post-war reconstruction of public health and hygiene. German racial hygienists were not invited. This provided an excellent opportunity for Czech eugenicists openly to declare their anti-German position and pro-French orientation.[59] During these years, moreover, Czech eugenicists followed and supported official Czechoslovak foreign policy.

With the official support of the Czechoslovak government, Czech eugenicists participated in the Second International Congress of Eugenics, held in New York in 1921. They gave papers in four sessions: "Human and Comparative Heredity"; "Eugenics and the Human Family"; "Human Racial Differences"; and "Eugenics and the State."[60] In New York, Czech eugenicists met not only racial theorists like Madison Grant (1865–1937), but also leading eugenicists such as Charles B. Davenport (1866–1944), Jon Alfred Mjöen (1860–1939), Harry H. Laughlin (1880–1943) and Henry F. Osborn (1857–1935), as well as prominent biologists like Thomas H. Morgan (1866–1945), Jacques Loeb (1859–1924), Alfred H. Sturtevant (1891–1970), and the psychologist Robert M. Yerkes (1876–1956). The American anthropologist of Czech origin, and curator of the National Museum of Natural History, Aleš Hrdlička (1869–1943), represented a fusion of the "new" and the "old" world.[61] The official representatives of the Czechoslovak Eugenics Society were Haškovec, Brožek and Hrdlička; while the Czechs representing the Medical Faculty of Charles University were Růžička and Hrdlička. Other officials included Antonín Sum (1877–1947) from the Czechoslovak Embassy in Washington D.C., and Bohumil Feierabend (1895–1933) from the International Health Bureau of the Rockefeller Foundation in New York.[62]

Besides their participation in the international eugenic network, Czech eugenicists also contributed to the furthering of scientific cooperation in the field of genetic research. On the occasion of the 100th anniversary of the birth of Gregor J. Mendel (1822–1884), the Czechoslovak Eugenics Society arranged several meetings in Brno and Prague marking the occasion. Many distinguished international and local participants and guests convened in the Augustinian Monastery in Brno, and at the Charles University in Prague, to celebrate the "person and the spirit of the father of genetics."[63] Both groups within the Czech

TSCHECHOSLOVAK. EUGENISCHE GESELLSCHAFT
IN PRAG.

EINLADUNG

zur

FESTVERSAMMLUNG

UND ZU VORTRÄGEN,

welche die

Tschechoslovakische Eugenische
Gesellschaft in Prag

unter dem Protektorate S. M. des Rectors der Karlsuniversität
Prof. Dr. B. NĚMEC
zur Erinnerung an die 100 Rückkehr des
Geburtstages von J. G. Mendel veranstaltet.

PROGRAMM:

I. FESTVERSAMMLUNG.

1. Eröffnung der Versammlung durch den Präsident der T. E. G. Universitätsprofessor Dr. Lad. Haškovec und Begrüssung der erschienenen Gäste und Delegate.
2. Ansprachen.
3. Festrede S. M. des Rectors der Karlsuniversität Prof. Dr. B. Němec: Die Bedeutung Mendels in der Biologie.
4. Festrede des Prof. Dr. Vlad. Růžička: Über Mendelismus und Genetik.

Diese Festversammlung wird Donnerstag am 19. Oktober 1922 9 Uhr vormittags im grossen Hörsaale des naturwiss. Institutes der Karlsuniversität, Prag-II., Albertov 6, abgehalten.

Fig. 3. The invitation to the 100th anniversary of Gregor Mendel's birth

eugenics movement (that is, pro-Mendelian and pro-Lamarckian) were present at these discussions. Němec, for example, underlined the importance of Mendelism for contemporary eugenics: "Mendel's [...] scientific work, duly appreciated, dispersed many fantastical conjectures and in their place established Natural Law. In this lies his greatest significance. Mendel ranks among those of us who lay down simple and strict laws, free of any embellishments, mysticisms and fancies, so that we might build upon them a system of natural science of the world. They prepare the breeding of the strong, and therefore let us be thankful to them!"[64] Moreover, speaking about the future of eugenics, Němec predicted: "The Czechoslovak Eugenics Society has good reason to celebrate the 100th anniversary of Mendel's birth, for it has its object in the science of heredity of which Mendel is the coryphaeus. The eugenic movement might well adopt as its maxim words from Horace: 'Fortes creantur fortibus et bonis!' Even in eugenics it is necessary to start with a genetic analysis in order to identify the 'fortes et boni', and then take measures adequate to secure a generation of the strong and the good."[65]

The influence of Mendelian genetics within the Czech eugenic movement was quite clear during the 1920s. In 1924, Němec—the contact person of the "Life Science Division" of the Rockefeller Foundation in Czechoslovakia—proposed Brožek for a six-month fellowship in the United States. During this period Brožek intended to study the "methodology of genetics and eugenics" with Thomas H. Morgan at Columbia University in New York, as well as eugenics with Charles Davenport, particularly the "correlations between feeblemindedness and other physical as well as psychological traits," and the "correlations between different physical states, both normal and abnormal, which accompany and characterize families with one or more suicides."[66] In June 1924, the fellowship was granted for six months, commencing on 1 September 1924. Fascinated by his American experience, Brožek stated upon his return: "I am sure that I have brought home many new ideas, methods, and plans for scientific work from America, all of which might be well adapted and conformed to methods used in Europe and especially in my country."[67] In the same year, he proposed the most ambitious eugenic project undertaken by Czech eugenicists during the first half of the twentieth century: the genetic screening of Czechoslovak citizens, known as the "Genetic Cadastre of the Czechoslovak Population" (*Vlohový soupis obyvatelstva československého*). The screening was to be carried out under the auspices of the Eugenics Commission of the Masaryk Work Academy. With this project, Czech eugenics completed the transition to applied genetics.

Conclusions

In the Czech—and later Czechoslovak—case, eugenic ideas penetrated intellectual circles from 1900 onwards, largely due to efforts to avoid the negative effects of "degeneration" and "deterioration." Eugenics was closely linked to other scientific fields, and it was in the main academically active scholars who played a major role in its reception. "Getting to know" eugenics usually took the form of deliberations and conceptual clarifications about its theoretical bases and about eugenics as "applied genetics" (*aplikovaná genetika*) and "social genetics" *(sociální genetika)*.

Thus both "Mendelian" and "Lamarckian" concepts of heredity and inheritance played an important role in the development of eugenics in Bohemia and Moravia between 1900 and 1925. Furthermore, the Czech case demonstrates that acceptance of the Mendelian principles of heredity was not an absolute prerequisite for those whose goal was the "improvement," or manipulation, of human nature on the basis of biological arguments. The First World War constituted an important turning point in the evolution of eugenics in the Czech Lands. War losses, fears of depopulation, and changes in social stratification made eugenics popular. Eugenics was connected to a broader range of medical and health reform programs emerging in the interwar years.

Between 1917 and 1925, a rough concept of Czechoslovak "national eugenics" was also developed. Following Galton and Pearson, this was declared to be a "national" program, one distinguished from the German concept of "racial hygiene" (*Rassenhygiene*) that often identified the "vital race" with the "Nordic race." "National eugenics" in the Czech lands was ultimately characterized by two goals: a) research in the field of the "progenic constitution," as defined by Růžička in his 1923 *Biological Foundations of Eugenics*; and, b) focusing on the hereditary and biological constitution of the population living in Czechoslovakia after 1918.

Endnotes:

[1] A look at the current discussion on the issue of the European Constitution, where eugenics—or eugenic practices—is explicitly mentioned, indicates that further understanding of what eugenics was historically seems to be of great importance.

[2] Mark B. Adams, "Toward a Comparative History of Eugenics," in Mark B. Adams, ed., *The Wellborn Science: Eugenics in Germany, France, Brazil and Russia* (Oxford: Oxford University Press 1990), 222–225.

[3] On the period of German occupation (1939–1945) see Michal Šimůnek, "Ein Neues Fach: Die NS-Erb- und Rassenhygiene an der Medizinischen Fakultät der Deutschen Karls-Universität Prag 1939–1945," in Kostlán Antonín, ed., *Die Wissenschaft in den böhmischen Ländern 1939–1945* (Prague: Výzkumné centrum pro dějiny vědy, 2004), 190–317.

[4] See Kai Buchholz, Rita Latocha, Hilke Peckmann, Klaus Wolbert, *Die Lebensreform. Entwürfe zur Neugestaltung von Leben und Kunst um 1900* (Darmstadt: Verlag Häusser, 2001).

[5] On the general history of medicine in this period see Ludmila Hlaváčková and Petr Svobodný, *Dějiny lékařství v českých zemích* (Prague: Triton, 2004); and Ladislav Niklíček, *Přehled dějin českého lékařství a zdravotnictví*, vol. 1 (Brno: Institut pro další vzdělávání stř. zdrav. pracovníků, 1989).

[6] Ladislav Haškovec, "Opening Speech of the Chairman," in Vladislav Růžička, ed., *Memorial Volume in Honor of the 100th Birthday of J. G. Mendel* (Prague: Fr. Borový, 1925), 3. See also Jarmila Veselá, *Sterilisace: Problém populační, sociální a kriminální politiky* (Prague: L. Mazáč, 1938), 121–122.

[7] Ladislav Haškovec, *Snahy veřejného zdravotnictví v otázce smlouvy manželské* (Prague: n.p., 1902). See also Ladislav Haškovec, *Snahy eugenické* (Třeboň: K. Brandejs, 1912).

[8] Ladislav Haškovec, "The Eugenics Movement in the Czechoslovak Republic," in Charles Davenport, et. al., eds., *Eugenics in Race and State: Scientific Papers of the Second International Congress of Eugenics* (Baltimore: Williams & Wilkins Company, 1923), 435, 438; also Haškovec, "The Importance of Mendelism for Medicine and Eugenics," in Růžička, ed., *Memorial Volume*, 107–120, especially 111.

[9] František Lašek, *O dědičnosti a jejím významu pro úpadek a zachování lidstva* (Prague: J. Otto, 1910); and František Lašek, *Zušlechtění lidstva (eugenika)* (Prague: J. R. Vilímek, 1916).

[10] See, for example, Artur Brožek, "A case of non-Mendelian inheritance of the white striped races (formae albomaculatae) of the Mimulus quinquevulnerus," in *Věstník I. sjezdu čs. botaniků v Praze* (Prague: Výbo sjezdu, 1923), 1–3.

[11] Artur Brožek, *Zušlechtění lidstva (Eugenika)* (Prague: F. Topič, 1914); and Artur Brožek, "C. B. Davenport: Eugenika, nauka o ušlechtění lidstva dokonalejším křížením," *Živa* 22, 1 (1912), 8–10; 2 (1912), 44–47; and 3 (1912), 78–80. See also Artur Brožek, *K nauce o mendelismu* (Třeboň: K. Brandejs, 1916).

[12] See Géza von Hoffmann, *Die Rassenhygiene in den Vereinigten Staaten von Nordamerika* (Munich: J. F. Lehmanns Verlag, 1913).

[13] Artur Brožek, "Eugenika, nauka o zušlechtění a ozdravění lidu, založená na pravidlech o dědičnosti," *Pražská lidová revue* 8, 6 (1912), 21–22.

[14] Brožek, "Eugenika," 22.

[15] See Karel Herfort, "Zpráva o činnosti v Ernestinu," *Revue* 15, 4–5–6 (1919), 140–151.

[16] Artur Brožek and Karel Herfort, "Die eugenische Zentrale des Ernestinums," *Eos* 10, 3 (1914), 1–13; Artur Brožek and Karel Herfort, "O dědičnosti s ohledem k dítěti mravně úchylnému," *Přednášky z nauky o mravně vadných – Knihovna úchylné* 3, 2 (1928), 68–86; Karel Herfort, "Rodokmeny slabomyslných I," *Anthropologie* 2, 3–4 (1924), 165–182; Karel Herfort, "Rodokmeny slabomyslných II," *Anthropologie* 3, 1 (1925), 1–44; and František Dolenský, "Institute pour l'étude des enfants et des adolescents," *Anthropologie* 2, 3–4 (1924), 16–21.

[17] Vladislav Růžička, *Biologické základy eugeniky* (Prague: Fr. Borový, 1923), 258–262.

[18] See Jan Sapp, *Genesis: The Evolution of Biology* (Oxford: Oxford University Press, 2003), 117–140.

[19] Hans Stubbe, "Kurze Geschichte der Genetik der Genetik bis zur Wiederentdeckung der Vererbungsregeln Gregor Mendels," *Genetik: Grundlagen, Ergebnisse und Probleme in Einzeldarstellungen* 3 (1965), 218–238.

[20] Wilhelm Johannsen, *Exakte Elemente der Erblichkeitslehre. Mit Grundzügen der biologischen Statistik* (Jena: Gustav Fischer Verlag, 1926), 134–174.

[21] See, for example, Paul Kammerer, *Allgemeine Biologie* (Berlin: Deutsche Verlags-Anstalt, 1915).

[22] Haškovec, *Snahy eugenické*, 12.

[23] Vladislav Růžička, "Diagnostika a eugenika," *ČLČ* 59, 20 (1920), 223–251. See also Jan Janko, *Vznik experimentální biologie v Čechách 1882–1918* (Prague: Academia, 1982), 61–92.

[24] Růžička, *Biologické základy eugeniky*, 387–391.

[25] Břetislav Foustka, *Die Abstinenz als Kulturproblem mit besonderer Berücksichtigung der österreichischen Völkerstämme* (Vienna: Verlag von Brüder Suschitzky, 1908), 5–6.

[26] See Jindřich Matiegka, *Plemeno a národ. Poměr mezi plemenem a národem zvláště s ohledem na československý stát* (Prague: B. Kočí, 1919).

[27] Haškovec, "The Eugenics Movement in the Czechoslovak Republic," in Davenport, et. al., eds., *Eugenics in Race and State*, 438.

[28] Haškovec, "Opening Speech of the Chairman," in Růžička, ed., *Memorial Volume*, 2.

[29] Karl Pearson, *The Life, Letters and Labours of Francis Galton*, vol. III (Cambridge: Cambridge University Press, 1930), 415.

[30] František Kulhavý, "Aktuálnost a význam eugenických problémů," *Revue* 16, 1–2–3 (1919), 19–22, 20.

[31] Statutes of the Czech Eugenics Society in Prague, 1915. Pražské policejní

ředitelství – spolkový katastr, IX/287–357, Archives of the Capital City of Prague.

[32] Haškovec, "The Eugenics Movement in the Czechoslovak Republic," in Davenport et. al., eds., *Eugenics in Race and State*, 440.

[33] Popular lectures and enlightenment activities, 1915–1925. Republikánská liga pro mravní obrodu národa, ANM Prague (Archives of the National Museum), box 4.

[34] See Michal Šimůnek, "Eugenika a Velká válka," in Kostlán Antonín, ed., *Inter arma... scientia* (Prague: Výzkumné centrum pro dějiny vědy, 2002), 53–73.

[35] "Memorandum České eugenické společnosti o základních potřebách eugeniky c. k. ministru veřejného zdravotnictví dr. Janu Horbaczewskému," *Revue* 14, 11–12 (1917), 434–437.

[36] "Zpráva o činnosti české eugenické společnosti za rok 1918," *Revue* 14, 1–2 (1919), 143; see also Vladislav Růžička, "Institute of National Eugenics," *Národ* 3, 2 (1919), 17–18.

[37] Jaroslav Kříženecký, "Organisace vědy – Česká eugenika," *Nové Atheneum* 1, (1919), 209–212; Jaroslav Kříženecký, "Cíle a program snah eugenických," *Osvětová škola sokolských žup pražských* 1, 2 (1919), 3–28.

[38] See note 29.

[39] Růžička, *Biologické základy eugeniky*, 707.

[40] However, as soon as external factors of development were recognized as having the same significance in shaping heredity as internal factors, adaptive, altruistic eugenics became possible. Růžička argued that the main principles of genetics, heredity and adaptation should become the principal means of eugenics; and that "selective" and "adaptive" eugenics should be distinguished from each other. See Růžička, *Biologické základy eugeniky*, 742–743.

[41] Ladislav Haškovec, "Česká účast na mezispojeneckém kongresu pro sociální hygienu," *Revue* 15, 18 (1919), 116.

[42] Jaroslav Kříženecký, "Zákonná úprava příbuzenských sňatků s hlediska biologicko-lékařského," *Právník* 57, 10 (1918), 325–331.

[43] Haškovec, "The Eugenics Movement in the Czechoslovak Republic," in Davenport et. al., eds., *Eugenics in Race and State*, 435.

[44] Břetislav Foustka, "Address for the Department of Sociology of the Faculty of Science in Prague," in Růžička, ed., *Memorial Volume*, 13; and Břetislav Foustka, "Ethics and Eugenics," in Růžička, ed., *Memorial Volume*, 121–129.

[45] Růžička, *Biologické základy eugeniky*, 724.

[46] Růžička, *Biologické základy eugeniky*, 672. For the opposite opinion see Jan Šebek, "Význam pohlavního výběru pro rassu v přírodě au člověka," *Revue* 15, 11–12–13 (1919), 290–294; and Břetislav Foustka, "Vyšší národ," *Vyšší národ* 1, 1 (1921), 2.

[47] Růžička, *Biologické základy eugeniky*, 673; and Vladislav Růžička, "K biologické definici pojmu národa a 'národní eugeniky'," *Revue* 15, 1–2–3 (1919), 32–35.

48 Růžička, *Biologické základy eugeniky*, 612.

49 Růžička, *Biologické základy eugeniky*, 612.

50 He was not related to Vladislav Růžička. Stanislav served between 1914 and 1919 as a physician of the Czechoslovak Legions in Russia, and following the war he became the first professor of hygiene at the medical faculty of the newly established Comenius University in the Slovak capital city of Bratislava.

51 Stanislav Růžička, *Ein Grundriss der sozialen Eubiotik. Die Familie im Zusammenleben mit der Natur ist der verlorene und vergessene einzige natürliche Urgrundpfeiler der volksgesundheitlichen Kultur* (Prague: Verlag von A. Haase, 1922).

52 Haškovec, "Opening Speech of the Chairman," in Růžička, ed., *Memorial Volume*, 2.

53 See, for example, Alois Hajn, *Ženská otázka v letech 1900–1920* (Prague: Pokrok, 1939).

54 Růžička, *Biologické základy eugeniky*, 737–739.

55 Vladislav Růžička, "Československý ústav pro národní eugeniku," *Čas* 32, 305 (1922), 1–2. See also Růžička's letter to the Ministry of Education and National Education of the Public, 4 October 1924. LF – osobní spisy (Růžička Vladislav), AUK (Archives of Charles University) Prague, box 44.

56 Statutes of the Czechoslovak Eugenics Society in Prague, 1922. Pražské policejní ředitelství—spolkový katastr, IX/287–357, AMP. See also Ludmila Cuřínová, "Ústav pro národní eugeniku," in Janko Jan and Emilie Těínská, eds. *Technokracie věd českých zemích (1900–1950). Studia historiae academia scientiarum Bohemicae*, vol. 3 (Prague: Archiv Akademie věd, 1999), 151–157.

57 Vladislav Růžička, "Proč potřebuje a žádá česká eugenika samostatné zastoupení v Masarykově akademii práce?" *Národ* 3, 19 (1919), 319–320.

58 Handwritten "Remarks on Eugenics" (1916), AAV (Archives of the Academy of Sciences) Prague, Čáda František – box 17/no. 155. See also Růžička, *Biologické základy eugeniky*, 599.

59 Haškovec, "Česká účast na mezispojeneckém kongresu pro sociální hygienu," 116–117; see also William H. Schneider, *Quality and Quantity: The Quest for Biological Regeneration in Twentieth-Century France* (Cambridge: Cambridge University Press, 1900), 116–140.

60 Charles Davenport, *Report of the Second International Congress of Eugenics: Membership, Organization, General Programme, and Business Proceedings* (Baltimore: Williams & Wilkins Company, 1923), 21–22.

61 Aleš Hrdlička, "Physical Anthropology: Its Scope and Aims; Its History and Present Status in America," *American Journal of Physical Anthropology* 1, 1 (1918), 3–23; and Aleš Hrdlička, "Potřeby a úkoly anthropologie zvláště v Československu," *Anthropologie* 1, 1 (1923), 3–8.

62 Davenport, *Report of the Second International Congress of Eugenics*, 15, 18–19.

63 Haškovec, "Opening Speech of the Chairman," in Růžička, ed., *Memorial Volume*, 1.

[64] Bohumil Němec, "Official Speech," in Růžička, ed., *Memorial Volume*, 27.

[65] Bohumil Němec, "Official Speech," in Růžička, ed., *Memorial Volume*, 29.

[66] A letter from Rose to Morgan, 30 April 1924. Rockefeller Archive Centre, New York, International Education Board (hereafter IEB); 9I-3 B46#665 (Artur Brožek). See also a letter from Rose to Davenport, 7 May 1924, RAC New York, IEB – 9I-3 B46#665 (Artur Brožek).

[67] An evaluation letter from Brožek to the International Education Board, 16 April 1925, RAC New York, IEB – 9I-3 B46#665 (Artur Brožek).

Progressivism and Eugenic Thinking in Poland, 1905–1939

Magdalena Gawin

The essential characteristic of the Polish eugenic movement, which developed from 1905 until the outbreak of the Second World War, was its progressivism. The term "progressivism" is used here to denote a certain perspective, founded on the belief that history is a process of conscious dynamic evolution and that man is responsible for his own destiny.[1] Progressivism equates scientific, technological, and ethical development. Polish eugenicists believed that it was possible to build a harmonious and advanced society, free from social problems such as alcoholism and prostitution, as well as physical disabilities and diseases. Eugenicists such as Leon Wernic (1870–1953), Tomasz Janiszewski (1867–1939), and Wiktor Grzywo-Dąbrowski (1885–1868) proposed the radical measure of compulsory sterilization during the 1930s. They insisted that their decision was motivated by a desire to reduce the scale of human suffering. After the experiences of the Second World War, the hybrid language of eugenics—combining social sensitivity with repulsion and contempt for the sick and the weak—was almost completely forgotten or wrongly identified with the Nazi regime and its extermination policies, whereas few remembered that, in fact, Polish eugenics was dominated by left-wing and liberal advocates of state welfare, thus resembling the Scandinavian model of state-sponsored eugenics.[2]

Political Determinants

Poland lost its independence in 1794, after defeat in the uprising against Russia. One year later, the three neighboring powers, Russia, Austria and Prussia, ratified a treaty under which they annexed Polish territories. From that time until the Treaty of Versailles at the end of the First

World War, Poland was absent from the map of Europe. Former territories including the regions of Belarus, Ukraine and Lithuania were incorporated into the Russian Empire. Following the Congress of Vienna in 1814–1815, the Congress Kingdom of Poland came into being, which was an independent state with a government, parliament and constitution of its own, connected to Russia by personal union. After suffering defeat in another uprising against Russia in 1831, the tsarist authorities abolished the independent status of the Congress Kingdom of Poland. Following the next anti-Russian uprising of 1863 (known as the January Insurrection), a number of repressive measures were imposed on the Congress Kingdom. Russian became the official language of instruction in all types of schools; insurgents' estates were confiscated, and many rebels were imprisoned or exiled to Siberia. Censorship was imposed and political parties and associations were banned. The tsarist administration was not interested in modernizing the Congress Kingdom of Poland. No health policy was developed, thus health care consisted of the private provision of medical assistance. Consequently, a sizeable proportion of the Kingdom's population was not covered by this policy. The number of hospitals remained insufficient and infant mortality remained high. Following the Industrial Revolution and increased social mobility from the 1870s onwards, venereal diseases spread at an alarming rate.

Similarly, there was no consistent health policy following the German partition. After the unification of Germany in 1871, centralist tendencies intensified, resulting in efforts to subordinate the Roman Catholic Church to the state. During the 1880s, the political activities pursued within the framework of the *Kulturkampf* campaign were transformed into a policy of socioeconomic discrimination against the Poles. German became the official language in schools and administration. As in the Russian partition, Polish political parties were banned. Politically speaking, the situation was by far the most advantageous for Poles in the region of Galicia, which had an autonomous government and a Diet in Lvov (L'viv, Lwów, Lemberg) from the late 1860s and early 1870s. This region pursued its own educational and economic policies. The autonomous authorities restored the Polish language to all schools, including the universities of Lvov and Krakow, and established other institutions of higher education, most notably the Academy of Sciences, the Academy of Fine Arts (Krakow), and the Technological University (Lvov).

Overall, historians of the region agree that in addition to language, culture and religion, the intelligentsia was a key element throughout the territories of the Polish partitions during the nineteenth century. It was only between the intelligentsia of the Russian partition and the Austrian partition (Galicia), however, that close co-operation developed. The territory of the German partition remained the most isolated. The nascent eugenic movement in the Polish territories reflected these divisions. There were two independent circles of social activists, scientists, and physicians that laid the ideological foundations for the development of Polish eugenics: one from the Congress Kingdom of Poland and Galicia (which was the result of the Austrian partition), and the other resulting from the German partition. It was only during the interwar period that these two movements were unified.

From an Ethical Revolution to Eugenics

Rising interest in the theories of inheritance and Darwinism is reflected in the Polish press from the 1880s onwards. Natural scientists and intellectuals organized debates about the ideas of August Weismann (1834–1914) and Ernst Haeckel (1834–1919). Study groups were arranged in which the works of Cesare Lombroso (1835–1909) and Francis Galton (1822–1911) were read. Polish interest closely followed developments in the field of race anthropology, particularly in Germany and France. The fashionable theme of "degeneration" dominated public discussion in the wake of the publication in 1886 of data relating to the health of Polish conscripts enlisted in the tsarist army.[3]

Yet the history of the Polish eugenic movement dates back only so far as the revolution of 1905, which spread rapidly from Russia to the Congress Kingdom of Poland. On the rising tide of revolutionary turmoil—which raised hopes for the liberalization of the political system and, especially, the abolition of the tsarist autocracy—medical circles aiming to combat prostitution, alcoholism and venereal diseases were mobilized. Among physicians committed to social causes, the thirty-five-year-old venereologist Leon Wernic, a graduate of the Imperial University of Warsaw, is notable. In Warsaw he socialized with members of the progressive intelligentsia. In addition to contributing to the feminist and socialist journals *Ster* (Helm)[4] and *Ogniwo* (Link),[5] Wernic wrote for various medical journals, including *Medycyna* (Medicine) and *Kronika Lekarska* (Medical Chronicle).[6] Wernic also lectured on hygiene at the *Mikołaj Rej* secondary school in Warsaw. Wernic was later employed at the St Lazarus Hospital, located in the heart of Warsaw's working-class district of Powiśle, first as a doctor in charge of a ward, then, from 1913, as the hospital's director. Between 1905 and 1907, Wernic worked as editor of *Zdrowie* (Health), the most important Polish journal to specialize in medical and social questions, published under the auspices of the Warsaw Hygienic Society *(Warszawskie Towarzystwo Higieniczne)*. The society, founded in 1898 in Warsaw, was one of the very few associations whose registration was successfully negotiated with the tsarist authorities prior to the revolution of 1905. In a 1907 article, Wernic clearly linked the revolution to ethical reform: "The great social and political revolution that has ensued for the past three years in the Kingdom of Poland has economic and political roots. The first wave of social and political upheaval was followed by a second, an ethical revolution. The most pressing of ethical issues is the issue of child rearing and the explanation of the facts of life to young people."[7] In 1907, Wernic also began contributing to *Czystość* (Cleanliness),[8] a journal established by the anarcho-syndicalist, scientist and chemist Augustyn Wróblewski (1866–?). Wróblewski fled from Galicia to Warsaw in 1905, following his conflict with the conservative academic milieu in Krakow. Revolutionary Warsaw seemed the perfect place to begin his publishing activities. Wróblewski brought together several different circles and organizations from the Congress Kingdom and Galicia devoted to promoting temperance and combating prostitution and venereal disease.[9]

Wernic, who in 1907 founded the "Society for Combating Venereal Diseases and Promoting the Principles of Abolitionism" *(Towarzystwo do Zwalczania Chorób Wenerycznych i Szerzenia Zasad Abolicjonizmu)*, soon became a regular contributor to *Czystość*. Well-known feminists, like Justyna Budzińska-Tylicka (1867–1936), the biologist Benedykt Dybowski (1833–1930), and the physician Wacław Miklaszewski (1868–1950), were invited to contribute. All contributors to the journal were devoted followers of Darwin, and the Swiss psychiatrist and eugenicist August Forel (1848–1931), whose lectures and works were translated into Polish. Wróblewski founded an affiliate journal, *Przyszłość* (The Future), that followed Forel's anti-alcohol appeal presented in *To the Polish People: On Degeneration* (1907).

At the beginning of the twentieth century, "degeneration" was one of the key terms employed by social and medical reformers.[10] The main causes of physical degeneration in humans were perceived to be alcoholism and venereal diseases. Physicians argued that the smallest dose of alcohol, if consumed during sexual intercourse, adversely affected the body of the conceived child. The progeny not only of alcoholics, but also of people sporadically consuming alcohol, were born weak and degenerate. Leon Wernic spoke in favor of an absolute prohibition of marriage between people suffering from serious diseases: "The fate of the human race should not be governed by a passing impulse and short-lived affection or by the sex drive of a given individual but, rather, by a general concern of mankind to exclude dwarfish types from among its ranks. Mankind must seek to create generations strong in spirit and in body. (...) The ultimate goal of marriage is to preserve and perfect the species."[11] It was thought that moral purity was threatened by promiscuity, which was in turn encouraged by prostitution and pornography. For this reason, eugenicists developed anti-alcohol and anti-pornography campaigns, promoted sex education among the young, and advocated premarital sexual abstinence. The editors of *Czystość* regarded industrialization and capitalism as the underlying cause of prostitution. In response, the zoologist and physician Benedykt Dybowski developed a eugenic ethical program.[12] In a series of articles published in *Czystość*, Dybowski highlighted the obstacles preventing social progress: alcoholism, nationalism and religious devotion. For Dybowski, religion—the bastion of superstition and ignorance—and prejudice were the greatest obstacles hindering social progress and the most difficult to overcome. He contrasted the harmful consequences of religious beliefs—including, most notably, intolerance, hatred and contempt—with the beneficial consequences of science, which "recommends love, unity and harmony."[13]

Conceived in these terms, the basis of the eugenic ethical revolution was a system of beliefs drawing upon Darwinism and, more broadly, the natural sciences. Science was to become the lasting foundation stone of the new ethics. In accordance with Max Weber's theory of the dual nature of secular rationalization, science was the tool with which the contributors of *Czystość* constructed their visions of a "disenchanted," perfectly rational world, one free from religious superstition. Accordingly, these authors considered religious dogma regarding the resurrection and immortality of the human soul to be "prejudicial."[14]

In contrast, they were concerned with the social and "racial" impact of alcoholics, whose progeny were described as "idiots," "neurasthenics," "natural-born killers" or, at best, "epileptics," and this was complemented by the "scientifically" proven thesis that prostitutes were on average twice as fertile as other women.[15]

Underlying this vision of social existence was the belief that progress was possible, but only if all forms of social injustice and national antagonism were eliminated from society, thus ensuring greater potential for individual happiness. In the case of conflict between individual and public interest, priority should be given to the collective: "We demand scientific human morality, that is, the understanding that each act is moral as long as it sustains and develops the existence of the individual and mankind as a whole."[16] This group of eugenicists were convinced that social pathologies, including prostitution and alcoholism, would disappear as soon as people learned how to live according to the principles of their new social ethics. At that stage in the formulation of eugenic thinking, eugenicists had not as yet considered the question of using coercive measures against those seen to be hampering progress (especially the sick and socially unacceptable misfits). Wernic's views about the prohibition of marriage between "socially unacceptable" individuals were therefore an exception at that time.

The Evolution of the Polish Eugenic Movement

The Society for Combating Sexually Transmitted Diseases *(Towarzystwo ku Zwalczaniu Zakaźnych Chorób Płciowych)* was established in 1903 as a pseudo-eugenic organization in Poznań (Pozen under German rule) by a group of physicians and social activists. Franciszek Chłapowski (1846–1923), a physician and social activist, headed the society. In the Congress Kingdom, the Bolesław Prus Society for Practical Hygiene *(Towarzystwo Higiny Praktycznej im Bolesława Prusa)* was the most dynamic. The patron of the society, Bolesław Prus (the pseudonym of Aleksander Głowacki, 1847–1912), an acclaimed Polish writer of the second half of the nineteenth century, belonged to an intellectual circle inspired by Western liberalism, called the "Warsaw Positivists" *(pozytywiści warszawscy)*.[17] Stress on the importance of public health, a key feature of modernization strategies, grew out of the evolutionary philosophy of the "Warsaw positivists."

Members of the Society for Practical Hygiene undoubtedly accomplished many good things. In Warsaw, for instance, their work led to the creation of public baths, supplies of clean clothing and living quarters free of charge to the poor. Exhibitions and lectures on hygiene in daily life were also organized. The society was particularly active during the First World War. It was at that time that the energetic and ambitious Leon Wernic distinguished himself as a leading eugenicist. In 1915, Wernic became head of the Department for Combating Prostitution and Venereal Diseases, established as a branch of the society. Two years later, Wernic transformed the department into a fully fledged Society for Combating Prostitution and Venereal Diseases (*Towarzystwo do Walki z Prostytucja i Chorobami Wenerycznymi*). In 1918, Wernic was appointed head of the "second section" at the Ministry of Public Health and Welfare, responsible for combating venereal diseases in the Second Republic sof Poland.

In 1917, a heated debate began in the medical press regarding the impact of the war on the size and health of the Polish population. In the same year a Congress of Polish Hygienists (*Drugi Zjazd Higienistów Polskich*) was held, where one of the speakers stated that it was the responsibility of the future Polish state to "breed a homogeneous and healthy type of Polish citizen."[18] Shortly before the Polish state was established in November 1918, Wernic, in collaboration with politicians and physicians of the new state, convened a congress on "The Depopulation of the Country" (*Zjazd w Sprawie Wyludnienia Kraju*). Women's rights activists, physicians and representatives of a number of non-governmental and government bodies discussed a wide range of issues concerning the system of health care and welfare of the Polish state. Other problems addressed by the participants included methods of birth control, the legal status of illegitimate offspring, the provision of care to neglected children and orphans, and the means by which to combat prostitution and venereal disease. But perhaps the most important question raised concerned the prevention of racial degeneration. As a result, the Executive Committee of the Congress filed a request to the state authorities for the compulsory sterilization of the incurably ill and criminals serving prison sentences, as well as the voluntary sterilization of individuals suffering from hereditary disease. Additionally, a request was made for the legal prohibition of marriage between those afflicted by hereditary diseases.[19]

These proposals were radical in nature. It was for the first time that the issue of sterilization was publicly discussed. The imminent prospect of the restoration of a Polish state was the decisive factor resulting in demands for eugenic measures, for proponents of eugenic "social correction" understood that social engineering required state infrastructure. In order to identify "individuals of little value" (people suffering from tuberculosis, alcoholics, vagrants, prostitutes, the mentally disabled), it was necessary to employ large numbers of clerical workers and, consequently, to allocate resources from the state budget. In addition to administration and funds, the state had special measures at its disposal, namely, the legal application of physical violence.[20] This final factor brought the nascent eugenic community and governmental bodies closer together.

Throughout the interwar period, eugenicists made repeated appeals to the state and finally succeeded in gaining support for the idea of a public health service. Partly owing to the persistence and determination of physicians such as Tomasz Janiszewski, a separate Ministry of Public Health and Welfare (*Ministerstwo Zdrowia Publicznego i Opieki Społecznej*) was created in 1918. Until its dissolution in 1924, a member of the Polish Eugenics Society headed the Ministry. Polish eugenics of the interwar period therefore considered the state as the sole institution empowered to pursue eugenic policies on a national scale.

The Polish Eugenics Society in the Interwar Period

Until the late 1920s, the Society for Combating Venereal Diseases (renamed the Polish Eugenics Society in 1922) remained an organization consisting of experts. At this time, Leon Wernic chaired the Society. Its members were physicians, dermatologists, psychiatrists, and pediatricians.[21] For example, practically every member of the editorial board of the quarterly *Zagadnienia rasy* (Questions of Race) had doctoral degrees, and some were academics. Many were physicians born during the 1870s and 1880s. The Polish Eugenics Society had numerous branches in towns and cities across the country. Consequently, the influence of the society expanded along with its membership. (On the eve of the outbreak of the Second World War the society had almost 10,000 members.) After 1927, prominent anthropologists from the Lvov School of Anthropology, as well as writers and journalists connected to the Polish Section of the World League for Sexual Reform established by the poet, translator and writer Tadeusz Boy-Żeleński (1874–1941), became involved in the work of the society.[22]

In many respects the development of the Polish eugenic movement resembled the Scandinavian model. Similar to Denmark, Norway and Sweden, Polish eugenics was supported by left-wing and liberal advocates of state welfare. The vast majority of eugenicists proposed raising benefits for citizens and state interventionism in the economy. Indeed, contributors to *Zagadnienia Rasy* did not conceal their aversion to free-market competition. In Janiszewski's opinion, "free-market egoism" leads to "human capital being wasted" and "the exploitation of the labor force."[23] Advocates of eugenics sought to introduce a rational system of economic management, one guaranteed by government administration, and to replace "free-market chaos." Such ideas stood in sharp contrast to nationalist visions, which aimed to combat *etatism* and promote free-market competition as a matter of principle. German racial hygiene also influenced its Polish counterpart in no small measure during the formative years of the movement's development. Books and pamphlets by Alfred Ploetz (1860–1940), Wilhelm Schallmayer (1857–1919), and Alfred Grotjahn (1869–1931) all played a decisive role in influencing Polish eugenics. The only handbook of eugenics published in Polish was translated from German.[24] Yet, as a result of the growing tide of racism in Germany after 1933, the title of *Zagadnienia Rasy* was changed in 1938 to *Eugenika Polska* (Polish Eugenics). Nevertheless, the closeness of these models was also signified in debates about sterilization in the draft eugenic legislation of the mid-1930s.

In the interwar years, eugenicists pursued community work on a large scale. The Polish Eugenics Society contributed to the reduction of venereal and other communicable diseases, as well as the international trafficking of women. To this end, Polish eugenicists co-operated with women's rights organizations, such as the Union for Women's Civic Work (*Związkiem Obywatelskiej Pracy Kobiet*), the Christian Society for the Protection of Catholic and Protestant Women (*Chrześ cijańskiego Towarzystwa Ochrony Kobiet Katolickich i Prostanckich*), and the Jewish Society for the Protection of Women (*Żydowskiego Towarzystwa Ochrony Kobiet*), as well as with representatives of the Polish Ministry of Labour and Welfare. In collaboration with these organizations, the society helped found the Polish Committee against the Trafficking of Women and Children (*Polski Komitet do Walki z Handlem Kobietami i Dziećmi*) in 1923.[25]

Nation, Class and Race

Eugenicists generally understood the concept of the "nation" according to historical and biological laws. The economist and social activist involved in fighting alcoholism, Zofia Daszyńska-Golińska (1866–1934), wrote: "The nation is an organization based on a bio-genetic community."[26] The historian and politician Apolinary Garlicki (1872–1940) believed that the nation "lasts only as long as it preserves a healthy and pure collective idioplasm."[27] In his numerous publications the most radical of Polish eugenicists, Tomasz Janiszewski, stressed that: "Health, and physical health alone determines the existence of nations and states."[28] He believed that the development of nations depended on the condition of race. However, they toned down the nationalist vision of an inevitable antagonism between nations and races, replacing it with an idea of "rivalry," in which Poland participated on an equal footing with other nations.[29] Ultimately, the protection of the race and the nation was a patriotic duty, a "dictate of reason and the heart."[30]

Throughout the interwar period Polish eugenicists used the word "race" in broad terms: as a synonym for community (nation, society, social group), designating genetic inheritance (inherited racial characteristics), or as an anthropological term denoting particular physical characteristics. Polish eugenicists tended to refrain from anti-Semitic and racist phraseology. This can be explained by the following factors: the "progressive," non-nationalistic nature of the society; the fact that at any given time its membership included many physicians and social activists of Jewish origin; and general disapproval of racist theories, in addition to the Nazi interpretation of physical anthropology. During the 1930s, anthropologists active in the eugenic movement deconstructed the main tenets of German racism, in articles for *Zagadnienia Rasy* and elsewhere. For example, in his works on physical anthropology, Stanisław Żejmo-Żejmis protested against the use of the term "race" in reference to nationality. He pointed to the harmful influence of Arthur de Gobineau's theory about the development of anthropology as a distinct discipline.[31]

The "Military Anthropological Photograph" (*wojskowe zdjęcie antropologiczne*), a large-scale project carried out by Polish anthropologists in the 1920s, underlined Polish rejection of the vulgarization of racial sciences. The project involved the "racial" examination of over

80,000 Polish soldiers (the measurement scheme comprised forty-five points, including hair, eye, and skin coloring, body proportions, and so on), concluding that the Nordic racial type prevailed in Poland.[32] Following the Nazi occupation of Poland in September 1939, the material compiled during the project was destroyed.

Of all Polish eugenicists, only Karol Stojanowski (1872–1947), a nationalist-leaning journalist, demanded the creation of an exclusivist Polish nation-state in which the civil rights of national minorities (especially the Jews) would be denied.[33] Unsurprisingly, his opinions were not published in *Zagadnienia Rasy*, for Wernic endeavored to ensure that the journal remained untainted by anti-Semitic prejudice. Yet Stojanowski did not belong to the eugenic mainstream; he did not, for instance, support the idea of compulsory sterilization pursued by the Eugenics Society during the 1930s.

The Polish eugenic movement was class- rather than race-oriented. In various draft versions of the laws on sterilization proposed by the society during the 1930s, the strictest eugenic legislation was targeted at the poor and the underclass. Eugenicists appealed to the state to regulate, by means of sterilization, the natural increase of "the handicapped, often degenerate; those that are a permanent burden on state and society: the degenerate, the retarded and the sick that constitute the majority of hospitalized patients and inmates of shelters and other specialized institutions."[34] However, the general prosperity of the state was to be promoted through the organization and selection of the labor force according to racial principles. It was thought that eugenic "breeding" would produce effective workers and good soldiers. Eugenicists stressed the importance of career guidance and published a cycle of pamphlets in 1925 under the title *O wyborze zawodu i wychowaniu* (On Career Choice and Upbringing). Furthermore, they recommended that employers introduce IQ tests and methods of psychological consultation when recruiting new workers. Improving working conditions would motivate workers to achieve higher productivity, while simultaneously increasing job satisfaction. Eugenicists declared that popularizing the principles of occupational hygiene would lead to an improvement in relations between private entrepreneurs and workers; therefore it was important to cultivate an environment in which new social and cultural attitudes could take root.

Furthermore, eugenicists aspired to assume a specific role in society, namely, that of experts on occupational hygiene.[35] However, when taking a closer look at eugenic discourse it becomes apparent that, despite the language of paternalism, eugenicists harbored a manipulative attitude towards workers, referred to as "human material," "human factor," "production material," "labor force," "production force" or "military material."[36] This type of thinking, in its attempts to rationalize and "technologize" daily life, was common during the interwar years. Still, this type of social engineering did have positive effects: the advent of the new architecture inspired by Charles Le Corbusier (1887–1965) helped raise the living standards of the labor force. That said, eugenicists added an extra dimension to this project: violence.

Politics and Eugenics

In the 1930s, eugenicists encountered obstacles in their efforts to introduce sterilization, not only in the form of the Roman Catholic Church but also, more importantly, by politicians, who remained indifferent to their arguments. When Józef Piłsudski (1867–1935) staged a military coup in May 1926, leading figures of the Polish Eugenics Society made proposals to Piłsudski. Piłsudski's circle of associates included socialists, liberals, and conservatives.[37] The ruling circle, known as the *Sanacja*, proposed a new "state ideology" that promoted loyalty to the Polish state. In turn, members of the Society conceived of eugenics as a form of state ideology. To this end they organized lecture series and exhibitions under the auspices of the Piłsudski camp (Piłsudski's portraits and excerpts from his speeches were used on the Society's publicity materials).[38] In the proposed eugenic legislation, the state was to offer special protection and assistance to "people from families with a good record of community work," "disinterested social activists," "model employees in all types of production," "healthy mothers, educators and housewives."[39]

These efforts notwithstanding, eugenics did not become part of Polish state ideology. Officials from the *Sanacja* camp sternly rejected all proposals for compulsory sterilization, obligatory prenuptial certificates, and any other form of eugenic regulation of the daily lives of citizens. The distinction between citizens of "greater" and "lesser" value was described by one official as "a caricature of censured society," and the relevant legislation as "a perfect arena for enormous abuse, nepotism and corruption."[40] The eugenicists' only tangible success in the 1930s was the establishment in 1935 of a Eugenics Section attached to the government-run National Health Council, whose eugenic legislation was debated and drafted independent of the Eugenics Society.[41] Clearly, eugenics was more popular in medical circles than in the political sphere. For various reasons, politicians from across the spectrum—and particularly socialists—were disinterested in eugenic legislation; socialists, in particular, were suspicious of the neo-Malthusian doctrine. Only a small proportion of socialists supported the idea of birth control. Likewise, Polish socialists coldly received eugenics, regarding it as yet another theory distracting the attention of the masses from more pressing problems.

Conclusions

The political isolation of physicians and eugenicists deepened from the mid-1920s onwards. The strong position of the Roman Catholic Church, aversion to theories of race—among both the political left and right—and the example of Nazi Germany all played an important role in diluting eugenic ideas in Poland. The idea of the "eugenic correction" of society was quashed as early as 1924.

The domination of the Ministry of the Interior over health policy increased the risk of the introduction of compulsory sterilization. In Poland, the dissolution of the Ministry of Public Health (the removal of physicians from policy making) averted the threat of artificial social selection. However, the lack of racist rhetoric among Polish eugenicists did not mitigate the grandeur of their utopian visions of a disease-free, happy, and vibrant society. The 1935 legislation on eugenics, the Eugenics Bill *(Projekt Ustawy Eugenicznej)*, identified epileptics, those afflicted by hereditary blindness or deafness, the insane (schizophrenics, manic-depressives), alcoholics, the mentally disabled and, finally, those with "grave hereditary physical defects" as cases for compulsory sterilization. Other proposed laws went even further. Wiktor Grzywo-Dąbrowski, for instance, suggested the compulsory sterilization of those suffering from tuberculosis and syphilis.[42] Other drafts included the sterilization of drug addicts and alcoholics.[43]

Eugenicists were motivated by a peculiar idea of the "common good." They imagined that sterilization, if applied in the long term, would eventually steer the biological evolution of man towards a final state of perfection. They were deaf to criticism and the warnings of opponents of sterilization.[44] The eugenic correction of society was what Karl Popper has termed "utopian social engineering"; that is, a large-scale project of change whose practical consequences are unpredictable.[45] For eugenicists, the introduction of sterilization as part of a wider social experiment was the only way to acquire knowledge about inheritance, thereby contributing to social progress.

Endnotes:

[1] Zdzisław Krasnodębski, *Upadek idei postępu* (Warsaw: Państwowy Instytut Wydawniczy, 1991), 8–9.

[2] Gunnar Broberg and Nils Roll-Hansen, eds., *Eugenics and the Welfare State: Sterilization Policy in Denmark, Sweden, Norway, and Finland* (East Lansing, MI: Michigan State University Press, 1996).

[3] See Magdalena Gawin, *Rasa i nowoczesność. Historia polskiego ruchu eugenicznego (1880–1952)* (Warsaw: Neriton, 2003).

[4] *Ster* was published between 1907 and 1914.

[5] *Ogniwo* was published between 1902 and 1905.

[6] *Medycyna* and *Kronika Lekarska* were published between 1908 and 1918.

[7] Leon Wernic, "O uświadomieniu płciowym młodzieży w okresie szkolnym i przedszkolnym," *Zdrowie* 6, 8 (1907), 455.

[8] *Czystość* was published between 1905 and 1909.

[9] For example the Academic Society of Ethos (*Akademickie Towarzystwo Ethos*); the Society for the Purity of Morals (*Towarzystwo Czystości Obyczajów*); and the Society for Combating Venereal Diseases and Promoting the Principles of Abolitionism (*Towarzystwo Dla Walki z Chorobami Wenerycznymi i Szerzenia Zasad Abolicjonizmu*).

[10] See Daniel Pick, *Faces of Degeneration. A European Disorder, c. 1848–c. 1918* (Cambridge: Cambridge University Press, 1996).

[11] Leon Wernic, "Małżeństwo z punktu widzenia higieny społecznej i seksualnej," *Czystość* 6 (1907), 85–89.

[12] Benedykt Dybowski, "O nadwyznaniowości i religii," *Czystość* 33–36 (1909), 513, 543.

[13] Dybowski, "O nadwyznaniowości i religii," 545.

[14] (Anonymous), "Nieśmiertelność," *Czystość* 41, 43 (1909), 625, 643; and (Anonymous), "Dusza," *Czystość* 43 (1909), 666.

[15] Justyna Budzińska-Tylicka, "O potomstwie alkoholików," *Czystóść* 1 (1909), 4.

[16] Augustyn Wróblewski, "Nasz program życia i czynu," *Czystóść* 26 (1909), 401–402.

[17] Prus was appointed the patron of the society because of his continual support of public awareness campaigns run by physicians to promote vaccination during the 1880s and 1890s, and of the construction of waterworks in urban areas and general access to clean water.

[18] Jan Boguszewski, "Zadania eugeniczne w Polsce," *Walka o zdrowie* 1, 6–7 (1918), 239.

[19] Zygmunt Zakrzewski, "Walka ze zwyrodnieniem," *Zagadnienia Rasy* 1, 11 (1921), 8–9.

[20] As Max Weber stated: "Specifically, at the present time, the right to use physical force is ascribed to other institutions or to individuals only to the extent to which the state permits it. The state is considered the sole source of the 'right' to use violence." See Max Weber, "Politics as a Vocation," in Max

Weber, *Essays in Sociology*, ed. H. H. Gerth and C. Wright Mills (New York: Oxford University Press, 1946), 77.

[21] The physicians included Witold Chodźko, Tomasz Janiszewski, Ludwik Hirszfeld and Wiktor Grzywo-Dąbrowski; and women rights activists included Zofia Daszyńska-Golińska, Teodora Męczkowska and Eugenia Waśniewska.

[22] Leading figures in the field of anthroplogy included Jan Mydlarski and Stanisław Żejmo-Zejmis.

[23] Janiszewski, *Polskie Ministerstwo Zdrowia Publicznego* (Krakow: Odbitka ze Zdrowia, 1917), 7.

[24] Alfred Grotjahn, *Higiena praktyczna* (trans. by T. Janiszewski) (Krakow: Wydawnictwo Polskiego Towarzystwa Eugenicznego, 1930).

[25] See Teodora Męczkowska, "Stosunek Towarzystwa Eugenicznego do towarzystw pokrewnych," *Zagadnienia Rasy* 2, 1 (1924), 17; and Wacław Borkowski, "Projekt organizacji misji kolejowych," *Zagadnienia Rasy* 2, 11 (1924), 11.

[26] Zofia Daszyńska-Golińska, *Polityka populacyjna* (Warsaw: Księgarnia Ferdynanda Hoesica, 1927), 15.

[27] Apolinary Garlicki, *Zagadnienia biologiczno-społeczne* (Warsaw: Książnica Naukowa, 1924), 258.

[28] Janiszewski, *Polskie Ministerstwo Zdrowia Publicznego*, 10.

[29] Wacław Wesołowski, "Walka z chorobami wenerycznymi a obrona rasy," *Zagadnienia Rasy* 1, 1 (1918), 7.

[30] Wesołowski, "Walka z chorobami wenerycznymi a obrona rasy," 7. See also Henryk Nussbaum, "Mens sana in corpore sano," *Zagadnienia Rasy* 1, 1 (1918), 4.

[31] Stanisław Żejmo-Żejmis, "Doktryna rasizmu," *Prosto z mostu* 6, 41 (1937), 3; and Stanisław Żejmo-Żejmis, "O rasie, rasach i rasizmie," *Zagadnienia Rasy* 4, 5 (1929), 79.

[32] Jan Mydlarski, "Sprawozdanie z wojskowego zdjęcia antropologicznego," *Kosmos* 50, 2–3 (1925), 530.

[33] See Karol Stojanowski, K. *Rasizm przeciw Słowiańszczyźnie* (Poznań: Wydawnictwo Głos, 1934); and Karol Stojanowski, *Polsko-niemieckie zagadnienie rasy* (Poznań: Księgarnia Katolicka, 1939).

[34] Ludwig Witowiecki, "Zagadnienie populacyjne z punktu widzenia eugeniki praktycznej," *Życie Młodych* 2 (1939), 40.

[35] Marcin Kacprzak, "Państwowa Szkoła Higieny," *Lekarz Polski* 6 (1927), 14; and Bruno Nowakowski, "Higiena pracy a zdrowie publiczne," *Lekarz Polski* 6 (1927), 8.

[36] See Kazimierz Karaffa-Korbutt, *Praca i odpoczynek* (Krakow: Okrągowy Związek Kas Chorych, 1929).

[37] See Jerzy Lukowski and Hubert Zawadzki, *A Concise History of Poland* (Cambridge: Cambridge University Press, 2004), 212–214.

[38] Leon Wernic, "Eugenika—jej zadania społeczne w Polsce oraz stosunek do medycyny i innych nauk," *Zagadnienia Rasy* 2, 2 (1925), 37; and Leon Wernic, "Wymieranie narodów w przeszłości i narodów współczesnych oraz

rola prawodawstwa i organizacji eugenicznych w chwili bieżącej," *Zagadnienia Rasy* 3, 10 (1927), 87.

[39] Leon Wernic, "*O ustawach eugenicznych w Polsce*," *Zagadnienia Rasy* 9, 1 (1935), 59.

[40] A Eugenics Bill (*Projekt Ustwy Eugenicznej*) 1935, New Records Archives, Warsaw (*Archiwum Akt Nowych., Warszawa*), Welfare Ministry Files (*Ministerstwo Opieki Społecznej*), signature 532, 25–26.

[41] See Wiktor Grzywo-Dąbrowski, *Postulaty w sprawie sterylizacji i kastracji*; Bohdan Ostromęcki, *Projekt ustawy eugenicznej*; Witold Łuniewski, *O hamowaniu rozrodu niepożąnego;* Leon Wernic, *O hamowaniu rozrodu osobników dysgenicznych*; *Sekcja Eugeniczna Państwowej Naczelnej Rady Zdrowia* (1936), Special Collection of the Chief Medical Library, Warsaw (Zbiory Specjalne Głównej Biblioteki Lekarskiej), signature D/501.MOS, no pagination.

[42] Zbiory Specjalne (Special Collection), Główna Biblioteka Lekarska w Warszawie (Central Medical Library, Warsaw), signature DI/501.MOS. Sekcja Eugeniczna Państwowej Naczelnej Rady Zdrowia (Ministry of Welfare, Eugenics Section of the National Health Council).

[43] See "A Eugenics Bill," Archiwum Akt Nowych, Warsawa, signature 532.

[44] See Stanisław Podoleński, "Eugenika dzisiejsza, jej drogi i bezdroża," *Przegląd Powszechny* 193–196, 581 (1932), 171; Stanisław Podoleński, "Eugenika i ruch eugeniczny," *Przegląd Powszechny* 193–196, 579 (1932), 319; Maria Kępínska, *Świadome macierzyństwo* (Poznań: Katolickie Zjednoczenie Związków Polek, 1934); and Stefan Dąbrowski, "Sterylizacja w świetle etyki," *Życie Medyczne* 5, 18–19 (1938), 6.

[45] See Karl Popper, *The Open Society and Its Enemies* (London: Routledge and Kegan Paul, 1966), 157–168.

The First Debates on Eugenics in Hungary, 1910–1918

Marius Turda

The history of eugenics in Hungary remains a neglected area in contemporary scholarship. Although studies dealing with German racial hygiene and eugenics during the interwar period record the eugenic ideals professed by various Hungarian political and intellectual figures, to date no scholarly discussion of the eugenic movement in Hungary has been undertaken.[1] One would have expected Hungarian scholarship to compensate for such a historiographic lacuna. However, in most Hungarian scholarship, eugenics is either marginalized as an insignificant historical detail, or treated indistinguishably from other subjects like bio-medical racism.[2]

Such historical and academic neglect has, however, no justification. Like elsewhere, Hungarian eugenicists addressed a wide range of medical, social, and political issues, from social hygiene and mental care to forced sterilization and serological research of ethnic groups. Hungarian eugenicists deemed resolution of these issues essential to the progress of Hungarian society at the time. Moreover, the fascination with eugenics in Hungary knew no ideological restrictions. Socialist and fascist supporters alike favoured it, and religious groups, such as Roman Catholics, offered some of the most sophisticated interpretations of the relationship between eugenics and religion of the interwar period in Hungary. The time has come for the history of eugenics in Hungary to receive the attention it deserves, and for Hungarian eugenicists to be integrated within the international scholarship on racial hygiene and eugenics.

This chapter discusses the first phase in the history of eugenics in Hungary, between 1910 and 1918. During this formative period, two schools of eugenic thought were formed: the first, internal, group was represented, most prominently, by István Apáthy, József Madzsar, Lajos Dienes, Zsigmond Fülöp, János Bársony and Mihály Lenhossék. This group believed in the pre-eminence of Hungarian eugenics over competing eugenic ideas from abroad. The second, external, group was dominated by the activity of one individual, Géza von Hoffmann, although eugenicists associated with the internal group, such as János Bársony and Ernő Tomor, also contributed to its development. The external group of Hungarian eugenicists endeavoured to accommodate ideas associated with American and German eugenics, especially racial hygiene, to Hungarian realities. The interaction between these two groups of eugenicists provided a propitious environment for the emergence of some of the most interesting theoretical debates on the meaning and practice of eugenics in early-twentieth-century Europe. Scholars concur that during the first two decades of the twentieth century, British, American and German eugenic movements were well established; yet no study has hitherto documented that between 1910 and 1918 Hungary was, in fact, in the vanguard of eugenic thinking in Europe.

The chapter is divided into four sections. First, I review some of the Hungarian reactions to the European debate about racial hygiene and eugenics during the first decade of the twentieth century. Interestingly, the development of British eugenics and German racial hygiene preoccupied not only established medical doctors such as Lajos Dienes and József Madzsar, or progressive sociologists like Oszkár Jászi, but also prominent Hungarian politicians like Count Pál Teleki. In the second section I examine the first Hungarian debate on eugenics, the so-called *Eugenika vita*, which was organized in 1911 by the Hungarian Academy of Sciences, with an impressive number of Hungarian intellectuals from various disciplines, including biology, anthropology and psychiatry, attending. In the third section I discuss the creation of the Eugenics Society in 1914. A lively dialogue on the role and importance of eugenics and racial hygiene accompanied the establishment of the Eugenics Society. Indeed, István Apáthy and Géza von Hoffmann published some of their most important studies relating to eugenics during this period. The publication of these studies notwithstanding, there was no agreement between Apáthy and Hoffmann with respect to their role within the Eugenics Society. The external group did not become an integral component of the eugenic movement in Hungary as Hoffmann had hoped, and the methodological rupture between his theory of eugenics and that proposed by Apáthy persisted until 1917.

In the final section I consider the evolution of eugenics during the First World War. After 1914, the social and national transformations caused by war became prime eugenic concerns, as illustrated by the work of Lajos Méhely, János Bársony, József Madzsar and Géza von Hoffmann. Equally importantly, the war reconfigured the relationship between the internal and external groups of Hungarian eugenicists. If, before the war, the preponderance of the internal group was unquestionable, after the outbreak of the war the influence of the external group notably increased. Between 1914 and 1917, Géza von Hoffmann, arguably the most internationally recognized Hungarian eugenicist, contributed to European and American debates on racial hygiene, eugenics and sterilization. He published both in Hungary and abroad, thus integrating Hungarian eugenics within a broader European and North American context.

In the climate of heightened nationalism and concerns about the declining birthrate, advocates of social hygiene like János Bársony and József Madzsar designed innovative schemes for promoting the health of the family. Although eugenicists saluted the introduction of these schemes in preventive medicine, they disagreed on its ideological message. Eugenicists associated with the conservative Right were concerned that socialists, and the Left generally, should not be seen to take the initiative in health education. As in Germany and Austria, conservative eugenicists in Hungary developed their own hygienic strategy, which promoted racial values and a new idea of nationalist morality. Eugenics became a useful instrument for supporters of state welfare and for those, like Ernő Tomor and Géza von Hoffmann, who urged the introduction of a population policy to support the quality rather than quantity of births. Both groups, however, agreed on one important subject, namely, that eugenic policies should improve the racial qualities of the nation. This was the programmatic vision upon which the Hungarian Society for Racial Hygiene and Population Policy was created in 1917. The establishment of this Society placed Hungarian eugenics firmly on the intellectual map of European eugenic movements.

I. The Roots of the Eugenic Movement (1904–1910)

The emergence of the eugenic movement in Hungary required the acceptance of an entirely different discipline by established academics and political actors of the day. Yet eugenics, however different, was not entirely new in Hungary. Preoccupations with the racial characteristics of peoples and their distinct somatic and psychological features had preoccupied generations of Hungarian anthropologists, naturalists, and practitioners of medical sciences during the course of the nineteenth century.[3] Attuned to the evolutionary theories of Jean Baptiste Lamarck (1744–1829), Herbert Spencer (1820–1893), Thomas Malthus (1766–1834), Charles Darwin (1809–1882), and August Weismann (1834–1914), Hungarian supporters of the theory of natural evolution advocated the revolutionary importance of new theories of heredity for the development of Hungarian science.[4] In addition to anthropology, sociology and medical hygiene were the other scientific disciplines to cultivate a Darwinist interpretation of human nature, a view that facilitated the transmission of eugenic ideas.

To be sure, the fascination with evolution and heredity transcended the confines of scientific disciplines. Count Pál Teleki (1879–1941), a prominent Hungarian aristocrat and prime minister of Hungary (1920–1921 and 1939–1941), for example, was fascinated by the relationship between science and politics and excelled in his knowledge of human geography. In 1904, in the journal established by progressive Hungarian intellectuals *Huszadik Század* (Twentieth Century), Teleki welcomed the publication of the first issue of the *Archiv für Rassen- und Gesellschaftsbiologie* (Journal of Racial and Social Biology)—a journal edited by Alfred Ploetz (1860–1940), the German racial hygienist. Teleki made no explicit attempt to discuss the concept of "racial hygiene" but he reiterated his support for Ploetz's eugenic endeavours and concurred with Ploetz that biology was the true venue for exploring the dynamics of both human societies and the evolution of individuals. "True social sciences," Teleki wrote, are dependent on biology in order to reveal the "social fabric of mankind."[5] The "father of eugenics," Francis Galton (1822–1911), was also reviewed favourably in *Huszadik Század*. In 1906, József Madzsar (1876–1940), a leading physician and social hygienist, briefly examined Galton's definition of eugenics, as presented in "Eugenics: Its Definition, Scope and Aims" (1904).[6] Madzsar's review was shortly followed by a more substantial translation of one of Galton's most important articles dealing with the relationship between eugenics and biometry, "Probability, The Foundation of Eugenics" (1907).[7] With this article Galton not only made it clear that he supported Pearson's biometrical studies, but he also validated the embryonic Hungarian discourse on biometry.

Magyar Társadalomtudományi Szemle (Hungarian Review of Social Sciences) was the other important journal helping to galvanize the dissemination of eugenic ideas in Hungary during the first decade of the twentieth century. In 1909, a new section on "Közegészségügy" (public health) appeared in the journal, quickly becoming a forum for debate about Darwinism and evolution, as well as feminism and biology, for authors of various political and intellectual orientations, including István Apáthy (1863–1922), the eminent Hungarian zoologist and dean of the Medical Faculty in Kolozsvár (today Cluj, Romania), and the lawyer and journalist Géza Kenedi (1853–1920).[8]

Scientific discussions on evolution and heredity also influenced attitudes towards practical experimentation in the field of traditional hygiene. A new biological ethos was gradually formed, one that refashioned general ideas about hygiene into topics connected to racial hygiene. In 1906, the Hungarian statistician Géza Vitéz (1871–1931) produced an analysis of how social conditions determined reproductive behaviour. In *Születési és termékenységi statistika* (Statistics on Births and Fecundity), Vitéz explained fluctuating birthrates as a consequence of fertility change and the "rationalization" of reproduction by wealthy families. Moreover, Vitéz envisioned an interventionist programme aiming simultaneously to protect, regulate, and emancipate Hungarian women in accordance with new ideas of social hygiene.[9]

Towards the end of the first decade of the twentieth century, ideas of social and racial hygiene were appropriated by an eclectic segment of the Hungarian scientific elite. The strongest support came, however, from medicine. Growing interest in the relationship between socioeconomic development and medical pathology transformed the national responsibility bestowed upon the medical profession. Physicians increasingly considered themselves responsible for promoting the social and national health of the nation. In a series of studies published between 1904 and 1910, József Madzsar outlined his conception about the social and medical transformation of Hungarian society. He was engaged in various medical projects, including combating "social plagues" such as alcoholism, as well as the popularization of Darwinism as a social theory.[10] According to Mária Kovács, Madzsar should be credited with introducing the "eugenic gospel" into Hungary.[11] Indeed, Madzsar was a respected physician and an enthusiastic supporter of Darwinism and evolution, but the dissemination of eugenic ideas in Hungary was not the achievement of one individual; instead, it was the result of concerted work by a number of physicians, sociologists, and anthropologists. However, it is indisputable that Madzsar played a significant role in stimulating interest in the benefits of the new science of eugenics among the scientific and political elite in early-twentieth-century Hungary.

A prelude to the debate about eugenics came in 1910 with the publication of three seminal articles in *Huszadik Század*. Physician and biologist Lajos Dienes (1885–1974) authored the first. Under the title "Biometrika" (Biometrics), the article analysed Francis Galton's contribution to the study of measurable biological characteristics.[12] József Madzsar wrote the second article, "Gyakorlati eugenika" (Practical Eugenics), in which he described the historical achievements of Mendelism and argued that the new *eugenika vallása* (eugenic religion) presented further venues for the understanding of the individual and society.[13] Finally, Zsigmond Fülöp (1882–1948), a naturalist and the editor of Darwin's works in Hungary, published "Eugenika" (Eugenics),[14] in which he discussed various definitions of eugenics and the relationship between eugenics and biometry. Biometrics and biometry became part of the European scientific vocabulary after the publication of the journal *Biometrika*, founded in 1901 by Francis Galton, Walter F. R. Weldon (1860–1906) and Karl Pearson (1857–1936). According to Pearson, biometry denoted the application of modern statistical methods to biology. It was assumed that Pearson's mathematical model was opposed to the newly discovered Mendelian laws of inherited characteristics (the law of segregation and the law of independent assortment). Familiar with the debate between biometricians and Mendelians in Britain, Dienes, Madzsar and Fülöp chose, however, to synthesize their own conflicting views on heredity and natural selection rather than create two separate schools of eugenic thought.

The publication of these articles in *Huszadik Század* thus had two immediate consequences. First, it demonstrated that Hungarian eugenicists were eager to unify various interpretations of the new science of eugenics and to initiate a discussion about eugenics within the larger scientific community. Second, it openly challenged Hungarian supporters of Lamarckism to accommodate Mendel's laws of heredity, particularly the notion of the transmission of human characteristics. Once eugenics had succeeded in attracting sufficient supporters, two dominant viewpoints formed with respect to its application to Hungarian society.

One category of eugenicists argued that economic reform and the

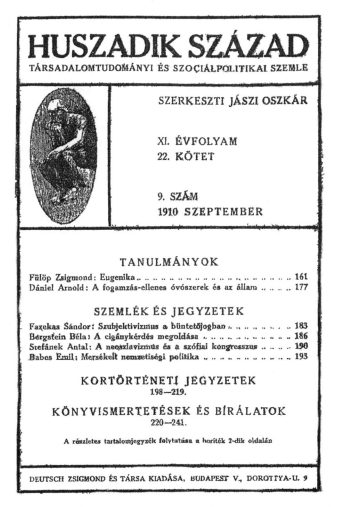

Fig. 1. Zsigmond Fülöp's article on eugenics announced in the progressive journal *Huszadik Század*

improvement of social conditions should be directed by the precepts of racial hygiene. The advocates of this view of social reform underlined the necessity for selective breeding policies designed to prevent those individuals with "negative" characteristics from social interaction and, ultimately, reproduction. Yet such a radical position did not go unchallenged. There were supporters of eugenics who, instead, perceived their actions primarily in terms of social and medical reforms. For this category, eugenics was part of a broader hygienic *Weltanschauung*, which included diverse policies, such as improving the living conditions of the urban underprivileged classes in the suburbs of Budapest, the prevention of venereal diseases, and the social reintegration of prostitutes. These two interpretations of the role of eugenics in society were to be tested during the first public debate on eugenics in Hungary, the *Eugenika vita* of 1911.

II. The First Debate on Eugenics in Hungary (1911)

The transformation of eugenics from a specialized scientific discourse involving a few physicians interested in heredity, into an organized discipline and public association occurred gradually and involved a number of crucial steps. The scientific jargon of eugenics had to be popularized, so that it appeared to address the general problems of Hungarian society. The debate between biometricians and Mendelians might have interested Dienes and Madzsar, but such a subject hardly ensured the favourable reception of eugenic rhetoric among the less scientifically oriented members of the Hungarian political and social elite. In Hungary, similar to other parts of Europe, eugenics was redesigned according to immediate social and national concerns.

In 1911, *Huszadik Század* published three articles devoted to eugenics. In the first, "Fajromlás és fajnemesítés" (Racial Degeneration and Racial Improvement), József Madzsar presented a conceptual synthesis of various works on eugenics including Francis Galton's *Natural Inheritance* (1899), Karl Pearson's *The Scope and Importance to the State of the Science of National Eugenics* (1909), and Leonard Doncaster's *Heredity in the Light of Recent Research* (1910). Agreeing with these authors, Madzsar argued that the new science of eugenics should be directed towards the eradication of medical problems facing Hungarian society, such as tuberculosis and malaria, the level of "fertility in diseased and healthy families," as well as the impact of alcoholism on these families. More importantly, however, Madzsar imbued eugenics with a social mission. Fears of social degeneration characterized much of Madzsar's previous work on alcoholism and fertility. Transposing these anxieties to the perceived decline of the racial qualities of the population confirmed Madzsar's belief: the proportion of the constitutionally weak and mentally disabled was on the increase in Hungary. Ultimately, "the institution of marriage" needed to be reformed. Echoing familiar tropes of British discourse on eugenics, Madzsar suggested that marriage between the mentally disabled caused profound social instability; eugenics should in turn consider regulating such marriages.

The wider intellectual objectives of this article largely mirrored Madzsar's own ideological goals: to overcome divided attitudes about the implementation of social policies in Hungary, most notably in the spheres of medical care and social assistance; to educate a responsive audience in the virtues of eugenics, especially its heuristic and ideological potential as a source of individual liberation; and, finally, to develop and clarify eugenic reform by purging elitist notions from its social doctrine. To create a "biological aristocracy"—one based not on social class but upon hereditary qualities—was perhaps a eugenic utopia, but one that could nevertheless be achieved by the next generation.[15]

Contributions by Francis Galton and Karl Pearson to the consolidation of eugenics as a scientific discipline were further analysed by Lajos Dienes. In "A fajnemesítés biometrikai alapjai" (The Biometrical Basis of Racial Improvement), he explored the relationship between eugenics and biometrics. On the one hand, Dienes acknowledged Madzsar's innovative contribution to the dissemination of eugenic ideas in Hungary; on the other, he maintained the importance of Pearson's approach in understanding racial qualities through statistical techniques, in particular. According to Dienes, differences in physical traits, health, and intelligence could be explained by the statistical study of natural selection within the population. Dienes thus combined Pearson's population approach to Darwinist variation with Galton's hereditary conception of society in order to justify his own concept of hereditary social policies. The introduction of such policies would, according to Dienes, improve the social condition of the nation.[16]

Zsigmond Fülöp added a new dimension to this discussion. In "Az eugenetika követelései és korunk társadalmi viszonyai" (The Claims of Eugenics and the Social Conditions of Our Age), Fülöp attempted to analyse eugenics from a comparative perspective. This was undertaken through an assessment of Madzsar's arguments about the social role of eugenics, as well as the works of Francis Galton and Wilhelm Schallmayer on heredity. Eugenics simultaneously aimed to improve (through "positive" eugenics) and impair (through "negative" eugenics) the racial qualities in a population. The differentiation of eugenic methods, as Fülöp acknowledged in his article, followed Galton's original definition of eugenics, posited in 1883. In addition to its potential for channelling social transformation, eugenics should also prompt the creation of a new "national ethics," namely, an evolved form of social solidarity ensuring the racial continuity of future generations.[17]

These articles catalysed a lengthy debate about eugenics in Hungary. Coordinated by the *Társadalomtudományi Társaság* (The Sociological Society), the *Eugenika vita* had a clear aim: to clarify the scientific and social challenges posed by *eugenika, eugenetika* (eugenics), *fajromlás* (racial degeneration), and *fajnemesítés* (racial improvement). This multiple terminology suggests that doubt may have existed among Hungarian supporters of eugenics as to whether the term "eugenics" was sufficient to encompass the competing viewpoints expressed during the debate. Under the provocative title "A fajnemesítés (eugenika) problémái" (The Question of Racial Improvement—Eugenics), *Huszadik Század* published in two consecutive issues some of the speeches delivered on the occasion.[18]

Initial reporting of the debate contained summaries of the arguments made by one group of participants. This included Sándor Doktor (1864–1945), who discussed Galton's theory of eugenics; László Detre (1874–1939), who reviewed Mendel's laws of heredity; and István Apáthy, who suggested the adoption of a Hungarian term, *faj egészségtana* (racial hygiene), instead of either *eugenika* or *eugenetika* (eugenics).

Dezső Buday (1879–1919), a jurist, led *Huszadik Század*'s second reporting of the debate. Buday maintained that eugenics should address social problems from two perspectives: the first, "biological," concerning the laws of breeding and heredity; and the second, "sociological," embracing biometrical statistics. Buday placed his programme of biological rejuvenation at the intersection of these two perspectives: "Now," he suggested, "we may become more humane through positive selection and more powerful through the negative one. Now, we could advertise the luxury of eugenics: breeding *Übermensch* and geniuses."[19] Yet again, the Mendelian and biometrical views on heredity appeared in Hungarian debates about eugenics; but on this occasion an amendment was suggested. A few years before the English statistician and biologist Ronald A. Fisher (1890–1962) attempted to reconcile these two positions, Buday proposed that Mendelism could complement biometry, especially in the field of social hygiene.

The social implications of eugenics were further discussed by the psychologist László Epstein (1865–1923), and the neurologist and feminist René Berkovits (1882–?). The latter went a step further and, following Madzsar, linked social hygiene to concerns about racial degeneration. According to Berkovits, "eugenics should be pursued not only from an explanatory but also from a practical point of view: to determine the heredity of acquired characteristics and the causes of degeneration."[20] Berkovits also realized the importance of institutionalizing eugenics: she suggested the establishment of a "eugenic committee" under the auspices of *Társadalomtudományi Társaság*. Indeed, Berkovits was the first supporter of eugenics in Hungary to advocate its institutionalization as a component of state social policy.

Yet Berkovits was not the only participant urging a more practical application of eugenics. Tibor Péterfi (1883–1953), a histologist, maintained that "hitherto there has been a discrepancy between prevalent eugenics theories and the low results derived from practice."[21] Likewise, commenting on the practical results of eugenics, a certain Dr. Imre Káldor noted that "the future true eugenic movement" would "create an improved living social organism."[22]

Other contributors favoured a more theoretical debate on eugenics. For example, Zoltán Rónai (1880–1940), a lawyer, insisted that eugenics should aim at solving the scientific conundrum between "nature" and "nurture." According to Rónai, the main question to be addressed was whether characteristics acquired by organisms could be inherited or whether the environment proved crucial in shaping the biological formation of the individual. In this context, biometrics was revered for providing illuminating arguments in support of eugenics. Mathematical statistics could, Rónai believed, offer a more comprehensive explanation of the occurrences of "racial degeneration" (*faj romlása*), thus assisting "racial hygiene" in finding solutions to social and venereal diseases like alcoholism and syphilis. Considering the level of social transformation envisaged by eugenics, Rónai remarked that racial hygiene could only be achieved in a "socialist society."[23] A similar left-wing vision of social reform influenced Zsigmond Engel's notion of state intervention through the practice of eugenics. Engel recommended that both the general public and scientists actively address the problem of degenerating family health and welfare. These problems, Engel insisted, required protective measures from the state and the medical profession.[24]

The discussion continued with Leó Liebermann (1852–1926), a prominent social hygienist and immunologist, who explored the relationship between "racial improvement" (*fajnemesítés*) and the social fabric of society;[25] and Vilma Glücklich (1872–1927), a teacher and feminist activist, who reasserted the importance of a Mendelian perspective on the issue, arguing that from a eugenic point of view, *fajnemesítés* cannot be deleterious to the individual because, ultimately, the "cell nucleus" is encoded in the organism and immune from external intervention.[26] Finally, Dezső Hahn (1876–?), a psychologist, admitted his reluctance to embrace Mendelism, and condemned the evidence amassed by Mendelians as contradictory. Hahn disagreed with the idea that the "hereditary composition" of the individual corresponded to any particular "cell nucleus," able to determine a particular character, and instead suggested fusing the theory of heredity with the importance of environmental factors, "from a eugenic point of view."[27]

The debate inclined, however, towards supporting Mendelism and its connection to the germ-plasm theory of heredity elaborated by August Weismann (1834–1914). This view strongly rebuked the Lamarckian theory of the "inheritance of acquired characteristics," which did however—albeit subtly—resurface in some of the arguments put forward in the debate. In "Az alkalmazott eugenika" (Applied Eugenics), Jenő Vámos (1882–1950), a sociologist and veterinary doctor, supported Weismann's view that evolution resulted from the random variation of germ cells. He also appealed to eugenics in helping to explain the significance of inborn characteristics in determining the physical constitution and temperament of individuals. With the consolidation of eugenics as a scientific discipline, progress was made towards understanding human nature, so that, according to Vámos, "In today's society applied eugenics is not a utopia anymore, and it will be even less so in the society of the future."[28]

Nevertheless, the disagreement between biometricians and Mendelians in Hungary did not mean that there were irreconcilable methodological fissures, as was the case in Britain at the time. According to Hungarian supporters of heredity and eugenics, the usage of statistical and mathematical techniques in measuring continuous variations within the population could complement the Mendelian emphasis on the discontinuous variations transmissible to progeny. Lajos Dienes and Zsigmond Fülöp, two of the most enthusiastic supporters of a fusion between biometry and Mendelism, outlined these arguments in their conclusive remarks to the debate. Both reiterated their support for biometrics and endorsed Galton's arguments about the need for eugenics-based reform. Subsuming eugenics to biology was also strongly underlined: "Again, it is reiterated that the tentative achievements in eugenic research are provided by decisive factors in general biology."[29]

This reconciliation of biometry with Mendelism suggests that there was no fixed pattern for establishing which theory of heredity was to prevail during the debate. The tendency was, however, to accept Person's theory of biometry, Galton's concept of eugenics, and to reinforce Mendel's laws of heredity. For those eugenicists supporting this view, "nature" rather than "nurture" provided the key to understanding how genetic material was transmitted from generation to generation. Disagreement persisted, however, and the willingness of some of the most prominent participants in the debate—such as Fülöp, Madzsar and Dienes—to accept the Mendelian laws of human inheritance should not obfuscate the fact that other contributors adhered to various forms of Social Darwinism, Lamarckism, and neo-Malthusianism.

The 1911 debate on eugenics illustrates how determined Hungarian eugenicists were in their efforts to understand and explain to larger audiences both the idea of heredity and its application to society. To exemplify this, it is worth returning to one paper presented to the debate. Briefly mentioned in *Huszadik Század*, Apáthy's interpretation of eugenics was published in *Magyar Társadalomtudományi Szemle* under the title "A faj egészségtana" (Racial Hygiene).[30] Apáthy was an authority on the structure of the nervous system and one of the most important supporters of Darwinism and evolution in Hungary at the beginning of the twentieth century. Together with József Madzsar and Géza von Hoffmann, Apáthy played a decisive role in the dissemination of eugenics during this period.

From the outset, Apáthy advanced a personal definition of eugenics. "Racial hygiene *(faj egészségtana)*," he declared, "is practically what Galton means by *Eugenics*."[31] Thus, although other participants in the debate used the term *eugenika* (eugenics) extensively, Apáthy rejected it. In many ways, Apáthy's interpretation of *faj egészségtana* resembled Alfred Ploetz's conception of *Rassenhygiene*. In his 1895 *Grundlinien einer Rassen-Hygiene* (The Outlines of Racial Hygiene), Ploetz defined *Rassenhygiene* as a new type of social hygiene, one that considered the future of the race to be more important than the health of the individual.[32]

As the main representative of the internal group of Hungarian eugenicists, Apáthy selected the term *faj egészségtana* over *eugenika* in an attempt to distance his theories from racist ideas.[33] Moreover, he viewed preoccupations with the health of the race as a consequence of policies of public health and social medicine. Apáthy's emphasis on social medicine is also reinforced by his use of the phrase "the health of the race" (*fajegészség*), employed interchangeably with "racial hygiene" (*faj egészségtana*).

Apáthy aimed at producing a localized Hungarian version of eugenics, one reflecting the achievements of English and German eugenics and racial hygiene but imitating neither. Considering the novelty of eugenic theories at the time, and the general tendency of other Hungarian supporters of eugenics simply to enumerate the achievements of their English and German counterparts, Apáthy's adherence to a Hungarian definition of the term demonstrates his commitment to eugenic ideals and his belief that Hungarian eugenics should be recognized as a distinct movement.

The *Eugenika vita* of 1911 had two immediate consequences. On the practical level, it demonstrated the organizational efficiency of the internal group of Hungarian eugenicists. On a theoretical level, the debate on eugenics prompted two broader questions about the role of eugenics in shaping views on the biological development of Hungarian society: How could scientific paradigms like eugenics be made compatible with the social and national particularities of Hungarian society? And could biologists and physicians be trusted as a source of scientific enlightenment amidst profound social and national transformation? By simultaneously raising the question of legitimacy and demanding practical action in the name of science, the supporters of eugenics challenged the cultural and political establishment to react more resolutely to the changes that, they argued, had been troubling Hungarian society since the late nineteenth century.[34]

Ideologically, the second decade of the twentieth century was a period of fermentation for Hungarian eugenicists. Evaluating the *Eugenika vita*, the Hungarian diplomat Géza von Hoffmann (1885–1921) noted the socialist orientation of some of the participants (Apáthy and Madzsar in particular). For Hoffmann, these contributors understood eugenics from a "social and political" perspective rather than from a "biological" standpoint; a dimension, Hoffmann argued, insufficiently explored during the debate. Accordingly, it was this "biological perspective" that Hoffmann placed at the centre of his interpretation of eugenics and racial hygiene.[35]

Die Rassenhygiene

in den

Vereinigten Staaten von Nordamerika

Von

Géza von Hoffmann

k. u. k. österr.- ungar. Vizekonsul

Mit einer Figur im Text

J. F. Lehmanns Verlag, München

1913

Fig. 2. Géza von Hoffmann's *Racial Hygiene in the United States of America*

Hoffmann was one of the few European observers at the time to have comprehensively researched eugenic legislation in America, as evidenced by the publication in 1913 of *Die Rassenhygiene in den Vereinigten Staaten von Nordamerika* (Racial Hygiene in the United States of America).[36] In this book, Hoffmann presented evidence to highlight certain concurrences between eugenic theories and practical social policies, such as sterilization and immigration. Hoffmann endorsed "negative" eugenic policies on the practical level, while simultaneously emphasizing their theoretical value.

Die Rassenhygiene in den Vereinigten Staaten von Nordamerika
catapulted Hoffmann to the forefront of the European movement on
eugenics and racial hygiene, and confirmed his status as a leading
authority on American eugenics.[37] The publication of the book also
had a profound impact on the evolution of the Hungarian eugenic movement. With this book Hoffmann articulated an alternative model of
eugenic thinking than that championed by István Apáthy and others
during the 1911 debate on eugenics. However, this period of theoretical germination led to methodological conflicts rather than to the development of a common eugenic platform. One uncertainty dominated the
aftermath of this first debate on eugenics in Hungary: Would the internal and external group of Hungarian eugenicists be able to overcome
ideological disputes in order to form one unitary eugenic movement?

III. The Establishment of the Eugenics Society (1914)

One attempt at reconciliation between the two groups occurred in
1914. István Apáthy reviewed Hoffmann's book *Die Rassenhygiene in
den Vereinigten Staaten von Nordamerika* in a special section devoted
to eugenics in *Magyar Társadalomtudományi Szemle*. The review represented Apáthy's attempt to find common ground between the two
dominant interpretations of eugenics.[38] As with previous writings,
Apáthy commenced his analysis with a terminological debate about
eugenics; he then went on to discuss the origin of the term eugenics
(derived from the Greek ευγένια) and its usage by Galton, Ploetz, and
Schallmayer. In the second part of his review, Apáthy presented a
detailed description of Hoffmann's book and its main theses, and concluded by praising Hoffmann's "interesting and valuable book." The
time had come, Apáthy believed, for a professional treatment of the
"eugenics question in Hungary." A "eugenics committee" was needed to
provide institutional support for eugenics. According to Apáthy, it was
not only the eugenicists who advocated the creation of such a committee, but also medical institutions in Budapest, including the Medical
Association (*Budapesti Orvosegyesület*) and the Association of Public
Health (*Közegészségügyi Egyesület*).[39]

Apáthy's dedication to institutionalizing eugenics received strong
support from Hoffmann.[40] Indeed, Hoffmann's main objective was to
establish an effective eugenic organization in Hungary. Unlike many

Hungarian conservatives, he sought common ground with the progressive intelligentsia, for he perceived the two groups to have a common goal: instilling racial hygiene in society through public awareness and scientific institutionalization. It seemed that the idea of eugenics supporting a quantitative (democratic) social system, as championed by Apáthy and Madzsar, could be reconciled with racial hygiene schemes envisioned by Hoffmann no matter how qualitative, selective, or authoritarian were their political implications.

In the same issue of *Magyar Társadalomtudományi Szemle*, Hoffmann wrote his first article on eugenics in his native Hungarian language. The title of the article, "Eugenika" (Eugenics), is provocative, as Hoffmann preferred *eugenika* rather than *faj egészségtana* (the term used by Apáthy) in advancing his definition of eugenics. This was clearly an attempt to counteract Apáthy's influence within the eugenic movement in Hungary. Thus in the first part of his article, "A fajegészségtan alapja" (The Foundation of Racial Hygiene), Hoffmann analysed Darwin's theory of natural selection, Mendel's hereditary laws, and contributions to the development of eugenics thinking by luminaries such as Francis Galton, Alfred Ploetz, and Wilhelm Schallmayer. The second and the third part of the article dealt with the methodologies assumed by various schemes of racial hygiene; while in the last part, Hoffmann returned to the practical implications of his eugenic agenda: the creation of a section for racial hygiene within the Hungarian Society of Social Sciences.[41]

Hoffmann's article was a complex and contradictory ensemble of theory and practice. American eugenics was central to his thinking, but Hoffmann also took ideas from German racial hygiene and British eugenics. He thus explained his support for eugenics as an expression of his more general commitment to the principle of racial improvement, but he showed in his acceptance of Apáthy's concept of *faj egészségtana* that he was prepared to put this commitment to eugenics above partisan ideological considerations. That said, Hoffmann's interpretation of racial hygiene was not passive, for racial hygiene went hand in hand with public education on the importance of eugenics. And for such an education to occur, Hoffmann argued, Hungary's public institutions should promote a strong eugenic policy.

These initial efforts to popularize eugenics in Hungary did produce the outcome Apáthy and Hoffmann had anticipated. On 24 January 1914, in the festival hall of the Royal Hungarian Society for Natural

Sciences (*Királyi Magyar Természettudományi Társulat*), the Royal Medical Association of Budapest (*Budapesti Királyi Orvosegyesület*) and the National Society of Public Health (*Országos Közegészségügyi Egyesület*) met to discuss the issue of "racial hygiene" and the creation of a "eugenics committee."[42] The paper discussed on this occasion was Hoffmann's "Eugenika," which, due to the absence of its author, was read by a certain Győző Alapy. The main discussant of Hoffmann's paper was István Apáthy, who profited from the circumstance to reinforce his critique of Hoffmann's vision of eugenics, and to launch his programme of the eugenic movement in Hungary.

The institutionalization of eugenics in Hungary thus occurred at the time when the two groups of Hungarian eugenicists had found common ground: the belief that eugenics should be recognized as an independent scientific domain. In his address Apáthy outlined the tasks of the Committee of the Eugenics Society (*Egyesületközi Fajegészség-ügyi Bizottság*), and based them on two principles: to orchestrate social and medical reform, and to ensure the safeguarding of the racial qualities of the nation.

Apáthy also devised a laborious plan for eugenics in "action," one based on five strategies. He named the first "praeparativus eugenika (*előkészítő fajegészségtan*):" the Eugenics Society was supposed to popularize and "prepare" the understanding of eugenic principles. Following the activity of similar organizations in Europe and America, the propagandistic actions of the Hungarian Eugenics Society were to include public lectures and the distribution of pamphlets and materials to clubs, libraries and schools. Like Galton before him, Apáthy knew that before the eugenic transformation of Hungarian society could take place, the public would have to become acquainted with the main principles of eugenic thinking. Accordingly, the second strategy outlined the practical implications of eugenics. Apáthy referred to this direction as "praevent vus eugenika (*megelőző fajegészségtan*):" eugenics should scrutinize the social and medical history of individuals in endeavouring to prevent the spread of disease. If, however, such preventive measures were unsuccessful, a third strategy was to follow, "diagnostikus vagy taxativus eugenika (*selejtező fajegészségtan*)." Here, eugenicists faced their utmost challenge: to diagnose and "select" the eugenic method appropriate for solving social and medical predicaments. Once the problem was localized, a fourth strategy would follow, which Apáthy termed "normativus eugenika (*rendelkező fajegészségtan*)." This would be translated into prophylactic measures to be "prescribed" in order to ensure that medical problems confronting individuals were treated efficiently. Finally, the fifth strategy regrouped the goals of eugenics into an ideal-type, "idealis vagy prospectivus eugenika (*jövőt intéző fajegészségtan*)," placing eugenics firmly into the future organization of society.[43]

Apáthy's eugenic manifesto is remarkable in its attempt to synthesise different interpretations of eugenics into a coherent whole. Of the different factors helping to crystallize his commitment to eugenics, the most significant by far was his capability to connect eugenics with a conception of social responsibility. Apáthy offered a practical formulation for consolidating the social and political basis of the nation, one aimed at the improvement of the health of larger segments of the population.

Thereafter, Count Pál Teleki was nominated president of the Eugenics Society, while István Apáthy became its secretary. Although the two represented different ideological standpoints, they shared the same interest in eugenics. Other members included József Ajtay, Jenő Gaal, Emil Grosz and Benedek Jancsó. Rezső Bálint, Sándor Korányi, Leó Liebermann, Vilmos Tauffer and Lajos Török represented the Royal Medical Association of Budapest; while Zoltán Dalmady, Béla Fenyvessy, Ferencz Hutyra, Géza Lobmayer and Henrik Schuschny represented the National Association of Public Health. The Eugenics Society fell under the protective patronage of the Geographical, Ethnographic, Economic, Genealogical and Turanian Societies.[44]

Yet one name was missing from the list of members: Géza von Hoffmann. This is surprising, considering Hoffmann's involvement in the popularization of eugenics in Hungary and his acknowledged expertise in Germany and America. Few Hungarian politicians, like Teleki and Hoffmann, became involved with the eugenic movement before the First World War. However, Hoffmann was not nominated for any honorary position within the Eugenic Society, nor did he become a member. The reasons for his absence were neither ideological nor political (after all, a Magyar magnate, Pál Teleki, was the president of the Eugenics Society), but intellectual. In 1914, hopes for a fusion between the internal and external groups of Hungarian eugenicists were high, yet rarely had the methodological schism between them been so pronounced. In a note for the *Archiv für Rassen- und Gesellschaftsbiologie* announcing the establishment of the Committee of

Eugenics in Hungary, Hoffmann offered a possible explanation for the chasm between his interpretation of eugenics and Apáthy's: the Hungarian Eugenics Society advocated a form of *faj egészségtana* (racial hygiene) that was opposed to both *Eugenik* (eugenics) and *Rassenveredelung* (race selection)—two forms of racial hygiene Hoffmann strongly advocated.[45]

The outbreak of the First World War interrupted the activities of the Eugenics Society, but not of the eugenic movement in Hungary as a whole, which regrouped its forces. Eugenics increasingly represented the interest of a political elite, weary of the outcome of the war. Admittedly, physicians and biologists like Madzsar and Apáthy continued to dominate the movement, but the influence of politicians and diplomats such as Teleki and Hoffmann continued to grow.[46] A frustrated eugenicist in 1914, Hoffmann would return victoriously to Hungary in 1917, to contribute to the establishment of the Hungarian Society for Racial Hygiene and Population Policy.

IV. The Hungarian Society for Racial Hygiene and Population Policy (1917)

As throughout Europe, the First World War profoundly affected Hungarian society. The social and national changes brought about by war prompted the intellectual and political elite to look to eugenic principles as a source of hope in their disillusioning environment. The prestige of social hygiene increased during wartime, chiefly due to the failing health of the population. Birthrates were declining and social reformers anxiously alerted the government to the need for a stringent health policy. In addition to occasioning the introduction of social and medical policies, the war also generated a resurgence of nationalist concerns about the alleged deterioration of "Magyar racial qualities." As a result, a new category of eugenicists emerged. Fuelled by fears of biological deterioration and the decline of Hungarian political supremacy in the Carpathian Basin, this group introduced a new ethos into the eugenic movement.

Lajos Méhely (1862–1953), for instance, the most prolific racist anti-Semitic author in interwar Hungary, began his scientific career as a promising biologist. After the outbreak of the war in 1914, the Hungarian Ministry of War (*Magyar Királyi Honvédelmi Minisztérium*) commissioned Méhely to write a study on the effects of warfare on national and biological development. In 1915, Méhely published a pamphlet suggestively called *A háború biológiája* (The Biology of War).[47] A similar, albeit less racist, Darwinist perspective on the importance of the war as a racial conflict was offered in 1915 by the respected neurologist Mihály Lenhossék (1863–1937), in "A háború és a létért való küzdelem tétele" (War and the Theme of Positive Struggle for Life). Revealingly, both articles were published in *Természettudományi Közlöny* (Natural Sciences Bulletin), the most important journal to specialize in Hungarian biology and the natural sciences at the time.[48]

During the war, the biological health of the nation gradually emerged as a major eugenic problem. In addition, an emerging form of social hereditarianism became apparent through the desire of some segments of the Hungarian political elite to dominate groups in society viewed to be outside the "Magyar national community." Both the Jews and the emerging urban proletariat were considered to be of non-Magyar origin, and were held responsible for the social pathologies afflicting Hungary during the war. Demands for social regulation and political control by the government intensified. It is not surprising, therefore, that medical discourses about how best to protect society against sexually transmitted diseases were also intimately connected to the issue of "national protection and racial degeneration." In 1916, for example, the *Nemzetvédelmi Szövetség* (League for the Protection of the Nation) organized a congress on sexually transmitted diseases, followed in 1917 by a congress on social hygiene.[49]

The eugenic and dysgenic effect of war was a subject that had interested eugenicists since the late nineteenth century. In 1895, Alfred Ploetz pointed to the "dysgenic" effects of war and proposed that the "worst individuals" should be drafted for military service in order for "healthy" individuals to be saved.[50] In Hungary, the eugenic response to the social and national crises brought about by war was two-pronged. Advocates of the first viewpoint, such as János Bársony and József Madzsar, focused on the deterioration in health of future generations of Hungarians. This group campaigned for public discussion of a wide range of issues, including reproductive hygiene, the protection of mothers, and the low birthrate. The second perspective was represented by Géza von Hoffmann and Ernő Tomor, and was concerned with the immediate improvement of the biological quality of the population. Following similar developments in Germany—especially the work on racial hygiene and heredity by the dermatologist Hermann W. Siemens (1891–1969)—this group of eugenicists linked population growth to national struggles for competition and resources.[51] Both groups agreed, however, that new eugenic policies were needed in order to increase the hereditary protection of the population.

In 1915, the prominent Hungarian gynaecologist János Bársony (1860–1926) contributed to the *Archiv für Frauenkunde und Eugenetik* (Journal of Women's Studies and Eugenics) with the study "Eugenetik nach dem Kriege" (Eugenics after the War).[52] According to Bársony, medical research confirmed claims made by racial hygienists that the war had destroyed the "healthy and strong men of the nation." Racial fears were thus seemingly justified by statistical evidence about the increase of the "inferior individuals" (dysgenic elements) in the population. In turn, eugenics needed to respond efficiently to wartime challenges and traumas. As Bársony noted: "After the war, eugenics, the creed of race improvement, will step into the foreground with full strength."[53]

In order to achieve this strength, Bársony envisioned two techniques to ensure the "recovery of the race" (*Hebung der Rasse*). The first course of action was to increase the birthrate. "In Hungary, for example, the family with six children is regarded as a rarity in contrast to the past, and there are entire regions in which the 'one-child system' (*Einkindersystem*) dominates."[54] Some of the factors contributing to "the stagnation of the Magyar race," such as "birth-prevention, abortion and abortionists," were to be neutralized by preventive eugenic measures. The second approach underlined precisely this point: "The new generation should be not only large, numerically speaking, but also primarily healthy. The health of the parents is the first condition for this [the recovery of the race] to happen."[55] More generally, the reappraisal of the eugenic role of the mother resulted in a nuanced evaluation of the relationship between eugenics and maternity. There was thus a convergence of interest between the future of the nation and the protection of the mother. In order to raise the racial quality of future generations, Bársony advised the Hungarian government to "begin by protecting women."[56] The priority of the existing political elite should be to use eugenic propaganda to create a sense of social responsibility towards future generations.

The eugenic preoccupation with the health of future generations was clear in speculation about the declining birthrate, and the protection of mothers and infants. Already in 1912, the gynaecologist Henrik

Rotter analysed the relationship between reproduction and eugenics in his study *Eugenika és szülészet* (Eugenics and Obstetrics), in which he criticized reproductive techniques, like abortions, that could have disastrous effects on future generations.[57] József Madzsar, too, was particularly interested in these topics, as illustrated by his publications during the first years of war, such as *Az anya- és csecsemővédelem a háborúban* (The Protection of Mothers and Infants during the War); *Az anya- és csecsemővédelem országos szervezése* (The National Organization for the Protection of Mothers and Infants); and *A jövő nemzedék védelme és a háború* (The Protection of Future Generations and the War).[58] In 1916, Madzsar completed *A meddő Budapest* (Sterile Budapest), his demographic analysis of some of the causes contributing to declining birthrates in Budapest.[59] The study largely followed the economic analysis applied earlier to demographic trends in Berlin, *Das sterile Berlin* (Sterile Berlin), published in 1913 by the German public health reformer Felix Aaron Theilhaber (1884–1956).[60] Like Berlin, Budapest was in need of a new social policy able to encourage immigration from the countryside, in addition to introducing sexual education into state institutions. Madzsar's pessimistic conclusions on low fertility trends and birthrates complemented Bársony's demands for the mechanical control of reproduction through eugenic policies. Interest in reproduction, fertility and maternity eventually captured the attention of the Hungarian political elite towards the end of the war. In 1917, the *Országos Magyar Anya- és Csecsemővédő Központi Intézet* (National Institute for the Protection of Mothers and Infants), the first medical institution in Hungary to combine medical care with research on eugenics and heredity, was established.[61]

These practical efforts to tackle declining birthrates and protect mothers were paralleled by attempts to revitalize the eugenic movement. The year 1916 saw the publication of Hoffmann's third book, *Krieg und Rassenhygiene* (War and Racial Hygiene). Hoffmann lamented that the war had exposed the nations involved in combat to various forms of biological extinction. For Hoffmann, it became increasingly clear that, besides obstructing population growth, war had a dysgenic effect on the level of hereditary strength among European nations. Hoffmann further assessed the "racial burden" and "degeneration" caused by war on the "genetic stock" of the national population, arguing for the protection of marriage through the introduction of prophylactic measures against venereal diseases.[62] He referred to preventive measures adopted by the German Society for Racial Hygiene, including the "furtherance of inner colonization with privileges of succession in favour of large families;" the "abolition as far as possible of certain impediments to marriage;" the "legal regulation of procedure in all cases necessitating abortion or sterilization;" and "awakening a national mind ready to bring sacrifices, and a sense of duty towards coming generations."[63]

Although the book dealt largely with Germany, Hoffmann suggested that other nations should follow the German example in addressing concerns about the preservation of the nation through racial hygiene. Hoffmann's book was welcomed by racial hygienists in Western Europe and served further to consolidate his position within the eugenic movement in Hungary.[64] As a record of developments in Hungarian eugenic thinking, however, the book attempted to find a middle ground between those supporting racial hygiene and those in favour of population policy. In Germany, for example, the two camps were institutionally separated: the racial hygienists were grouped around the German Society for Racial Hygiene, while the supporters of population policy had founded the *Deutsche Gesellschaft für Bevölkerungspolitik* (German Society for Population Policy) in 1915. Thematically and practically, however, there was a constant overlapping between the two camps, and Hoffmann argued that the two needed not be separated, either institutionally or ideologically.

Shortly after the publication of *Krieg und Rassenhygiene*, Hoffmann was given the opportunity to put his ideas into practice. In 1917 he was appointed "population adviser" (*Bevölkerungspolitiker*) in the Hungarian Ministry of War. More importantly, his return to Hungary coincided with the reorganization of the eugenic movement there. On 24 November 1917, under the patronage of the Hungarian Academy of Sciences, the Eugenics Society was transformed into the Hungarian Society for Racial Hygiene and Population Policy (*Magyar Fajegészségtani és Népesedéspolitikai Társaság*). Again, Count Pál Teleki was appointed president, while Hoffmann became the vice-president.

In several respects the Hungarian Society for Racial Hygiene and Population Policy differed from similar societies in Europe. As Hoffmann remarked: "The double movement which divided the efforts of race regeneration in Germany was united in Hungary from the beginning."[65] The intersection of racial hygiene and population policy reflected the two preoccupations of its two principal founding members: racial hygiene (Hoffmann) and population policy (Teleki). It was important, however, Hoffmann argued, that the Hungarian Society for Racial Hygiene and Population Policy combined "racial hygiene" with "eugenics" and "population policy." According to Hoffmann, conceptual delineation between these terms was detrimental to the Hungarian eugenic movement. As a result, the Hungarian Society for Racial Hygiene and Population Policy had three main objectives. It campaigned for "1. The scientific exploration of those damages that threaten the body of the Hungarian nation, particularly the declining birthrate; 2. Establishing means and ways by which to increase the number of births; 3. The support of those endeavours whose purpose is the creation of an environment in which the Magyar race can prosper."[66]

"The central theme" (*Der Leitgedanke*), argued Hoffmann, was "that race consciousness, the consideration of future generations and the high estimation of proficient big families, was to be inculcated into all branches of the state, social, economic, political and moral life."[67] The positions occupied by Teleki and Hoffmann at the Ministry of War meant that the Hungarian Society for Racial Hygiene and Population Policy received the institutional support needed for eugenic propaganda.[68] Moreover, in response to the serious problems affecting Hungarian civilians during the war—particularly regarding contagious disease and mortality—the Ministry created special commissions to promote the well-being of the family, including marriage counselling, and medical assistance in the case of venereal infections.[69] A new journal was founded, *Nemzetvédelem* (The Protection of the Nation), which, although short-lived, was the first journal in Hungary whose sole purpose was the popularization of eugenics.[70]

The creation of the Hungarian Society for Racial Hygiene and Population Policy completed the complex process of the nationalization of eugenics initially commenced in 1911 with the *Eugenika vita*. It also marked the fusion between the internal and external groups of Hungarian eugenicists. As noted earlier, there was no apparent contradiction between racial hygiene, eugenics, and population policy in Hoffmann's thinking. A difference between racial hygiene and eugenics did exist, however, as Hoffmann declared: "Eugenics and race hygiene are not quite identical. (...) The motto of eugenics we may define as 'Quality, not quantity'. Race hygiene says: 'Quality *and* quantity'."[71] This was an important detail, for it was on the issue of the biological quality of the population that eugenics intersected with population policies. According to Hoffmann, eugenics should deal with qualitative population policies, and should serve racial and hereditary interests.

Although it clearly applied to the programme of the Society for Racial Hygiene and Population Policy, how important was Hoffmann's racial hygiene model within the eugenic movement in Hungary? Although the establishment of the Society for Racial Hygiene and Population Policy acknowledged Hoffmann as a leading eugenicist, his eugenic theories did not go unchallenged. In his 1918 *Neubegründung der Bevölkerungspolitik* (The New Foundation of Population Policy), Ernő Tomor (1884–?), a medical superintendent at the Metropolitan Institute for Tuberculosis in Budapest, endeavoured to demonstrate that "population policy" was, in fact, a "borderland between hygiene and social sciences."[72] Tomor devoted an entire chapter—provocatively called "Irrwege der Rassenhygiene" (The Questionable Paths of Racial Hygiene)—to a discussion of American eugenics.[73] According to Tomor, American eugenicists "transferred unfinished and unclear results of racial hygiene research hastily into practice in an overzealous and overheated manner."[74] This statement was, however, a rebuke only of American eugenics for, as Tomor acknowledged, in Europe racial hygiene had a different application in society. While in America, eugenic legislation was aimed primarily at controlling immigration and "racial mixing," these issues posed only theoretical challenges to forms of racial hygiene prevalent in Europe. "Bastardization" (*Bastardierung*) thus separated eugenic policies in Europe from those in America.

Moreover, in comparison to other countries, in the USA eugenics was firmly enmeshed in politics, making it difficult to have it analysed solely from a medical and biological perspective. Racial hygiene in Europe, in consequence, should not imitate American eugenics. Tomor detected such a precaution towards the application of eugenic principles to society within the German racial hygiene movement, which, according to him, "[did] not hasten to transfer unsolved scientific opinions into practice."[75] German racial hygiene had two aims: one moral, especially through the hygiene education of the young generation; the other medical, namely, the combat of venereal diseases and the elimination of "sickness" (*Geschlechtskrankheiten*) from the family.[76] Morality and dedication to the well-being of the family were complemented by a more general aim of racial hygiene: "the gradual biological regeneration of the entire national body (*eine allmähliche biologische Regeneration des ganzen Volkskörper*)" so that "damage to the nation" could be avoided.[77]

Fig. 3. Ernő Tomor's *The New Foundation of Population Policy*, a critical
attempt to combine social hygiene, eugenics and population policy

Tomor's iconoclastic critique of American eugenics resonated favourably with the internal group of Hungarian eugenicists, who were similarly attempting to dissociate racial hygiene from modern war and national degeneracy. In "*A fajegészségtan köre és feladatai*" (The Domain and Task of Racial Hygiene), István Apáthy refuted the argument that war exemplified the application of natural selection to society, and suggested a more rigorous methodology when dealing with the eugenics of war.[78]

Arguing that natural selection should enhance "social morality" and not generate conflict, Apáthy declared that "war is in all respects the very opposite of what we call natural selection." However, as a Social Darwinist, Apáthy acknowledged war to be an important element in human history: "I will even concede that human development required wars in the past. But let everyone open his eyes and he will see that at the present stage of human development, human culture, administration and technical knowledge, war is not necessity but sheer madness. This world war will not bring anything profitable for mankind in general, or for the development of individual nations in particular, but it will cause the destruction of numerous prerequisites for progress; it will result in the need to create new things from inferior remnants; things which would still exist if only they had not been destroyed."[79] War and eugenics were ultimately incompatible, for, as Apáthy suggested, "No step forward can be made in the field of racial hygiene until mankind is freed from the present war and the spectre of another war."[80]

A similar endorsement of racial hygiene and condemnation of war was offered by Mihály Lenhossék. In the 1918 article "A népfajok és az eugenika" (Races and Eugenics), Lenhossék reformulated some of the eugenic arguments put forward in "A háború és a létért való küzdelem tétele." On the one hand, Lenhossék combined eugenics with nationalism; on the other, he reinforced Hoffmann's claim that the utmost goal of eugenics was to preserve the "quality of race." Consistent to his training in anthropology, Lenhossék defined "race" in Darwinist terms; that is, as a group of human beings characterized by constitutional and psychological traits inherited from generation to generation. He believed in the generic unity of the "human race," but accepted that, within it, there was racial diversity. But such diversity should not, he cautioned, create the impression that some races were "superior" to other races.[81]

The debate about racial hygiene, eugenics, and population policy which occurred during the last years of the war is revealing for a number of reasons. On the one hand, the debate produced a diversity of interpretations of eugenics, as well as its immediate social and political purposes, that illustrates, yet again, the importance racial science had acquired in national politics in Hungary between 1910 and 1918. Yet, on the other hand, the debate addressed issues of racial and national

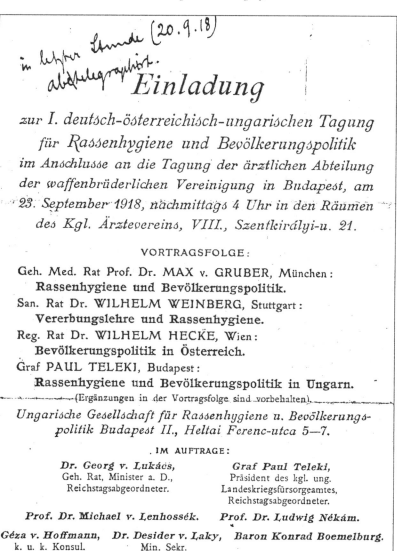

in lehter Stunde (20.9.18)
absfeliegraphist.

Einladung

*zur I. deutsch-österreichisch-ungarischen Tagung
für Rassenhygiene und Bevölkerungspolitik
im Anschlusse an die Tagung der ärztlichen Abteilung
der waffenbrüderlichen Vereinigung in Budapest, am
23. September 1918, nachmittags 4 Uhr in den Räumen
des Kgl. Ärztevereins, VIII., Szentkirályi-u. 21.*

VORTRAGSFOLGE:

Geh. Med. Rat Prof. Dr. MAX v. GRUBER, München:
Rassenhygiene und Bevölkerungspolitik.
San. Rat Dr. WILHELM WEINBERG, Stuttgart:
Vererbungslehre und Rassenhygiene.
Reg. Rat Dr. WILHELM HECKE, Wien:
Bevölkerungspolitik in Österreich.
Graf PAUL TELEKI, Budapest:
Rassenhygiene und Bevölkerungspolitik in Ungarn.
————————(Ergänzungen in der Vortragsfolge sind vorbehalten).————————

*Ungarische Gesellschaft für Rassenhygiene u. Bevölkerungs-
politik Budapest II., Heltai Ferenc-utca 5—7.*

IM AUFTRAGE:

Dr. Georg v. Lukács,
Geh. Rat, Minister a. D.,
Reichstagsabgeordneter.

Graf Paul Teleki,
Präsident des kgl. ung.
Landeskriegsfürsorgeamtes,
Reichstagsabgeordneter.

Prof. Dr. Michael v. Lenhossék.

Prof. Dr. Ludwig Nékám.

Géza v. Hoffmann,
k. u. k. Konsul.

Dr. Desider v. Laky,
Min. Sekr.

Baron Konrad Boemelburg.

Anfragen wegen Teilnahme an der Reise sind zu richten an
Prof. Dr. Adam, Berlin, Kaiserin Friedrich Haus, Luisenplatz 2—4.

*Anfragen betreffend Veranstaltung der Tagung sind
zu richten an die Ungarische Gesellschaft für Rassenhygiene
und Bevölkerungspolitik, Budapest II., Heltai Ferenc-u. 5—7.*

Fig. 4. The invitation to the "First German–Austrian–Hungarian Conference on Racial Hygiene and Population Policy". Alfred Ploetz's Archive. Courtesy of Paul J. Weindling. The invitation was also published in *Nemzetvédelem* I, 1–2 (1918).

survival; therefore, it was perhaps inevitable that political allegiances had infiltrated scientific debates—as illustrated by the conflict between conservative-nationalist and socialist supporters of eugenics between 1917 and 1918.

In February 1918, a congress on population policy was held in Berlin, under the auspices of the Society to Promote Friendship between the Central Powers (*Waffenbrüderliche Vereinigung*). Delegates from Germany, Austria-Hungary and Bulgaria attended the congress, but, according to Hoffmann, "the shadow of the great antagonism between conservative and radical thought which later led to the revolution already disturbed the discussion."[82] Hoffmann hoped to reconvene with the German and Austrian eugenicists on 23 September 1918, when the Hungarian Society for Racial Hygiene and Population Policy announced the "First German–Austrian–Hungarian Conference on Racial Hygiene and Population Policy" to take place in Budapest, in collaboration with the Austrian Institute for the Statistics of National Minorities (*Institut für Statistik der Minderheitsvölker*) and the Austrian Society for Population Policy (*Österreichische Gesellschaft für Bevölkerungspolitik*). Speakers were to include the Austrian bacteriologist and hygienist Max von Gruber (1853–1927), the German geneticist Wilhelm Weinberg (1862–1937), and Count Pál Teleki.[83] The conference was cancelled at the last minute, on the eve of the disintegration of the Austro-Hungarian Empire and the socio-political revolutions following it.

Conclusions

With the end of the war and the collapse of the Austro-Hungarian Empire, Hungary experienced major political transformations. On 11 November 1918, the revolutionary regime of Mihály Károlyi (1875–1955) proclaimed a Hungarian republic. Eugenicists affiliated with the Left during the war period, including István Apáthy and József Madzsar, became public figures during Károlyi's regime. A Chair of Social Medicine was established at the University of Budapest and was first taken up by Madzsar. Within a few months he was appointed to the post of state secretary in Count Károlyi's newly established Ministry of Public Welfare, and later became the director of the National Council of Public Health.[84]

Once in power, eugenicists and social hygienists attempted to implement many of the medical policies discussed immediately before and during the war. A new social doctrine was formulated in order to improve the health of both the urban and the rural masses. However, soon the constraints imposed by scarce economic resources, in addition to social conflicts generated by the philanthropic nature of some of the medical measures taken by the revolutionary regime, turned against eugenics. Indeed, so apparently defective were some of the medical policies of the new regime that the skeptical liberal Baron Sándor Korányi (1866–1944), a leading authority on internal medicine, noted: "Mere charity in medical care will no longer solve our national problems. The solution can only be a reversal of the role of the state and society: leadership has to be assumed by the state."[85]

And it was to the state that eugenicists now turned. Madzsar, for instance, suggested that mandatory sterilization and birth control should be introduced in order to prevent the reproduction of "genetically inferior" individuals. He also proposed the creation of medical agencies which would encourage the breeding of "healthy stock," and insisted that individuals obtain the consent of these agencies before marriage.[86] Madzsar's new eugenic programme resembled traditionalist measures previously advocated by the conservatives Hoffmann and Teleki, rather than the ideals of revolutionary democracy advocated by Apáthy. Eventually, the change of political regime in the early 1920s favoured those who opposed the social programmes advocated by left-wing eugenicists.

Before, and especially after, the communist experiment of Béla Kun (1886–1939) in 1919, Hungary experienced a resurgence of the conservative right. Racist societies, such as *Ébredő Magyarok Egyesülete* (The Society for Awakening Magyars) and *Magyar Országos Véderő Egylet* (The National Society of Magyar Defence), both founded in Budapest in 1918, came to dominate the political sphere. The resurgence of racial nationalism had significant consequences for the future evolution of the eugenic movement in Hungary. As Hoffmann explained in 1921: "Since Bolshevism was broken (...) the whole country needs 'race regeneration', not so much in the sense of eugenics, but sound morals, order and law, healthy family life, and regard for future generations. Everybody's whole time and energy is devoted to the reorganization of the country and to averting the consequences of a so-called peace. Later, when conditions change, the time will come to continue the work of eugenics."[87]

The early eugenic movement in Hungary was not a monolithic intellectual structure but a diverse blend of various orientations: liberal, progressive, socialist, and conservative. By the end of the First World War—and especially during the democratic and communist experiments of 1918 and 1919—left-wing groups dominated the eugenic movement and institutional centres of the Hungarian medical system. Apart from Soviet Russia, Hungary was the only country in Europe to attempt a large-scale reform of its social and medical system on the basis of socialist ideas. Leading Hungarian eugenicists like István Apáthy and József Madzsar occupied important positions within the political regimes that followed the collapse of the Austro-Hungarian Empire. However, while the democratic regime of 1918 and the communist experiment of 1919 both favoured the emergence of a new approach to social hygiene—one based on sanitary improvement and philanthropic welfare—the counter-revolution of 1920 represented an ideological turning point that markedly contributed to the resurgence of conservatism and racial nationalism within interwar Hungarian eugenics.

Endnotes:

[1] One such case is the reference to Géza von Hoffmann's book, *Rassenhygiene in den Vereinigten Stateen von Nordamerika* (1913). See Stefan Kühl, *The Nazi Connection. Eugenics, American Racism, and German National Socialism* (Oxford: Oxford University Press, 1994), 16; and Paul Crook, "American Eugenics and the Nazis: Recent Historiography," *The European Legacy* 7, 3 (2002), 377 (note 19). See also Paul Weindling, *Health, Race and German Politics between National Unification and Nazism, 1870–1945* (Cambridge: Cambridge University Press, 1989), 150, 240, 301, 303.

[2] See Endre Réti, "Darwinista humanizmus Apáthy és Lenhossék szemléletében", *Communicationes ex Biblioteca Historiae Medicae Hungarica*, 23 (1963), 111–116; Endre Réti, "Magyar darwinista orvosok (1945-ig)," *Communicationes ex Biblioteca Historiae Medicae Hungarica* 31 (1964), 117–313; Endre Réti, "Darwin's Influence on Hungarian Medical Thought," in *Medical History in Hungary* (Budapest: Hungarica, 1972), 157–167. The topic of eugenic sterilization has attracted the attention of one historian interested in the history of psychiatry in Hungary. See Béla Siró, "Eugenikai törekvések az ideg- és elmegyógyászatban Magyarországon a két világháború között," *Orvosi Hetilap* 144, 35 (2003), 1737–1742. Finally, political scientists and historians have briefly dealt with the eugenic movement in Hungary. See Mária M. Kovács, *Liberal Professions and Illiberal Politics. Hungary from the Habsburgs to the Holocaust* (Washington: Woodrow Wilson Center Press, 1994); and Balázs Ablonczy, "Az eugenikai vonzásában. A társadalom biológïai tervezése," *Rubicon* 2 (2004), 15–18. Ablonczy also edited *Teleki Pál. Válogatott politikai írások és beszédek* (Budapest: Osiris kiadó, 2000), and recently published a monograph about the Hungarian politician. See Balázs Ablonczy, *Teleki Pál* (Budapest: Osiris, 2005).

[3] See Béla Balogh, "Die Geschichte der ungarischen Anthropologie", *Ungarische Jahrbücher* 19, 2–3 (1939), 141–181.

[4] Erzsébet Ladányiné, *A magyar filozófia és darwinizmus XIX. századi történetéből, 1850–1875* (Budapest: Akadémiai kiadó, 1986).

[5] See Pál Teleki, "Társadalomtudomány biológiai alapon," *Huszadik Század* 5, 9 (1904), 318–322.

[6] József Madzsar, "A szaporodás higienéje (F. M. Galton. Eugenics)," *Huszadik Század*, 7, 4 (1906), 366–367.

[7] Francis Galton, "A valószínűség, mint az eugenetika alapja," *Huszadik Század*, 8, 22 (1907), 1013–1029.

[8] István Apáthy, "A darwinizmus birálata és a társadalomtan," *Magyar Társadalomtudományi Szemle* 1 (1909), 309–339; and Géza Kenedi, "Feminizmus és biológia", *Magyar Társadalomtudományi Szemle* 1 (1909), 218–234.

[9] Géza Vitéz, *Születési és termékenységi statistika* (Budapest: Grill, 1906).

[10] See József Madzsar, "Darwinizmus és szocialismus," in Endre Kárpáti, ed., *Madzsar József válogatott írásai* (Budapest: Akadémiai kiadó, 1967), 141–145; and József Madzsar, "Larmark és Darwin," in Kárpáti, ed., *Madzsar József válogatott írásai*, 165–171. The latter article became the book *Darwinizmus és Lamarckismus* (Budapest: Deutsch Mátkus, 1909).

218   *"Blood and Homeland"*

11 Kovács, *Liberal Professions and Illiberal Politics*, 33.
12 Lajos Dienes, "Biometrika," *Huszadik Század* 21, 1 (1910), 50–51. See also Marius Turda, "Heredity and Eugenic Thought in Early Twentieth-Century Hungary," *Orvostörténeti Közlemények*, 194–195 (2006), 101–118.
13 József Madzsar, "Gyakorlati eugenika," *Huszadik Század* 21, 2 (1910), 115–117.
14 Zsigmond Fülöp, "Eugenika," *Huszadik Század*, 22, 9 (1910), 161–175.
15 József Madzsar, "Fajromlás és fajnemesítés," *Huszadik Század* 23, 12 (1911), 145–160
16 Lajos Dienes, "A fajnemesítés biometrikai alapjai," *Huszadik Század* 23, 12 (1911), 291–307
17 Zsigmond Fülöp, "Az eugenetika követelései és korunk társadalmi viszonyai," *Huszadik Század* 23, 12 (1911), 308–319.
18 (Anonymous), "A fajnemesítés (eugenika) problémái," *Huszadik Század* 23, 12 (1911), 694–709; and (Anonymous), "A fajnemesítés (eugenika) problémái," *Huszadik Század* 23, 12 (1911), 29–44; 157–170; 322–336. See also Marius Turda, "'A New Religion'? Eugenics and Racial Scientism in Pre-First World War Hungary," *Totalitarian Movements and Political Religions* 7, 3 (2006), 303–325.
19 (Anonymous), "A fajnemesítés (eugenika) problémái," 35.
20 (Anonymous), "A fajnemesítés (eugenika) problémái," 39–40.
21 (Anonymous), "A fajnemesítés (eugenika) problémái," 157.
22 (Anonymous), "A fajnemesítés (eugenika) problémái," 159.
23 (Anonymous), "A fajnemesítés (eugenika) problémái," 169.
24 (Anonymous), "A fajnemesítés (eugenika) problémái," 170.
25 (Anonymous), "A fajnemesítés (eugenika) problémái," 322–324.
26 (Anonymous), "A fajnemesítés (eugenika) problémái," 324.
27 (Anonymous), "A fajnemesítés (eugenika) problémái," 327–329.
28 Jenő Vámos, "Az alkalmazott eugenika," *Huszadik Század* 24, 12 (1911), 571–577, especially 576–577.
29 (Anonymous), "A fajnemesítés (eugenika) problémái," 336.
30 István Apáthy, "A faj egészségtana," *Magyar Társadalomtudományi Szemle* 4 (1911), 265–279.
31 Apáthy, "A faj egészségtana," 265.
32 See Alfred Ploetz, *Grundlinien einer Rassen-Hygiene* (Berlin: S. Fischer, 1895), 13–14.
33 This argument is also supported by Réti, "Darwin's Influence on Hungarian Medical Thought," 165.
34 Since the mid-nineteenth century, Budapest, in particular, has experienced a spectacular growth in size and population. Economic transformations fostered by urbanization and industrialization resulted in significant changes in the ethnic and social composition of the city. Social reformers and hygienists feared that such profound changes would have serious repercussions on the "natural evolution" of the Hungarian nation.
35 Géza von Hoffmann, "Rassenhygiene in Ungarn," *Archiv für Rassen und Gesellschaftsbiologie* 13, 1 (1918), 55–67.

36 Géza von Hoffmann, *Rassenhygiene in den Vereinigten Staaten von Nordamerika* (Munich: J. F. Lehmanns Verlag, 1913).

37 See, for example, Fritz Lenz, "Hoffmann, Géza von, Die Rassenhygiene in den Vereinigten Staaten von Nordamerika," *Archiv für Rassen und Gesellschaftsbiologie* 10, 1–2 (1913), 249–252; Edgar Schuster, "Von Hoffmann, Geza. Die Rassenhygiene in den Vereinigten Staaten von Nordamerika," *The Eugenics Review* 5, 3 (1913), 279; and Amey Eaton Watson, "Von Hoffmann, Geza. Die Rassenhygiene in den Vereinigten Staaten von Nordamerika," *The Journal of Heredity* 5, 9 (1914), 373.

38 István Apáthy, "Fajegészségügy és fajegészségtan," *Magyar Társadalomtudományi Szemle* 7 (1914), 52–65.

39 Apáthy, "Fajegészségügy és fajegészségtan," 64–65.

40 Apparently, the creation of a Hungarian Society for Eugenics, had preoccupied Hoffmann since 1913. See also Hoffmann, "Rassenhygiene in Ungarn," 57.

41 Géza Hoffmann, "Eugenika," *Magyar Társadalomtudományi Szemle* 7 (1914), 91–106.

42 The *Királyi Magyar Természettudományi Társulat* sent the invitation on 2 December 1913. I found the letter amongst the papers belonging to the *Orvosegyesület* (Medical Society) (1913–1914) at the *Semmelweis Orvostörténeti Levéltár* (Archive of the Semmelweis Museum of Medicine). I should like to thank Benedek Varga, the director of the archives, and Lajos Domján, the archivist, for their support.

43 (Anonymous), "A fajegészségügyi (eugenikai) szakosztály megalakulása," *Magyar Társadalomtudományi Szemle* 7 (1914), 165–172.

44 (Anonymous), "A fajegészségügyi (eugenikai) szakosztály megalakulása," *Magyar Társadalomtudományi Szemle* 7 (1914), 172.

45 Géza Hoffmann, "Ausschüsse für Rassenhygiene in Ungarn," *Archiv für Rassen und Gesellschaftsbiologie* 10, 6 (1913), 830–831. See also Géza von Hoffmann, "Eugenics in Hungary," *The Journal of Heredity* 7, 3 (1916), 105.

46 See Ablonczy, *Teleki Pál*, 119–125.

47 Lajos Méhely, *A háború biológiája* (Budapest: Pallas Irodalmi és nyomdai részvénytársaság, 1915). The pamphlet was also published in *Természettudományi Közlöny* (1915).

48 Mihály Lenhossék, "A háború és a létért való küzdelem tétele," *Természettudományi Közlöny* 47, 619–620 (1915), 91–95.

49 Géza von Hoffmann, "Eugenics in the Central Empires," *Social Hygiene* 7, 3 (1921), 294.

50 Ploetz, *Grundlinien einer Rassen-Hygiene*, 147.

51 Hermann W. Siemens, *Die biologischen Grundlagen der Rassenhygiene und der Bevölkerungspolitik* (Munich: J. F. Lehmanns Verlag, 1917)

52 János Bársony, "Eugenetik nach dem Kriege," *Archiv für Frauenkunde und Eugenetik* 2, 2 (1915), 267–275.

53 Bársony, "Eugenetik nach dem Kriege," 267.

54 Bársony, "Eugenetik nach dem Kriege," 268.

55 Bársony, "Eugenetik nach dem Kriege," 272.

[56] Bársony, "Eugenetik nach dem Kriege," 275.

[57] Henrik Rotter, *Eugenika és szülészet* (Budapest: Franklin, 1912).

[58] See József Madzsar, *Az anya- és csecsemővédelem a háborúban* (Budapest: Stefánia Szövetség kiadó, 1914); József Madzsar, *Az anya- és csecsemővédelem országos szervezése. A Stefánia Szövetség alapszabályainak tervezete* (Budapest: Stefánia Szövetség kiadó, 1915); and József Madzsar, *A jövő nemzedék védelme és a háború* (Budapest: Politzer, 1916).

[59] József Madzsar, *A meddő Budapest* (Budapest: Pfeiffer, 1916).

[60] Felix A. Theilhaber, *Das sterile Berlin. Eine volkswirtschaftliche Studie* (Berlin: Eugen Marquardt Verlag 1913).

[61] See Kárpáti, ed., *Madzsar József válogatott írásai*, 41–43.

[62] Géza von Hoffmann, *Krieg und Rassenhygiene. Die bevölkerungspolitischen Aufgaben nach dem kriege* (Munich: J. F.Lehmanns Verlag, 1916).

[63] Hoffmann, *Krieg und Rassenhygiene*, 21–23.

[64] See Fritz Lenz, "G. von Hoffmann, Krieg und Rassenhygiene," *Archiv für Rassen und Gesellschaftsbiologie* 12, 5–6 (1918), 510–511.

[65] Hoffmann, "Eugenics in the Central Empires," 291.

[66] "A Magyar Fajegészségtani és Népesedéspolitikai Társaság memoranduma a be-, ki-, és visszavándorlás kritériumairól," Széchényi Library, Manuscripts Collection, Quart. Hung. 2454/II, 10–14.

[67] Hoffmann, "Eugenics in the Central Empires," 291–294.

[68] See the speech delivered by Teleki in 1917 in the Hungarian Parliament. The speech is included in Ablonczy, *Teleki Pál. Válogatott politikai írások és beszédek*, 27–48.

[69] Hoffmann, "Rassenhygiene in Ungarn," 59–66.

[70] Only three issues were published, between 1918 and 1919. The editor-in-chief was György Lukács; the editorial committee included Elemér Fischer (war widows and orphans); Géza Hoffmann (racial hygiene and population policy); Bela Kolarits (sexually transmitted diseases); and Fülöp Rottenbiller (protection of children).

[71] Hoffmann, "Eugenics in the Central Empires," 286.

[72] Ernst Tomor, *Neubegründung der Bevölkerungspolitik* (Würzburg: Verlag Curt Kabitzch, 1918), 7. Tomor is also the author of *A socialis egészségtan biológiai alapjai* (Budapest: Singer és Wolfner, 1915), one of the most sophisticated interpretations of social hygiene in early-twentieth-century Hungary.

[73] Tomor, *Neubegründung der Bevölkerungspolitik*, 47–68.

[74] Tomor, *Neubegründung der Bevölkerungspolitik*, 47–48.

[75] Tomor, *Neubegründung der Bevölkerungspolitik*, 55.

[76] Tomor, *Neubegründung der Bevölkerungspolitik*, 57.

[77] Tomor, *Neubegründung der Bevölkerungspolitik*, 60.

[78] István Apáthy, "A fajegészségtan köre és feladatai," part I, *Természettudományi Közlöny* 50, 689–690 (1918), 6–21; and István Apáthy, "A fajegészségtan köre és feladatai," part II, *Természettudományi Közlöny* 50, 691–692 (1918), 81–101.

[79] Apáthy, "A fajegészségtan köre és feladatai," part I, 8–9.

[80] Apáthy, "A fajegészségtan köre és feladatai," part II, 101.

[81] Mihály Lenhossék, "A népfajok és az eugenika," *Természettudományi Közlöny* 50, 695–696 (1918), 212–241.

[82] Hoffmann, "Eugenics in the Central Empires," 288.

[83] Alfred Ploetz papers. I should like to thank Paul Weindling for making the invitation to the conference available. The invitation was also published in Nemzetvédelem I. 1–2 (1918).

[84] See Hoffmann, "Eugenics in the Central Empires," 294; and Kovács, *Liberal Professions & Illiberal Politics*, 41–42.

[85] Quoted in Kovács, *Liberal Professions & Illiberal Politics*, 41.

[86] Kárpáti, ed., *Madzsar József válogatott írásai*, 180.

[87] Hoffmann, "Eugenics in the Central Empires," 294.

Taking Care of the National Body: Eugenic Visions in Interwar Bulgaria, 1905–1940*

Christian Promitzer

In the 1942 issue of the *German–Bulgarian Society Yearbook* (*Deutsch-Bulgarische Gesellschaft*), the Bulgarian zoologist Stefan Konsulov (1885–1954) contributed a piece on the "nature of the Bulgarian." One page was devoted to Bulgarian attitudes towards racial hygiene:

> In the past, the selection of bride and groom was made by the aged, by those who were well experienced [...]. In its essence, the nature of this responsibility was definitely racial hygiene. As in many cases the Bulgarian people expressed their experience of the past in coarse proverbs and popular sayings. To quote one of the many racial hygien- ic proverbs: 'Take dogs and women from a good tribe!' The relatives of the mate carefully investigate the tribe and the descent of the bride and the groom: whether the members of the family are economically active, quarrelsome, alcoholics or mentally disabled and so on; and after having finished preparations for marriage, the women of the kin of the groom accompany the bride to a bathing place, while the men of the bride's kin accompany the groom in the same process. Therewith each group should take a close look at the body of the future mate and ascertain whether there are any defects, which may have been hidden by clothing.[1]

Stefan Konsulov was well aware that he was writing for a German audience; he had come into contact with German racial hygiene during his stay at the University of Breslau in 1920.[2] In Bulgaria, Konsulov hoped, the "preparation and execution of a programme of racial hygiene will meet fewer obstacles than in other countries," largely due to a strong eugenic tradition in the Bulgarian Orthodox Church, "the only church with racial hygiene laws."[3] According to Konsulov, in 1871 in the statutes of the Exarchate, the autocephalous Bulgarian Church adopted a clear stance on the principles of racial theory and conferred legal legitimacy on the principles of racial hygiene. In the interest of healthy descendants this ecclesiastical legislation did not grant permission to marry if one of the suitors was suffering from psychic or venereal diseases or alcoholism.[4]

Two questions arise from this situation: one is whether "mating behaviour," as outlined by Konsulov, was a forerunner of modern eugenics; the second concerns Konsulov's reference to the past as a source of legitimacy in applying the principles of racial hygiene to Bulgaria. This chapter discusses the latter point, arguing that Bulgarian preoccupation with racial hygiene following the First World War arose from a specifically Bulgarian understanding of the process of modernization and urbanization and their consequences for Bulgarian society.[5] The recourse to an alleged tradition in Bulgarian eugenic thinking, however, may have helped integrate racial hygiene into familiar thought patterns in bringing about a moral authorization of the eugenic approach—one not completely rooted in Bulgarian society, as its supporters were inclined to believe. This chapter will further discuss the failure of eugenic societies; the dispute about racism at the end of the 1930s; and, finally, the 1943 "Law on Families with Many Children."

This outline of the history of eugenics in Bulgaria is also intended to contribute to the historical understanding of bio-politics in Bulgaria. Bio-politics is defined here as a set of various disciplines and technologies whose purpose is to examine and control the biological aspects of the nation, especially with respect to public health and reproduction. The sphere of bio-politics includes, among other aspects, racial and criminal anthropology, population policy, and classifying instruments of social work.[6] In the Bulgarian case, however, it seems appropriate to start with eugenics, for, to date, there exists no historical account of racial hygiene and eugenics in Bulgaria.[7]

The Origins of Racial Hygiene in Bulgaria

German racial hygiene, and German medicine in general, greatly influenced Bulgarian eugenic thinking. Prior to the founding of a medical faculty at the University of Sofia in 1918, students of medicine attended foreign universities, particularly German institutions.[8] At the time, the first Bulgarian attracted to the new science of heredity was Konstantin Pašev (1873–1961). As a young ophthalmologist, he broadened his knowledge of the subject in Berlin. During his stay there (1903–1904) he wrote *Higiena na okoto* (The Hygiene of the Eye).[9] In 1935, Pašev recalled that he developed his interest in heredity some five years after the Mendelian laws of heredity had been rediscovered.[10] In his book, for instance, Pašev mentioned the findings of the German ophthalmologist, Richard Heinrich Deutschmann (1852–1935), in reference to the role of heredity in affecting eyesight.[11] These findings led Pašev to formulate ideas about eugenics: "The consequences and the meaning of heredity must be popularized to our nation; the idea of a greater sense of responsibility should be instilled in the population [...] nobody has the right to override natural laws and grant existence to weak and suffering creatures. In foreign countries the term 'heredity' has already spread beyond the doors of medical academies, and has become a part of daily reading material in private households."[12] Pašev illustrated his arguments with the saying "The sins of the fathers are visited upon the children," in order to illustrate the negative role of hereditary diseases in the next generation. Therefore, Pašev argued, those contaminated with hereditary eye diseases, as well as people with eye defects, the blind and those with myopia, should not marry.[13]

The publication of Pašev's book in 1905 did not have immediate consequences for the development of Bulgarian eugenics. At the turn of the twentieth century, 1.33 per cent of the population in Bulgaria suffered from blindness, higher than in most Western and East-Central European countries (with 0.76 per cent in Switzerland and Italy, and 0.90 per cent in France).[14] However, most eye diseases and forms of blindness were not hereditary; they were the result of infection and precarious sanitary conditions. Supporters of hygiene laid emphasis on the improvement of poor sanitary conditions in the growing cities and towns. They attempted to raise awareness about hygiene first "with the pen" in medical journals, and later "by organizational means," through the founding of hygienic societies. Thus in 1909, the state-run Hygienic Institute of the Direction for the Protection of National Health *(Higienen institut pri Direkcijata za opazvane narodnoto zdrave)* was established.[15] However, with the exception of Pašev no one at that time connected the principles of hygiene—and social hygiene in particular—with heredity. Epidemic diseases, like cholera and tuberculosis, represented a serious threat during the two Balkan Wars (1912–1913 and 1913) and the First World War. Fighting against their dissemination in Bulgaria was far more important than tackling the long-term issue of improving the "racial qualities" of the nation.[16] Even Stefan Vatev (1866–1946) and Krum Drončilov (1889–1925), the leading Bulgarian supporters of physical and racial anthropology, were sceptical of eugenic ideas at the time, although Vatev was among the first to propagate hygiene in Bulgaria and later, during the 1930s, became a staunch supporter of eugenics.[17]

As such, hygienic measures in Bulgaria did not play the significant role as antecedents of racial hygiene that advocates of the latter hoped to demonstrate. Stefan Konsulov, a promoter of racial hygiene in the early 1920s, ranked the different branches of hygiene as if they were different stages of human development. For Konsulov, individual hygiene constituted the lowest level of physical and mental health; social hygiene, which fought against infection and contamination, was graded on the second level; yet racial hygiene, the highest level, aimed "to lift the health and strength of the nation."[18] Konsulov had gained experience as a hygienist during the First World War, when he was engaged in the preventive campaign against malaria. But this experience was not apparent in the evolution of his ideas about racial hygiene. As a student and young natural scientist, Konsulov had already analysed heredity and evolution in the popular scientific journal *Priroda* (Nature), where he discussed the achievements of illustrious biologists like Charles Darwin (1809–1882), August Weismann (1834–1914), Gregor Mendel (1822–1884), and Ernst Haeckel (1834–1919).[19] In discussing the possible consequences of intermarriage, he adopted a strict anti-Lamarckian position in rejecting the idea of the inheritance of acquired traits.[20]

In 1922, in one of his many articles on the Mendelian laws, Konsulov discussed the role of recessive factors in the dissemination of hereditary diseases among human beings (diabetes, albinism, abnormal small stature, epilepsy, "feeblemindedness," and so on). Moreover, he insisted that such individuals were obliged to fulfil their moral responsibility to refrain from reproduction in order to save their children from similarly miserable fates. Healthy people, on the other hand, should be made aware of the dangers of marrying individuals with hereditary diseases, for such negative phenomena could result in widespread social degeneration. The general application of birth control (Konsulov did not specify whether he meant sterilization or other forms of contraception) would strengthen the nation and allow for the development of individuals with positive qualities: "In this way the bad will be removed and the good will be saved; the [Bulgarian] race will develop in a positive direction and will be capable of a higher level of culture."[21]

However, Konsulov had to create an appropriate Bulgarian terminology for the application of methods of breeding to humans. After considering the terms *eugenika, evgenika,* and *rasova služba* (racial service) he decided in favour of *rasova higiena* (racial hygiene).[22] With the term *rasova higiena,* Konsulov expressed his appreciation of the German tradition of *Rassenhygiene* as opposed to the British and American eugenics, or French *eugénique.* The short-lived term *rasova služba* had a German equivalent, *Rassedienst,* which was used by the German racial hygienist, Wilhelm Schallmayer (1857–1919), in his 1918 *Grundriß der Gesellschaftsbiologie und der Lehre vom Rassedienst* (Outline of Social Biology and The Theory of Race Service).[23] Thanks to Konsulov, the terms *rasova higiena* and *evgenika* would be used over the next two decades in Bulgaria. Those that preferred the term *rasova higiena* indicated their support for contemporary German eugenics, while the term *evgenika* encompassed eugenic ideas from other countries.

In Germany, racial hygiene provided a possible way out of the general crisis arising from the disastrous experience of the First World War. As a result of precarious socioeconomic and political circumstances, the long-term consequences of urbanization and declining birthrates were discussed in *Grundriß der menschlichen Erblichkeitslehre und Rassenhygiene* (Outline of Human Genetics and Racial Hygiene), published in 1921 by Erwin Baur (1875–1933), Eugen Fischer (1874–1967), and Fritz Lenz (1887–1976).[24] In some respects, the situation in Bulgaria was similar to that in Germany: Bulgaria also belonged to the group of defeated powers. The 1919 Peace Treaty of Neuilles sealed Bulgaria's loss of Southern Dobrudja, the Aegean shore, and a major part of Macedonia; the country was over-crowded with refugees from these regions, and it had to pay reparations; additionally, cities suffered from inflation and unemployment. Bulgarian intellectuals considered the aftermath of recent wars to be a "national catastrophe." The failure to achieve national aims, and the unstable political situation throughout the 1920s and 1930s, fostered various in-depth analyses about the Bulgarian "national character," in which the essence of a nation in the midst of a deep existential crisis was the dominant theme. Expressing this anxiety, Bulgarian intellectuals affirmatively adopted the arguments put forward by Oswald Spengler (1880–1936) in his 1918 *Der Unter-gang des Abendlandes* (The Decline of the West).[25]

Konsulov's interpretation of racial hygiene implicitly referred to the eugenic arguments developed by Baur, Fischer, and Lenz. For Konsulov, culture and material wealth were responsible for the degeneration of the nation.[26] Wealth offered the weak and inept, who otherwise would not find a mate, the possibility of marrying and having children. In turn, degeneration resulted from a decrease in the general level of physical and mental qualities intrinsic to the nation. This process was an urban phenomenon, visible among the "white races" of Northern America and Western and East-Central Europe. Even in Bulgaria, "the peasants have physically healthier descendants than is the case with the residents of the cities."[27]

For Konsulov, changing gender roles in modern society were also responsible for declining birthrates and genetic degeneration: "Talented women represent the genetic treasure of a nation, but instead of bearing many children they often prefer to work in science and to stay childless. They squander their valuable genes; the number of intellectually gifted will decline in the long run."[28] Bulgarian feminists, unsurprisingly, denounced Konsulov. They insisted that the issue he raised was not about saving the "genetic treasure of the nation," but about preserving the monopoly of men in academia and in high social positions.[29] Such criticism notwithstanding, clearly Konsulov called for the application of the principles of racial hygiene in Bulgaria; it alone could save the nation from decline and misfortune. This was the argument developed in numerous articles published in the journal *Priroda* and the newspaper *Slovo*. Konsulov, who assumed the Chair of Zoology at the University of Sofia in 1923, subsequently started recruiting like-minded individuals to support his form of eugenics.

The Bulgarian Society for Racial Hygiene

In 1926, a small group of intellectuals in Sofia, eager to discuss eugenics, gathered around Konsulov, forming an informal "Circle for the Study of Racial Hygiene" *(Obštestvo za rasovo-higienični proučavanija)*. During their weekly meetings they discussed the decreasing birthrate in Bulgarian society, alcoholism, venereal diseases and similar "social evils," as well as the spread of hereditary diseases.[30] Damjan Ivanov, Ljubomir Ivan Rusev, Petăr Penčev, Ivan Germanov Kinkel (1883–1945) and Konsulov formed the core of the inner circle of the group.[31]

In 1928, the group established the Bulgarian Society for Racial Hygiene *(Bǎlgarsko družestvo za rasova higiena)*. The aims of the Society were as follows: "1) to inform members of the Society about scientific achievements in the field; 2) to investigate problems connected to the fight against the degeneration of the Bulgarian nation; 3) to inform Bulgarian society and state about these issues; 4) to organize a special library and to edit a bulletin about its activities, and later also a scientific journal series."[32] Damjan Ivanov eventually became the president, and Konsulov the vice-president of the Society. Members of the executive committee included Asen Ivanov Hadžiolov (1903–1994), later a leading histologist and cytologist in Bulgaria.[33] The Bulgarian Society for Racial Hygiene was an exclusive circle. Konstantin Pašev, who later became a prominent representative of eugenics, did not play any role in the Society, nor did Toško Petrov (1872–1942), a leading Bulgarian hygienist and professor of hygiene at the Medical Faculty of the University of Sofia. Petrov sympathized with the work of the Austrian hygienist, Julius Tandler (1869–1936), a representative of Social Democratic eugenics and head of welfare and health in Vienna during the 1920s.[34]

Konsulov was responsible for writing a programmatic introduction to eugenics, pointing at cattle-breeding and the cultivation of plants; he insisted that man, by continuous selection of the best breed and stock, has consistently applied the laws of heredity for the purpose of genetic improvement. However, due to sporadic care, degeneration ensued: "Compare oxen in some parts of the Balkans with the Iskăr race [a special Bulgarian breed], and one is convinced of the degree to which the first group has degenerated. Also our cereals have degenerated (...) Why has this happened? Because we do not practice selection."[35] The laws of heredity regulated mankind in the same way; mankind was either doomed to degeneration or would become the subject of systematic selection. Konsulov envisaged a strange eugenic utopia:

> If mankind fell into the hands of a superior being which uses these methods which we apply to our plants and cattle, it would result in the creation of humans with positive qualities (...) the difference between them [the new men] and contemporary man would be so marked that it would be greater than that between the beautiful apple from Kjustendil [a city and region in western Bulgaria] and the tart type from the mountains. It would be possible to create a mankind which has not known disease; [which is] highly intelligent, musically gifted and moral; for these people legal regulations and prisons would be needless. From a biological point of view they would truly constitute a superman of sound, gifted and moral character.[36]

Despite such far-reaching visions, the Bulgarian Society for Racial Hygiene organized only a modest lecture series on racial hygiene. The first cycle of lectures began in the summer of 1928 at the Institute of Biology at the University of Sofia. Damjan Ivanov delivered the introductory lecture about the history of racial hygiene, and Konsulov was assigned three lectures about the theories of heredity; scientific experimentation and Mendelism; and the theory of evolution. Ljubomir Ivan Rusev was scheduled to deliver three lectures (on human genetics, human hereditary diseases, and the basic principles of applied racial hygiene); Petăr Penčev's topic was the evolution of civilized nations. Further lectures about heredity and crime, heredity and education, as well as alcoholism and venereal diseases, were planned.[37] The anatomist and physical anthropologist Milko Balan (1888–1973) also addressed the Society.[38]

In 1928, Konsulov published two booklets outlining the activities of the Society.[39] The first discussed the laws of heredity—especially regarding human heredity—while the second addressed the "degeneration of civilized mankind." In the first booklet, Konsulov declared that the progress of biology during the previous two decades, in the fields of evolution and heredity, had led him to present his new findings to a broader Bulgarian audience.[40] He rejected Lamarckism once again and assessed the theories of Weismann and Mendel, in addition to considering chromosomal theory. By addressing the fatal effects of hereditary diseases he underlined the necessity of preventing marriage in such cases, or, at the very least, preventing the possibility of reproduction: "We remain uneducated in this spirit, and such measures may be considered strange. But they will certainly impose themselves on the society of the future."[41] In the second booklet, Konsulov followed Schallmayer, Baur, Fischer, and Lenz. First, he described the characteristics of degeneration, including the elimination of natural selection and the reversal of declining birthrates. Thereafter, he discussed racial hygiene as a means by which to prevent decline and extinction and to improve the physical and mental qualities of mankind. Dissemination of the principles of racial hygiene could thus aid in educating Bulgarians about "racial ethics." However, these ideas could only be achieved through the implementation of specific legislation. Racial ethics "asks the individual to be prepared to make sacrifices (...) Racial ethics asks from a human being who is suffering from dangerous hereditary weaknesses to refrain from marriage, to go without the joy of having a family (...) Why? So that, after many years, when this individual is buried, society may enjoy better health."[42]

Among the racial hygiene measures, Konsulov recommended family certificates that listed hereditary weaknesses. The certificates were to be used by marriage councils to allow or prevent proposed marriages. Society should have the right to intervene in the private life of the individual, especially in the case of reproduction, which Konsulov regarded as the sole purpose of marriage. Therefore, marriage should be prohibited only in those cases where the danger of degeneration was high. The mentally disabled, criminals, incurable alcoholics, and disfigured individuals should all be sterilized.[43] Konsulov furthermore demanded the idolization of the family in order to ensure the fertility of "valuable" members of society. Special legislation would grant that each "valuable" couple would give birth to at least four children in order to compensate for the mortality rate.[44] In 1928, following the publication of various booklets and the organization of lecture series, the activities of the Society were stopped for "different reasons."[45]

The first collective venture of Bulgarian racial hygienists did not succeed because they did not capture the interest of the general public. Eugenics and related questions were too difficult to grasp for a country that had much more pressing health problems to tackle. Moreover, Bulgarian racial hygiene lacked the basis of an elaborate population policy, as was the case in Germany. Until the late 1920s, Bulgarian racial hygienists only suspected that a drop in the birthrate among the Bulgarian population was underway, for they still lacked concrete data.

Whether the activities of the Bulgarian Society for Racial Hygiene affected the "Law for National Health" passed by the Bulgarian Parliament in 1929 remains a matter of further investigation. To be sure, the law contained passages about fighting tuberculosis and alcoholism. In its preamble, it addressed the necessity of an increase in the physically and mentally healthy members of the Bulgarian population, and the importance of avoiding reproduction among those suffering from hereditary diseases. The law also stated that individuals entering into marriage must receive printed instructions about the importance of health and the potentially harmful consequences of hereditary diseases on spouses and descendants. Municipalities had the right to establish a bureau for premarital medical counselling. With respect to demographic policies, the law made abortion and the distribution of contraceptive devices liable to prosecution.[46] However, by the time the law came into force the Bulgarian Society for Racial Hygiene was no longer active.

In 1930, a certain Meri (Mary) Kričeva translated and edited the book *Le Haras Humain* (The Human Stud-Farm) by the French psychopathologist Charles Binet-Sanglé (1868–1941), originally published in 1918. The book demanded the application of euthanasia as well as the foundation of a eugenic institute for the improvement of the human race. Kričeva's edition also included a translation of the standing rules of the British Eugenics Society, and the Norwegian Program for Race Hygiene, as presented in 1908 by the Norwegian eugenicist Jon Alfred Mjøen (1860–1939).[47] But this endeavor to revive the Bulgarian eugenics movement met with resistance.

Meanwhile, Stefan Konsulov continued to popularize racial hygiene both in popular articles and his two textbooks on biology.[48] In 1929, the statistician Petăr Penčev published an article on the declining birthrate in Bulgaria; he also assessed the differential birthrate between cities and the countryside, as well as that between different professions. For him, the reasons for the declining birthrate were not only social but also "national" and "biological."[49] With his extensive *Izsledvanija vărhu demografijata na Bălgarija* (Studies on the Demography of Bulgaria), published in 1931, the jurist and economist Georgi Danailov (1872–1939) emulated Penčev's views. Danailov demonstrated that the national and "racial" differences between the Bulgarian population and national minorities like the Turks and Muslim Pomaks would find their expression in different demographic behavior, especially with respect to fertility. After the First World War, the Muslim population increased, and subsequently the Muslim population grew faster than the Bulgarians. However, Danailov insisted that the Bulgarians were more physically virile and more culturally advanced than the Muslims living in Bulgaria.[50]

The Renewal and Demise of the Bulgarian Society for Racial Hygiene

After 1932, eugenics grew in popularity among the Bulgarian public. Several factors account for this changed attitude. The first was a heightened state awareness about the health of the population. Malaria was widespread in the Bulgarian countryside. During the 1930s, frequent cases of typhus and diphtheria were recorded in small cities and some quarters of Sofia. Every year approximately 150,000 to 200,000 people were infected by tuberculosis, with 10 to 20 per cent fatalities ensuing. Increased awareness of these illnesses resulted in a general consensus that the health of the population was getting worse.[51] In 1931, the journal *Bǎlgarski higienen pregled* (Bulgarian Review of Hygiene) was launched as a medium of information and discussion about hygienic measures. It was followed in July 1932 by the foundation of the Society for Hygiene and Preventive Medicine in Bulgaria *(Družestvo za higiena i predpazna medicina v Bǎlgarija)*. The journal and Society were directed by Toško Petrov. Both addressed eugenics in a more moderate way than the Bulgarian Society for Racial Hygiene.[52] Finally, Konstantin Pašev demanded the introduction of Racial Eye Hygiene *(Rasova očna higiena)*, on account of the high rate of blindness among the Bulgarian population.[53]

The debate about the statutes of the Bulgarian Exarchate also contributed to the revival of eugenics in the early 1930s. As already mentioned, Bulgarian Orthodox statutes prohibited marriage if one spouse suffered from syphilis, epilepsy or a mental disorder. This statute was theoretically binding on all newly married couples; however, there are no known examples of its implementation in practice.[54] When, in 1932, the Supreme Secular Exarchate Council discussed new amendments to the statute, the Society of Hygiene and Preventive Medicine asked for recognition of the principles of eugenics, to allow for the exclusion from marriage of carriers of incurable venereal diseases, as well as those inflicted with pulmonary and laryngeal tuberculosis.[55] In turn, Konstantin Pašev and his Bulgarian Ophthalmologic Society additionally demanded the further exclusion of all individuals born blind.[56] The reformulated statute was passed in 1937; however, it did not contain any of these amendments.[57]

The third factor related to the stimulation of Bulgarian racial hygiene was the impact of Hitler's rise to power and the Nazi laws on sterilization. Bulgarian intellectuals disgusted by the parliamentary system, and political scandals in particular, were attracted by the Nazi model, which demonstrated that a fascist state was able to implement effective eugenic legislation. Adherents of eugenics who had been politically neutral now began to show their right-wing political affiliation. In 1932, a year before Hitler's takeover, Pašev demanded the introduction of "National Solidarism" in Bulgaria; its proximity to the term National Socialism was not accidental.[58] Pašev was also an enthusiastic observer of the progress of racial hygiene in Germany. His approval of the Nazi law on sterilization was shared by several colleagues, all of whom joined the reinvigorated eugenics society.[59]

The publication of the *Osnovni principi na Evgenikata* [*"Rasova higiena"*] [Basic Principles of Eugenics (Racial Hygiene)] in 1934 by Ivan Ljubomir Ivanov Rusev represented the fourth factor. For the first time, this book summarized the aims and visions of eugenics in Bulgaria, and encouraged the renewal of the inactive Bulgarian Society for Racial Hygiene (Rusev was its last secretary). Therefore, it seems reasonable to discuss the important passages of the book in some detail. Three-quarters of it dealt with the effects of the laws of heredity on human beings, and their connection to degeneration. The last quarter of the book discussed "applied racial hygiene."

Taking the ideas of the Austrian racial hygienist Max von Gruber (1853–1927) as a starting point, Rusev argued that all eugenic measures should be conducted in a situation whereby external conditions were equal for all people. On the other hand, in order to eliminate the negative effects of selection, the state had to ensure that its legislation was in accordance with the laws of nature.[60] To do so, Rusev distinguished between *qualitative* and *quantitative* racial hygiene. The latter related to the increase of progeny, while the former was divided into "eliminatory" racial hygiene (negative eugenics) and "selective" racial hygiene (positive eugenics). Eliminatory racial hygiene concerned the prevention of negative heredi-

Д.-ръ ЛЮБОМИРЪ ИВ. РУСЕВЪ

ОСНОВНИ
ПРИНЦИПИ
НА
ЕВГЕНИКАТА
(,,РАСОВА ХИГИЕНА")

СЪ ПРЕДГОВОРЪ ОТЪ ПРОФ. СТ. КОНСУЛОВЪ

СОФИЯ — Печатница ,ПОЛИГРАФИЯ' а. д. — 1934

Fig. 1 The cover of Ljubomir Ivanov Rusev's *Basic principles of Eugenics* (*Racial Hygiene*) (1934)

tary disorders like alcoholism, syphilis, and tuberculosis, in addition to the exclusion from reproduction of those with serious hereditary diseases. Selective racial hygiene, on the other hand, focused on the recovery and strengthening of the race by promoting positive selection and encouraging the reproduction of healthy and talented individuals.[61]

"Eliminatory racial hygiene" meant the elimination of alcoholism, venereal diseases (with the compulsory registration of those diagnosed with syphilis) and tuberculosis, whereby carriers of those diseases were not allowed to marry and procreate. This principle extended to all individuals with hereditary diseases, requiring that the state regulate marriages and births. Premarital medical examinations were therefore compulsory, together with the introduction of health certificates for newborns, as demanded by Wilhelm Schallmayer. The ideal method with respect to achieving the desired eugenic results was sterilization; Rusev did not consider it to be a more complicated procedure for women than men. He envisaged that a council of specialized medical doctors would assess each individual case for sterilization. Other effective means of eliminatory racial hygiene included the isolation of the mentally disabled in asylums and the internment of criminals in prisons.[62]

These measures still allowed a large number of individuals with an "average" hereditary disposition to reproduce to a greater extent than elite groups having "more valuable genes." Inciting the latter to have more children was therefore the task of "selective racial hygiene." Which stock of people satisfied the aims of selective racial hygiene? Rusev conceived a utopia of elite reign, one which was exclusively

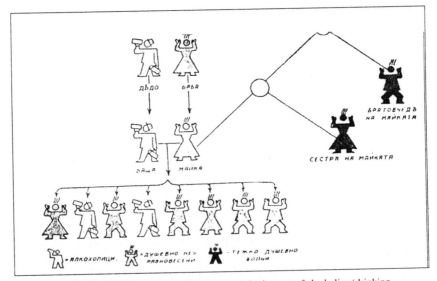

Fig. 2. Diagram showing how the Mendelian laws work in the case of alcoholics (drinking figures), mentally unbalanced (grey figures), and seriously mentally ill (black figures). Source: *Narod i potomstvo*, vol. 1 (1935)

male and disinterested in material wealth. Ironically, his description of the elite perfectly reproduced the social profile of Bulgarian eugenicists. For Rusev, contemporary society demanded intelligent individuals of a high moral standing rather than those having perfect physical attributes. But due to a corrupt education system, not all people with high-school and university degrees could fulfill the high standards of the intellectual elite in order to be selected for reproduction. Therefore, teachers should register the mental qualities and character of their pupils in health certificates. Those of higher talent should not be compelled to share the classroom with children of lesser intelligence, and should thus be instructed in particular schools. Long periods of instruction were not necessary for highly talented people; they only delayed reproduction. Therefore, the duration of instruction should be shortened by two or three years.[63]

Other factors impeding the birthrate of the "intelligent class" included socioeconomic conditions, which often prevented a couple from having children. From the eugenic viewpoint, the women's movement for equal rights was also an obstacle—not because women would not be able to replace men in society but, as Rusev put it, because men could not replace women as mothers: if a woman from the "intelligent class" occupied an important position otherwise given to a man, the latter would lose the material means necessary to start a family; the man would thus be childless. In turn, the working woman would not wish to bear children: "The damage for race is thus twofold," declared Rusev.[64]

In contrast, "quantitative racial hygiene" was related to the increasing rate of reproduction regardless of genetic quality. Some racial hygienists, like the German dermatologist Hermann Werner Siemens (1891–1969), rejected quantitative measures as senseless, in light of the fact that lower social classes reproduced at a higher rate than the upper classes. Rusev, on the other hand, counted himself among eugenicists believing that, with respect to declining birthrates, measures of quantitative racial hygiene (family policy via tax reforms and financial support; the fight against abortion, and so on) were necessary because the cultural and socioeconomic progress of a nation depended on a rising birthrate in general.[65]

The four factors outlined here (increased awareness of the role of the state in supporting national health; the statutes of the Bulgarian Exarchate; the example of Nazi Germany; and Rusev's book on eugen-

ics) created a propitious atmosphere for the regrouping of the supporters of eugenics in Bulgaria. On 28 November 1934, Stefan Konsulov
invited Stefan Vatev, Konstantin Pašev, Nikola Saranov (1895–1974)
and ten others (many of them members of the first Bulgarian Society
for Racial Hygiene) to the Institute of Zoology at the University of
Sofia. The time had come to revive the old eugenics society. Saranov,
the founder of Bulgarian criminology, and an adversary of eugenic
sterilization during the 1920s, declared that the refounding of the society was not only useful but also necessary: "The current government,
which has assumed responsibility for overseeing the strengthening of
the nation, must support the purpose of this Society."[66] Saranov was
referring to the government of Kimon Georgiev (1882–1969), whose
right-wing group *Zveno* (The Link) along with the so-called Military
League (*Voenen săjuz*), organized a coup d'état in May 1934; the new
government suppressed political parties and abolished parliamentary
rule. In 1935, King Boris III (1894–1943) removed Georgiev, but the
authoritarian regime remained in power.

During the 1934 meeting at the Institute of Zoology, Pašev suggested that the Society amend its original statutes to include the following
goal: "It (the Society) recommends concrete legislative measures in the
spirit of eugenics." Konsulov indicated that the term "racial hygiene"
should be replaced by "eugenics" in the official name of the Society.
Both suggestions were approved. The meeting elected an executive
committee, once again with Damjan Ivanov as president, and this time
with Nikola Saranov as vice-president and Ljubomir Ivan Rusev as
secretary.[67]

It took until February 1935, however, for the Ministry of the Interior and National Health (*Ministerstvo na vătrešnite raboti i narodnoto
zdrave*) to approve the existence of the new Bulgarian Eugenics Society
(*Bălgarsko družestvo za evgenika [rasova higiena]*).[68] Already in January 1935, under the editorship of Ljubomir Ivan Rusev, the group
began to publish the monthly *Narod i potomstvo* (Nation and Progeny).
This journal appealed to a wide readership, for the intended role of the
journal was educational and propagandistic, with the editors explicating the basics about heredity and eugenics. The editors justified the
importance of premarital medical examinations, eugenic sterilization,
and the general principles of eugenics. As such, the survival of nation
and race was closely connected to the introduction of eugenic measures by the "new state."[69] After five issues and a desperate appeal for

subscribers, the Eugenics Society finally ceased publication of the journal for financial reasons. Yet, until the end of the year, *Narod i potomstvo* appeared as a supplement in the medical journal *Zdravna prosveta* (Health Education). In early 1936, both the supplement and the Society ceased their activities altogether.[70]

Like its predecessor, the second Bulgarian Eugenics Society was short-lived. Many critics—largely communists—had serious doubts about the scientific basis of its ideology, its admiration for Nazi racial hygiene, and its rejection of environmental factors in the process of socialization.[71] Other critics, such as Toško Petrov, chairman of a rival organization, the Society for Hygiene and Preventive Medicine, deemed the introduction of radical eugenic measures like sterilization to be inappropriate: "In our country we are still at the beginning of our study of social hygiene and the biological conditions of life. We must first preserve and secure the general health of the population through social hygiene. A more urgent task is to oversee the preservation of health and the morality of the family. This can be achieved through the marriage of those capable of giving birth to socially suitable individuals. In this respect eugenic measures such as premarital medical examinations would be sufficient."[72]

One of the last activities of the Bulgarian Eugenics Society was to break the existing stalemate with the public. In collaboration with Toško Petrov and a physiologist from the Society for Hygiene and Preventive Medicine, Dragomir Mateev (1902–1971), members of the Society drew up an official statement relating to their stance on sterilization and premarital medical examinations. It was clear that the Eugenics Society had abandoned its previous support for the introduction of compulsory sterilization. Bulgarian eugenicists now accepted that eugenic measures should be applied where appropriate. In any case, they contended that individuals with hereditary diseases were less likely to reproduce, and therefore should be "socially sterilized." Both societies concluded that social eugenic instruction and systematic genealogical family research were immediately necessary. The latter research would form a base for assessing carriers of recessive hereditary diseases, in addition to facilitating premarital medical counseling. Moreover, both societies wanted to combine the introduction of a system of premarital medical counselling with the introduction of civil marriages. A common statement in this sense was sent to the Commission for Civil Marriage at the Ministry of Justice (*Komisija za graž-*

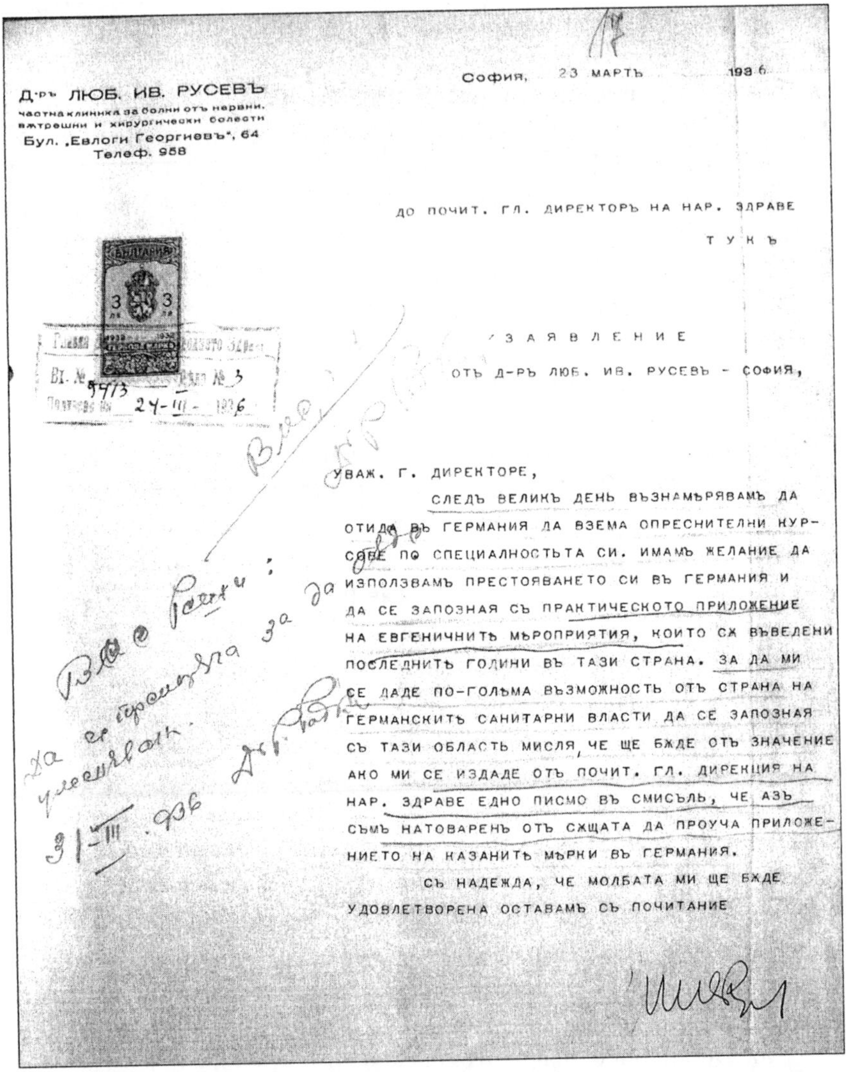

Д-ръ ЛЮБ. ИВ. РУСЕВЪ
частна клиника за болни отъ нервни, вътрешни и хирургически болести
Бул. „Евлоги Георгиевъ", 64
Телеф. 958

София, 23 мартъ 1936

ДО ПОЧИТ. ГЛ. ДИРЕКТОРЪ НА НАР. ЗДРАВЕ

ТУКЪ

ЗАЯВЛЕНИЕ

ОТЪ Д-РЪ ЛЮБ. ИВ. РУСЕВЪ – СОФИЯ,

УВАЖ. Г. ДИРЕКТОРЕ,

СЛЕДЪ ВЕЛИКЪ ДЕНЬ ВЪЗНАМѢРЯВАМЪ ДА ОТИДА ВЪ ГЕРМАНИЯ ДА ВЗЕМА ОПРЕСНИТЕЛНИ КУРСОВЕ ПО СПЕЦИАЛНОСТЬТА СИ. ИМАМЪ ЖЕЛАНИЕ ДА ИЗПОЛЗВАМЪ ПРЕСТОЯВАНЕТО СИ ВЪ ГЕРМАНИЯ И ДА СЕ ЗАПОЗНАЯ СЪ ПРАКТИЧЕСКОТО ПРИЛОЖЕНИЕ НА ЕВГЕНИЧНИТѢ МѢРОПРИЯТИЯ, КОИТО СѪ ВЪВЕДЕНИ ПОСЛЕДНИТѢ ГОДИНИ ВЪ ТАЗИ СТРАНА. ЗА ДА МИ СЕ ДАДЕ ПО-ГОЛѢМА ВЪЗМОЖНОСТЬ ОТЪ СТРАНА НА ГЕРМАНСКИТѢ САНИТАРНИ ВЛАСТИ ДА СЕ ЗАПОЗНАЯ СЪ ТАЗИ ОБЛАСТЬ МИСЛЯ, ЧЕ ЩЕ БѪДЕ ОТЪ ЗНАЧЕНИЕ АКО МИ СЕ ИЗДАДЕ ОТЪ ПОЧИТ. ГЛ. ДИРЕКЦИЯ НА НАР. ЗДРАВЕ ЕДНО ПИСМО ВЪ СМИСЪЛЪ, ЧЕ АЗЪ СЪМЪ НАТОВАРЕНЪ ОТЪ СѪЩАТА ДА ПРОУЧА ПРИЛОЖЕНИЕТО НА КАЗАНИТѢ МѢРКИ ВЪ ГЕРМАНИЯ.

СЪ НАДЕЖДА, ЧЕ МОЛБАТА МИ ЩЕ БѪДЕ УДОВЛЕТВОРЕНА ОСТАВАМЪ СЪ ПОЧИТАНИЕ

Fig. 3 Ljubomir Ivanov Rusev asks the Bulgarian Main Direction for Public Health for an official letter of recommendation in order to facilitate his intention 'to become acquainted with the application of eugenic measures' during his stay in Germany in 1936. Source: Central Historical State Archives of Bulgaria

danskija brak pri Ministerstvoto za pravosădieto), the Main Direction of National Health (*Glavna direkcija na narodnoto zdrave*), and the Bulgarian Women's Union (*Bălgarski ženski ssăjuz*).[73]

Eugenic Legislation without Eugenicists

Did the demise of the second Eugenics Society signal the demise of Bulgarian racial hygiene? To be sure, several members continued to propagate eugenic ideas in the public sphere during the late 1930s. The immediate reasons were an intense dispute on racism and growing public interest in the decline of the birthrate.

The dispute on racism commenced in late 1937, when an anonymous author—actually Stefan Konsulov—published *Rasovijat oblik na Bălgarite* (The Racial Countenance of the Bulgarians).[74] This account of physical traits and blood groups represented the continuation of a series of articles on the racial characteristics of Bulgarians published two years previously by Stefan Vatev in *Narod i potomstvo*.[75] Yet Konsulov went further. He claimed that Bulgarians, although Slav speakers, were of non-Slavic origin, and called for an adherence to the principles of racial hygiene. He concluded that no political ideology in Bulgaria had integrated racial hygiene into its agenda except for a rising movement, the Fighters for the Progress of Bulgarianhood, or Fighters for the Nation (*Ratnici za napredăka na bălgarština*).[76] The leader of this pro-Nazi, anti-Semitic group was the economist Asen Kantardžiev (1898–1981).[77] The *Ratnici* had, in fact, published the "Racial Characteristics of the Bulgarians" as the fifth volume in their series of pamphlets.[78]

In early 1938, the doyen of Bulgarian biology, Metodij Popov (1881–1954), delivered two public lectures *Bălgarskijat narod meždu evropejskite rasi i narodi* (The Bulgarian Nation and the European Races and Nations) and *Nasledstvenost, rasa i narod* (Heredity, Race and Nation).[79] Both lectures were critical of Konsulov's writings. According to Popov, Bulgarians belonged, without a doubt, to the Slavic nations. He rejected the idea of racial purity and the claim that there was a racial hierarchy with a "Master Race" on the top. As far as Popov was concerned, the decline in birthrates among the elites was part of a larger historical process: "After a period of stagnation and decline, a period of spiritual rebirth follows; radical change within the nation is brought about by mixing [the old elites], either with new social classes (...) or with closely related races."[80] With respect to the alleged biological degeneration of other European nations, as declared by various representatives of racial hygiene, Popov concluded that the elapsed period of observation was far too short for such an assessment: "Human evolution, however, does not take place and is not achieved within the bounds of limited and short historical periods. From a biological point of view ours is not a period of decline of either human civilization or Western civilization, as Spengler has predicted, but a period which signifies the emergence of civilization."[81]

As a result, the *Ratnici* disturbed Popov's public lectures, and Konsulov attacked his conception of "race." The philosopher Dimitǎr Mihalčev (1880–1967), on the other hand, reproached Popov for not going far enough in his rejection of "racial science." For Mihalčev, the "nation" was a group of people with a common historical fate. The nation was thus an entity that could not be understood simply by using the term "race," or through the discipline of biology.[82]

The debate on the racial origins of the Bulgarians had limited repercussions; this was not the case with another debate on the declining birthrate. In 1938, in the daily newspaper *Mir*, Konsulov warned that declining birthrates would lead to defeat in war by the far more "reproductive" neighboring countries: Romania, Yugoslavia, and Turkey.[83] Therefore, measures protecting families with many children, following the German example, would need to be introduced.[84] In late 1938, *Mir* conducted an opinion survey of experts on the reasons for declining birthrates in Bulgaria, as well as measures that could be introduced to reverse the trend. The Main Direction for National Health (*Glavna direkcija za narodnoto zdrave*), the central body of the Bulgarian Ministry of the Interior and National Health, appointed a commission with the task of drawing up legislation for "the preservation and increase of the Bulgarian nation."[85] In response, the Society for Hygiene and Preventive Medicine suggested various pro-natalist measures, the prosecution of illegal abortions, and the introduction of a special law for the protection of families. (The last point included obligatory premarital medical examinations, distribution of farmland, and tax privileges.[86]) During the early 1940s, the League of Parents with Many Children (*Sǎjuz na mnogodetnite v Bǎlgarija*) became an influential pressure group advocating the passing of laws like those suggested by the Society for Hygiene and Preventive Medicine.[87] Bulgaria's accession to the Axis Pact in 1941 clearly signaled the dominant influence of the Third Reich, particularly in the field of pronatalist population policy. In 1942, the Bulgarian Ministry of the Interior and National Health prepared the "Law on Families with Many Children," which was enacted in March 1943. The law granted tax relief, state credits and other privileges to young couples intending to produce more than two children. It also included eugenic regulations: it stipulated that the groom should not be over forty years of age; and in order to secure state credits, all spouses were obliged to undergo compulsory premarital medical examinations to ensure that neither had hereditary diseases.[88]

The law was formulated according to the Nazi system of marriage loans introduced into Germany in 1933, and was finalized with the addition of tax-free allowances in 1934 and 1939. In contrast to similar laws in Europe at that time, the German legislation excluded "inferior" and "racially alien" individuals like Jews, Gypsies, those who had been sterilized, and those with hereditary diseases.[89] Petăr Gabrovski (1898–1945), minister of the Interior and National Health, who had drafted the Bulgarian bill, initially restricted the scope of legislation to Bulgarian families "in order to maintain the purity of the Bulgarian race."[90] King Boris III, who was not of Bulgarian origin, was deeply offended by the new legislation, which only recognized ethnically pure Bulgarians. The Bulgarian prime minister, Bogdan Filov (1883–1945), apologized to the king, explaining that the regulation did not apply to the royal family, as it was aimed primarily at the Jews.[91] In its final draft, the law stipulated that "the men and women must be Bulgarians and subjects of the Bulgarian Kingdom."[92] Further research is necessary in order to reveal whether this passage excluded only the Jews, or extended to other Bulgarian ethnic minorities as well.[93]

Bulgarian eugenicists were not consulted in the process of drafting the new legislation passed in March 1943. Yet in June 1943, the Society for Hygiene and Preventive Medicine belatedly discussed the law and underlined its eugenic character.[94] In the statement sent by the Society to the Main Direction of National Health, members praised the law as an "enormous achievement," although they doubted it could "fully solve eugenic and health problems." They insisted that premarital medical examinations should focus on identifying heredity diseases: "It is the opinion of the Society that we consider the sterilization of the socially inferior and harmful in our country as a serious option. Sterilization should be organized in such a way that makes use of practical experience in countries where it has been introduced."[95]

Conclusions

Due to wartime circumstances there was little time in which to apply the "Law on Families with Many Children." In the late summer of 1944, the Communist-led Fatherland Front (*Otečestven front*) assumed power in Bulgaria. Due to his pro-Nazi orientation and racist ideas, the most prominent Bulgarian racial hygienist, Stefan Konsulov, was relieved of his position at the University of Sofia, and in 1945 was sentenced to seven years in prison.[96] However, other Bulgarian eugenicists, like Asen Ivanov Hadžiolov and Dragomir Mateev, who had adopted moderate positions, were able to continue their research in the postwar socialist Bulgaria. The "Law on Families with Many Children" was not abolished by the Communists, but merely modified. In 1951 it was replaced by a Decree for the Stimulation of the Birthrate and Abundance of Children (*Ukaz za nasărčenie na raždaemostta i mnogodetstvoto*). During the 1960s, premarital medical examinations became obligatory for all.[97]

The history of Bulgarian eugenics cannot be written without considering the influence of German racial hygiene. In both Germany and Bulgaria, a period of national optimism and romanticism ended with the First World War and was replaced by pessimistic "introspection," expressed in the Bulgarian case by critical examinations of "national character."[98] In contrast, racial hygiene provided a different perspective: its conceptualization of a glorious national future in the form of a eugenic utopia offered recompensation for the lost of the optimism characterizing the pre-1914 period.

As in the German case, Bulgarian racial hygienists also felt empowered to redefine gender roles. They did not derive their anti-feminist positions from the rules of the traditional patriarchal society but from a position of neo-traditionalism. This neo-traditionalism was connected to right-wing and fascist ideologies, and was nurtured by a particular interpretation of biology, ascribing to women the role of motherhood and of carrier of the "genetic treasure" of the nation.[99] The short-lived success of Bulgarian eugenics was ultimately encapsulated by the "Law on Families with Many Children," a law prepared without the input of eugenicists. They did, however, prepare the ground for the legislation of 1943.

German racial hygiene was initially designed as a scientific answer to the challenges of modernity. But in a largely agricultural country like Bulgaria, eugenics experienced far more difficulty in gaining a similar position. As the case of Romania shows, a high level of urbanization and an advanced health system were not necessarily prerequisites for the acceptance of eugenics in the long term. During the 1930s, Bulgarian eugenics finally managed to establish itself as the junior partner of hygiene. This was also a response to the eugenic legislation introduced by the Third Reich. From the late 1930s, increasing political and cultural domination by the Third Reich provided Bulgarian eugenics with a final push. The issue of Bulgarian racial hygiene was consequently reduced to a discussion of which items from the German eugenic agenda could be adopted in Bulgaria, given the low level of modernization within the country. In consequence, numerous initiatives—like compulsory sterilization—had to be postponed until a "more favorable moment."

Endnotes

* This paper is a result of the project MOEL 063 funded by the Austrian Research Association (ÖFG) in 2004.

1 S[tefan] Konsulov, "Das Wesen der Bulgaren," in Ewald von Massow, ed., *Bulgaria. Jahrbuch 1942 der Deutsch-Bulgarischen Gesellschaft Berlin* (Leipzig: Felix Meiner, n.d.), 1–11, especially 8.

2 "Konsulov – prof. Stefan Georgiev professor – biolog," opis 1 [inventory 1], fund 583k [private collection of Stefan Konsulov], Central State Archive of Bulgaria. See also Ljubomir Rusev, *Pedagogičeski rečnik* (Sofia: Dobromir Čilingirov, 1936), 170–171. On German–Bulgarian cultural and intellectual reciprocity following the First World War see Elena Kjuljumova-Bojadžieva, *Germanskata kulturna politika i Bălgarija 1919–1944* (Sofia: Sv. Kliment Ohridski, 1991); Atanas Natev, ed., *Usvojavane i emancipacija. Vstăpitelni izsledvanija vărhu nemska kultura v Bălgarija/Aneignung und Emanzipation. Einleitende Forschungen über deutsche Kultur in Bulgarien* (Sofia: K & M, 1997); and Elena Bojadžieva, "Istoričeskite srešti meždu bălgarskata i nemskata kultura. Mežduličnostni obštuvanija (1919–1939)," *Istoričeski pregled* 59, 1–2 (2003), 32–50.

3 Konsulov, "Das Wesen der Bulgaren," 9.

4 Konsulov, "Das Wesen der Bulgaren," 9.

5 See the comparative case of Romania, in Maria Bucur, *Eugenics and Modernization in Interwar Romania* (Pittsburgh: University of Pittsburgh Press, 2002).

6 For a critical assessment of bio-politics, see Ross Dickinson, "Biopolitics, Fascism, Democracy: Some Reflections on Our Discourse on Modernity," *Central European History* 37, 1 (2004), 1–48.

7 The only exception is Gergana Mirčeva, "Bălgarskijat evgeničen proekt ot 20-te i 30-te godini na minalija vek i normativnijat kod na 'rodnoto'," *Kritika i humanizăm* 17, 1 (2004), 207–221.

8 Zdravko Radonov, "Vrăzki meždu bălgarski i germanski učeni i muzei (1878–1944 g.)," in V[eselin] Hadžinikolov, ed., *Bălgarsko-germanski otnošenija i vrăzki. Izsledvanija i materiali*, vol. 2. (Sofia: BAN, 1979), 305–326.

9 See Asen Hadžiolov, "Akademik Konstantin Pašev kato učen i značenieto mu za bălgarskata i svetovna oftalmologija," in Asen Hadžiolov, ed., *V pamet na Konstantin Pašev. Sbornik ot naučnite trudove po oftalmologija* (Sofia: BAN, 1966), 5–19.

10 Konstantin Pašev, "Nasledstvoučenieto u nas," *Oftalmologičen pregled* 3, 1 (1935), 19–20.

11 Konstantin Pašev, *Higiena na okoto* (Sofia: Sv. Sofia, 1905), 40.

12 Pašev, *Higiena na okoto*, 42–43.

13 Pašev, *Higiena na okoto*, 43.

14 Pašev, *Higiena na okoto*, 145–146.

15 See Konstantin Pašev, "Modernata dăržavna sanitarna politika: lečenie ili opazvane," *Letopisi na Lekarski săjuz v Bălgaria* 13, 5 (1920), 181–187, especially 182–183.

[16] This was different in Germany. See Paul J.Weindling, *Health, Race and German Politics between National Unification and Nazism, 1870–1945* (Cambridge: Cambridge University Press, 1989), 184–88, 214–226.

[17] (Anonymous), "Pečatni trudove na d-r. Stefan Vatev," in *Sbornik v čest na profesor St. Vatev* (Sofia: P. Gluškov, 1936), 73–84, esp. 82–83; see also Christian Promitzer, "Vermessene Körper: 'Rassenkundliche' Grenzziehungen im südöstlichen Europa," in Karl Kaser, Dagmar Gramshammer-Hohl and Robert Pichler, eds., *Europa und die Grenzen im Kopf* (Klagenfurt: Wieser, 2003), 365–393, esp. 375–376.

[18] S[tefan] Konsulov, "Lična, obštestvena i rasova higiena," *Zdravna prosveta* 3, 20 (1924), 1–2.

[19] See S[tefan] Konsulov, "Nasledstvenostta i nejnite zakoni," *Priroda* 14, 5 (1908), 82–85; S[tefan] Konsulov, "Mehanizăm na nasledstenvostta," *Priroda* 18, 1 (1912), 12–16; S[tefan] Konsulov, "Mendelevijat zakon za nasledstvenostta," *Priroda* 20, 2 (1914), 35–36; and S[tefan] Konsulov, "Gregor Mendel," *Priroda* 23, 1 (1922), 1–2.

[20] S[tefan] Konsulov, "Krăvosrodnite brakove i tehnite posledstvija," *Priroda* 20, 4–5 (1915), 67–68; S[tefan] Konsulov, "Predavat li se po nasledstvo pridobitite osobenosti?" *Priroda* 22, 9 (1922), 131–132.

[21] S[tefan] Konsulov, "Mendelizăm," *Priroda* 22, 7–8 (1922), 98–101, esp. 101.

[22] S[tefan] Konsulov, "Biologičnoto razvitie na narodite," *Priroda* 24, 1 (1923), 1–2; and S[tefan] Konsulov, "Podobrenieto na čoveškite rasi," *Priroda* 24, 3 (1923), 33–34.

[23] This book was actually the third edition of Schallmayer's *Über die drohende körperliche Entartung der Kulturmenschheit und die Verstaatlichung des ärztlichen Standes* (1891). See Peter Weingart, Jürgen Kroll, and Kurt Bayertz, *Rasse, Blut und Gene. Geschichte der Rassenhygiene in Deutschland* (Frankfurt-am-Main: Suhrkamp, 1992), 196–198, 210.

[24] See Bernhard Brocke, "Die Förderung der institutionellen Bevölkerungsforschung in Deutschland zwischen Weltkrieg und Diktatur," in Rainer Mackensen, ed., *Bevölkerungslehre und Bevölkerungspolitik vor 1933* (Opladen: Leske & Budrich, 2002), 39–60, especially 41–42; and Jürgen Cromm, "Gesellschaft versus Individuum. Bevölkerungswissenschaftliche Standorte und Postulate in der Zeit vor dem Nationalsozialismus," in Mackensen, ed., *Bevölkerungslehre und Bevölkerungspolitik*, 77–102, especially 94–95.

[25] See Ivan Elenkov, *Rodno i djasno. Prinos kăm istorijata na nesădnatija "desen proekt" v Bălgarija ot vremeto meždu dvete svetovni vojni* (Sofia: LIK, 1998); and the collection of sources published in Ivan Elenkov and Roumen Daskalov, eds., *Zašto sme takiva? V tărsene na bălgarskata kulturna identičnost* (Sofia: Svetlostruj, 1994).

[26] See Konsulov, "Podobrenieto na čoveškite rasi," 33.

[27] Konsulov, "Biologičnoto razvitie na narodite," 2.

[28] S[tefan] Konsulov, "Ženata i nauka," *Slovo* 5, 1241 (1926) (newspaper cutting, no page number given).

29 B. Nikolova, "Ženata i nauka," *Slovo* 5, 1247 (1926) (newspaper cutting, no page number given). See also Krasimira Daskalova, "Bălgarskite ženi v socialni dviženija, zakoni i diskursi (1840–1940)," in Krasimira Daskalova, ed., *Ot sjankata na istorijata: Ženite v bălgarskoto obštestvo i kultura (1840–1940)* (Sofia: Bălgarska grupa za izsledvanija po istorija na ženite i pola – Dom za naukite za čoveka i obštestvoto, 1998), 11–41, especially 19–20.

30 (Anonymous), "Obštestvoto za rasovo-himični proučavanija," *Priroda* 26, 9 (1926), 142.

31 Damjan Ivanov was a medical doctor, employed in an insurance company for civil servants; Ljubomir Ivanov Rusev, also a medical doctor, worked in a private neurological hospital in Sofia; Petăr Penčev was head of the Economic Department of the state-run Central Direction for Statistics; and finally, Ivan Germanov Kinkel, one of the first sociologists in Bulgaria, was docent for political economy at the Law Faculty in Sofia. In the 1930s, Kinkel became notorious for his biologist interpretations of social phenomena. For biographical data on Ivanov and Rusev, albeit fragmentary, see (Anonymous), "Bălgarsko družestvo za rasova higiena," *Priroda* 28, 7 (1928), 110. On Penčev, see the cover of the journal *Trimesečno spisanie na Glavnata direkcija na statistika* 1, 1 (1929) (no page number), were he is mentioned as a member of the editorial board. On Kinkel, see the entry "Kinkel" in Rusev, *Pedagogičeski rečnik*, 164–65 and *Enciklopedija na Bălgarija*, vol. 3, ed. Vladimir Georgiev (Sofia: BAN, 1982), 398.

32 (Anonymous), "Bălgarsko družestvo za rasova higiena," 110.

33 (Anonymous), "Bălgarsko D-stvo za evgenika," *Narod i potomstvo* 1, 1 (1935), 3.

34 (Anonymous), "Pominal se e v Moskva Julius Tandler," *Bălgarski higienen pregled* 6, 4 (1936), 259. On Tandler and eugenics in Austria, see Kurt Sablik, *Julius Tandler. Mediziner und Sozialreformer. Eine Biographie* (Vienna: Schendl, 1983); Monika Löscher, "Zur Popularisierung von Eugenik und Rassenhygiene," in Mitchell G. Ash and Christian H. Stifter, eds, *Wien in Wissenschaft, Politik und Öffentlichkeit. Von der Wiener Moderne bis zur Gegenwart* (Vienna: WUV-Universitätsverlag, 2002), 233–265.

35 S[tefan] Konsulov, "Evgenika—nauka za podobrene na čoveškite rasi," *Priroda* 28, 8 (1928), 124–126.

36 Konsulov, "Evgenika—nauka za podobrene na čoveškite rasi," 125.

37 (Anonymous), "Kurs po rasova higiena," *Priroda* 28, 9–10 (1928), 156–157.

38 (Anonymous), "Bălgarsko D-stvo za evgenika," 3.

39 S[tefan] Konsulov, *Zakonite za nasledstvenostta. Nasledstvenost pri čoveka* (Sofia: Populjarna nauka, 1930);and S[tefan] Konsulov, *Izraždaneto na kulturnoto čovečestvo. Borba s izraždaneto* (Sofia: Populjarna nauka, 1928).

40 Konsulov, *Zakonite za nasledstvenostta*, 3.

41 Konsulov, *Zakonite za nasledstvenostta*, 94.

42 Konsulov, *Izraždaneto na kulturnoto čovečestvo*, 35.

43 Konsulov, *Izraždaneto na kulturnoto čovečestvo*, 37–39.

44 Konsulov, *Izraždaneto na kulturnoto čovečestvo*, 39–41.

[45] (Anonymous), "Bălgarsko D-stvo za evgenika," 3; Ljubomir Iv[anov] Rusev, *Osnovni principi na Evgenikata ("Rasova higiena") s predgovor ot prof. St. Konsulov* (Sofia: Poligrafija, 1934), 84.

[46] See (Anonymous), "Zakon za narodnoto zdrave," *Dăržaven vestnik* 277 (1929) (no page given); Rusev, *Osnovni principi na Evgenikata*, 73; Zahari Bočev, *ABV na zdravnoto delo. Azbučen spravočnik na zakonite, pravilnicite i naredbite po narodno zdrave v Bălgarija* (Sofia: Glavna Direkcija na narodnoto zdrave, 1943), 344; and Hristofor Stefanov Mihajlov, "Usilvane borbata na demokratičnata obštestvenost za narodno zdraveopazvane, zdravo zakonodatelstvo i progresivni napravlenija v medicinata prez perioda na fašistkoto upravlenie u nas (1924–1944)," in Vera Pavlova, ed., *Istorija na medicinata v Bălgarija* (Sofia: Medicina i Fizkultura, 1980), 187–211, especially 189–190.

[47] Meri (Mary) Kričeva, ed., *Rasovo izraždane i rasova higiena* (Sofia: Hudožnik, 1930).

[48] See Stefan Konsulov, *Učebnik po biologijata za VIII klas na realnite i poluklasičeski gimnazii i za pedagogičeskite učilišta* (Sofia: Hudožnik, 1929), 155–157; and Stefan Konsulov, *Obšta biologija* (Sofia: Gutenberg, 1931), 369–376.

[49] See P[etăr] Penčev, "Raždaemostta v Bălgarija," *Trimesečno spisanie na Glavnata direkcija na statistika* 1, 1 (1929), 13–32, especially 31.

[50] Georgi T. Danailov, *Izsledvanija vărhu demografijata na Bălgarija* (Sofia: BAN, 1931), 427–429. For a contemporary survey see Kristina Popova, *Nacionalnoto dete. Blagotvoritelnata i prosvetna dejnost na Săjuza za zakrila na decata v Bălgarija 1925–1944* (Sofia: LIK, 1999), 45–52, 62–63.

[51] See Konstantin Kantarov, "Tuberkulozata v T.-Pazardžik," *Bălgarski higienen pregled* 4, 2 (1933), 83–89; and Vera F. Davidova, *Istorija na zdraveopazvaneto v Bălgarija* (Sofia: Nauka i izkustvo, 1956), 143–145.

[52] See T[oško] Petrov, "Našata zadača," *Bălgarski higienen pregled* 1, 1 (1931), 1–9, especially 2; T[oško] Petrov, "Ustav na Družestvo za higiena i predpazna medicina v Bălgarija," *Bălgarski higienen pregled* 5, 5 (1935), 297–300.

[53] See K[onstantin] Pašev, *Problemite na slepotata* (Sofia: Hudožnik, 1932), 94–96, 101.

[54] See Rusev, *Osnovni principi na Evgenikata*, 69.

[55] See (Anonymous), "V stolicata zasedava arhierejski săbor," *Bălgarski higienen pregled* 2, 5 (1932), 291; and Bălgarsko družestvo za higiena i predpazna medicina, "Izloženie do Arhierejskoto săbranie po văprosa za zdravni uslovija za vstăpvane v brak," *Bălgarski higienen pregled* 3, 5 (1933), 293.

[56] (Anonymous), "XXIV zasedanie na Bălgarskoto oftalmologično družestvo," *Oftalmologičen pregled* 2, 2 (1934), 59.

[57] See Z[ahari] Bočev, "Zakonodatelstvo po zdravnata zakrila na majčinstvoto i detstvoto u nas," *Narod i zdrave* 2, 8–9 (1942), 241–244.

[58] K. Silvenov K. (Konstantin Pašev), *Novite socialni dviženija v Evropa (Nacionalen solidarizăm)* (Sofia: Hudožnik, 1932).

[59] See K[onstantin] Pašev, "Nasledstvenata slepota i germanskija zakon za evgeničnoto obezplodavane," *Oftalmologičen pregled* 2, 3 (1934), 69–80.

⁶⁰ Rusev, *Osnovni principi na Evgenikata*, 63–65.

⁶¹ Rusev, *Osnovni principi na Evgenikata*, 63.

⁶² Rusev, *Osnovni principi na Evgenikata*, 73–75.

⁶³ Rusev, *Osnovni principi na Evgenikata*, 75–80.

⁶⁴ Rusev, *Osnovni principi na Evgenikata*, 81–82.

⁶⁵ Rusev, *Osnovni principi na Evgenikata*, 82–84.

⁶⁶ "Protokol na văzobnovitelnoto săbranie na d-voto za rasova higiena," 28 November 1934, arhivska edinica 38 (archival unit 38), opis 1 (inventory 1), fund 3ᵏ, State Archive of the City and District of Sofia (I am grateful to Kristina Popova for making this document available to me). On Saranov's earlier critical view on eugenics see N[ikola] Saranov, "Izpolzuvane na r'ontgenovite lăči v borbata s prestăpnicite," in N[ikola] Saranov, *Borbata s prestăpnicite* (Sofia: Pečatnica na Armejskija voenno-izdatelski fond, 1929), 1–34, especially 28–29.

⁶⁷ (Anonymous), "Protokol na văzobnovitelnoto săbranie na d-voto za rasova higiena," 28 November 1934.

⁶⁸ (Anonymous), "Ustavăt na Bălg. d-vo za evgenika," *Narod i potomstvo* 1, 3 (1935), 4.

⁶⁹ A sub-section of the journal was devoted to eugenic measures and the activities of related organizations in other countries, especially in Germany. The authors reviewed books and articles on eugenics published in Bulgarian. Anthropological articles discussed the racial characteristics of the Bulgarians.

⁷⁰ No. 7 of *Narod i potomstvo* was published in *Zdravna prosveta* 14, 22 (1935) without page reference; the other supplements were published under the title *Evgenični văprosi* (Eugenic issues) in *Zdravna prosveta* 14, 24 (1935), 3–4; 26 (1935), 3–4; and 28 (1935), 3–4.

⁷¹ Kamen Ivanov, "Rasova ili socialna higiena?" *Misăl* 2, 3 (1936), 12–15.

⁷² See (Anonymous), "Evgenika i polovo sterilizirane," *Bălgarski higienen pregled* 5, 1 (1935), 41.

⁷³ See (Anonymous), "Graždanskija brak i rasovoto podobrenie na našija narod," *Zora* 17 (28 March 1936), 5; and (Anonymous), Bălgarsko družestvo za higiena i predpazna medicina, "Izloženie po văprosite za evgeničnija podbor, predbračnoto osvitetelstvuvane i graždanskija brak," *Bălgarski higienen pregled* 6, 3 (1936), 186–188.

⁷⁴ Stefan Konsulov, *Rasovijat oblik na Bălgarite* (n.d.): for the year of publication see H. Barten, "Rassenkundliche und bevölkerungspolitische Fragen in Bulgarien," *Zeitschrift für Rassenkunde* 5, 9 (1939), 64–67, esp. 64–65; on the authorship of Konsulov see Nikola Altănkov, *Narekoha gi fašisti* (Sofia: TANGRA TanNakRa, 2004), 266–267.

⁷⁵ St[efan] Vatev, "Rasovi osobenosti na Bălgarite," *Narod i potomstvo* 1, 1 (1935), 2.

⁷⁶ (Anonymous), *Rasovijat oblik na Bălgarite*, 27–42, 45–46, 53–55.

⁷⁷ See Nikolaj Poppetrov, "Idejno-političeskite shvaštanija na 'Săjuz na bălgarskite nacionialni legioni' i 'Ratnici za napredăka na Bălgarštinata' v godinite na Vtorata svetovna vojna," *Istoričeski pregled* 47, 6 (1991), 53–67; and Altănkov, *Narekoha gi fašisti*, 260–327.

[78] Two thousand copies of the pamphlet were published. See Altănkov, *Narekoha gi fašisti*, 266–67.

[79] Metodij Popov, *Bălgarskijat narod meždu evropejskite rasi i narodi* (Sofia: Pridvorna pečatnica, 1938); and Metodij Popov, *Nasledstvenost, rasa i narod* (Sofia: Pridvorna pečatnica, 1938).

[80] Popov, *Nasledstvenost, rasa i narod*, 82.

[81] Popov, *Nasledstvenost, rasa i narod*, 82–83.

[82] See Ivan Elenkov, "Rasovite belezi na bălgarskoto v tărsene na bălgarskata kulturna identičnost meždu dvete vojni," in Natev, ed.,*Usvojavane i emancipacija*, 305–318; Georgi Kapriev, "Bălgarskijat spor za rasizma prez 30-te godini na XX vek: filosofski i obštokulturni perspektivi," in Natev, ed., *Usvojavane i emancipacija*, 329–342; Dimităr Denkov, "Kăm istorijata na ideologičeskite modi v Bălgarija: Rasizmăt," in Natev, ed., *Usvojavane i emancipacija*, 343–354; and Aleksandăr K'osev, "Rasizăm in potentia?" in Natev, ed., *Usvojavane i emancipacija*, 355–368.

[83] See S[tefan] Konsulov, "Stopjavaneto na bălgarskija narod," *Mir* (17 May 1938). See also S[tefan] Konsulov, "Naj-goljamata opasnost za bălgarskija narod—negovoto stopjavane," *Prosveta* 3, 7 (1938), 794–808.

[84] See S[tefan] Konsulov, "Kak se bori sreštu namaljavašta se raždaemost u nas?" *Mir* (1938) (newspaper cutting, no page number given).

[85] I.,"Kakvo e nužno za zapazvaneto i razrastvaneto na bălg. narod," *Mir* (1938), 3.

[86] (Anonymous), "Iz naučnite družestva," *Bălgarski higienen pregled* 9, 1 (1939), 55–67; and (Anonymous), "Družestvo za higiena i predpazna medicina," *Bălgarski higienen pregled* 11, 1 (1941), 65–72, especially 69.

[87] (Anonymous), "Săjuz na mnogodetnite v Bălgarija," *Rodna probuda* 3, 22 (1941), 2; (Anonymous), "Po primera na kulturnite narodi," *Rodna probuda* 3, 23 (1941), 1; (Anonymous), "Otgovarjame," *Rodna probuda* 3, 29 (1942), 2; and (Anonymous), "Pred navremenna dălžnost," *Rodna probuda* 3, 30 (1942), 1.

[88] See "Zakon za mnogodetnite bălgarski semejstva," *Izvestija na Glavnata direkcija na narodnoto zdrave* 28, 260 (1943), 211–219.

[89] See Gisela Bock, *Zwangssterilisation im Nationalsozialismus. Studien zur Rassenpolitik und Frauenpolitik* (Opladen: Westdeutscher Verlag, 1986), 146–153, 175–177.

[90] See Z[ahari] Bočev, "Novo zakonodatelstvo za zakrila na bălgarskite pokolenija," *Narod i zdrave* 2, 8–9 (1942), 224–231, especially 226.

[91] See Bogdan Filov, *Dnevnik* (Sofia: Izdatelstvo na Otečestven front, 1990), 544.

[92] (Anonymous), "Zakon za mnogodetnite bălgarski semejstva," 211.

[93] On the situation of Jews in Bulgaria in that period see, in general, Michael Bar-Zohar, *Beyond Hitler's Grasp: The Heroic Rescue of Bulgaria's Jews* (Holbrook, Mass.: Adams Media Corporation, 1998).

[94] See Mirčo Sirakov, "Načala na rasova higiena i organizacija na truda kato evgeničen problem," *Izvestija na Glavnata direkcija na truda i Instituta za obštestveno osigurjavane* 3, 5–6 (1943), 11–15, especially 11; Mirčo Sirakov,

"Načala na rasovata higiena," *Mediko-pedagogičesko spisanie* 10, 1 (1943), 28–31, especially 29, 2 (1943), 80–87.

[95] Bălgarsko družestvo za higiena i predpazna medicina, "Izloženie," *Bălgarski higienen pregled* 12, 3 (1943), 88–90. The statement shows that after the death of Toško Petrov in 1942, a more radical current replaced the moderate eugenic orientation of the Society for Hygiene and Preventive Medicine.

[96] See Petăr Semerdžiev, *Narodnijat săd v Bălgarija 1944–1945. Komu i zašto e bil neobhodim* (Sofia: Makedonija Pres, 1997), 283–284.

[97] See Hr[isto] Maksimov, *Spravočnik po zakonodatelstvoto na Narodnija Republika Bălgarija ot 9 septemvri 1944 g. do 30 juni 1957 g.* (Sofia: Nauka i izkustvo, n.d.), 292; "Zapoved No. 2247 za zdravnoto osvidetelstvuvane na vstăpvaštite b brak," *Dăržaven vestnik* (6 August 1968), 1. On pro-natal policies in socialist Bulgaria see Ulf Brunnbauer and Karin Taylor, "Creating a Socialist Way of Life: Family and Reproduction Policies in Bulgaria, 1944–1989," *Continuity and Change* 19, 2 (2004), 283–312.

[98] See Maria Todorova, "The Course and Discourses of Bulgarian Nationalism," in Peter Sugar, ed., *Eastern European Nationalism in the Twentieth Century* (Washington: American University Press, 1995), 55–102, especially 80, 83–88, 100-101; and Roumen Daskalov, *Meždu iztoka i zapada. Bălgarski kulturni dilemi* (Sofia: LIK, 1998), 217, 220.

[99] Daskalova, "Bălgarskite ženi v socialni dviženija, zakoni i diskursi (1840–1940)," in Daskalova, ed., *Ot sjankata na istorijata*, 35–38.

The Self-Perception of a Small Nation: The Reception of Eugenics in Interwar Estonia

Ken Kalling

Contemporary history has shown that the ideology of eugenics is more diverse as a body of knowledge than as a practical application. The approval of eugenic legislation, especially laws relating to sterilization, provides good criteria for testing the eugenic movements in different countries. The pervasive influence of eugenics in Scandinavian countries, the US, and Nazi Germany are the most known cases. Less well known is the case of the Baltic states, particularly Latvia and Estonia, and the passing of legislation during the 1930s according to the eugenic principles of obligatory sterilization and abortion. Besides shedding light on the emergence and development of eugenics and its ideology in Estonia, this chapter will explore specific features related to the dissemination and acceptance by Estonian eugenicists of so-called positive eugenics, especially in its pro-natalist forms. Eugenics can be divided into "positive" and "negative" branches. "Positive eugenics" emphasized the need to increase the ratio of racially "superior" members of society; "negative eugenics" was motivated by inhibiting the fertility of those deemed "biologically unfit." In interwar Estonia, the two branches intersected.

The Eugenic Movement in Estonia

Discussions about eugenics in the Estonian language can be traced back to the turn of the twentieth century. Estonians were perceived by non-Estonian members of society, and sometimes even by some Estonian politicians, as a peasant nation "without history," and thus, seemingly, with no future. However, during the Estonian national awakening (commencing around 1850 and culminating in 1918, when independence was declared), nationalist scholars and politicians seized upon arguments from the natural sciences with which they could justify the evolution of the Estonian nation. The biological laws of nature present-ed nationalists with a challenge: if humans evolved from apes, why, then, should the Estonians not be able to escape the limitations of a peasant existence and evolve into a cultivated society?[1] The leading Estonian eugenicist, Juhan Luiga (1873–1926), for example, argued that nationality should be defined according to natural laws rather than to theories relating to history.[2]

The fin-de-siècle witnessed increasing debates on supposed racial traits within the Estonian population. Similar to other Baltic Finns, Estonians were described as "Mongols" and, therefore, inferior to other "White Europeans" in the "Great Chain of Being."[3] Unlike in neigh-boring Finland, where race theories became bogged down in political arguments between Finnish nationalists and the local Swedish-speak-ing aristocracy, Estonian discussions about race remained within the academic realm.

Before 1918, in the light of Russian disapproval of political and cultural Estonian nationalism, the conception of the nation as a biolog-ical entity became a core feature in the justification of the idea of "Estonianness." The "biologization" of Estonian nation-building en-compassed both qualitative and quantitative characteristics held to comprise the population. From the outset, eugenics was concerned with demography due to high rates of emigration (which became common in the second half of the nineteenth century) and the denationalization of ethnic Estonians, who adopted the German or Russian language as their own.

At the beginning of the twentieth century, the abstinence movement (*karskusliikumine*), which included several Estonian public figures, voiced its concerns about the threat of social and biological degenera-tion. Estonian eugenics was thus born. The Estonian Eugenics Society (*Eesti Eugeenikaselts "Tõutervis"*) was founded in 1924, institutional-izing ideas that were already relatively widespread and accepted in educated circles.

The political orientation of leading eugenicists in Estonia was largely centrist or left of centre, with several belonging to the Estonian Labor Party (*Tööerakond*). Here Juhan Vilms (1883–1952) deserves special mention; his writings from the mid-1930s included strong criticisms of the socialist idea of equality. Besides being a leading advocate of corporatism, he was also an anti-Semite.[4] His final monograph on the subject of demography, *Rahvaste edasielamise alustest* (On the Successful Future Existence of Nations) nevertheless reveals that towards the end of this life, Vilms had returned to his original social democratic principle.[5]

The first major achievements of the eugenic movement in Estonia came in 1927, when the first marriage counseling service was established, along with the Congress of National Education (*Rahvusliku Kasvatuse Kongress*) and a public poll on the demographic changes in Estonia. A lecture series on eugenics was introduced at the University of Tartu in 1928, followed by the opening of the Institute of Eugenics (*Eugeenika Instituut*) at the same university in 1938, under the direction of Hans Madissoon (1887–1956). Although Juhan Aul (1897–1994), a well-known Estonian physical anthropologist, remained in close contact with the eugenic movement, the fields of eugenics and racial studies remained institutionally distinct in Estonia. A separate Institute of Anthropology and Racial Studies (*Antropoloogia ja Rassiteaduste Instituut*) was created only in 1943, during the German occupation.

Despite the fact that leading physical anthropologists in Estonia expressed sympathy for eugenics, the eugenic movement was generally devoid of racist thinking. Some scholars encouraged marriages between Estonians and minority ethnic groups (Swedes, Finns, Jews, and so on); however, marriages with Germans and Russians were generally discouraged, resulting from Estonian fears of cultural domination by their powerful neighbors.

In 1940, after the Soviet invasion, the Communist ban on public organizations included the Eugenics Society, which simultaneously also resulted in the abolition of the "Sterilization Law" on the grounds that it contradicted the moral norms of socialism.[6] Indeed, Estonians and Latvians were the first in Europe to declare the abolishment of national sterilization policies. (Latvia had introduced sterilization and abortion in the Public Health Act of 1938; however, the possibility of abortion on the grounds of racial hygiene was first mentioned in the 1933 Latvian Criminal Code.[7])

Estonian Sterilization Policy

On 1 April 1937, the Estonian "Law on Sterilization" (a document previously signed by the head of state in 1936) came into effect.[8] It proposed obligatory sterilization or abortion for a limited group of mentally disabled or "feebleminded individuals" ("those suffering severe
epilepsy, but also those incurably physically handicapped on hereditary grounds").[9] It stressed, on the grounds of heredity, that these individuals posed a threat to society because they could potentially breed
handicapped descendants. The right to propose sterilization or abortion
was limited to (besides the subjects themselves) medical doctors, cus-
todians, directors of residential care establishments and medical institutions. The final decision was to be taken by one of the two sterilization commissions (*steriliseerimise ringkonnakomisjon*). These commissions were composed of several medical doctors and officials nominated by the communities involved.

   Yet sterilization practices in Estonia remained modest when compared to other countries with similar legislation.[10] Official data on sterilization in Estonia only exist for the years 1937 and 1938; no data are
available for 1939 and 1940 due to the Soviet occupation and the limitations this placed on the compilation of statistics. In 1937, a total of
thirty-two patients (including twenty-six women) appeared before
Estonian sterilization commissions. Sterilizations were authorized in
twenty-one cases, of which nineteen were women.[11] Of the twenty-two
individuals brought before the commissions in 1938, twenty were sterilized (and six abortions were carried out); again, eighteen of those
sterilized were women. Thus, during the first two years of this legislation, forty-one people were sterilized, of whom only four (ten per cent)
were males. (The potential candidates for sterilization, according to the
criteria for mental disability and feeblemindedness, amounted to 3,358,
of whom 1,220 individuals were not held in psychiatric wards.)[12]

   The legislation regulating sterilization in Estonia was probably passed
due to the political need to legalize an already existing practice.[13] This
is confirmed by various sources, including documents from marriage
counseling centers, which reveal that sterilization and abortion were
both medically considered before the promulgation of the law.[14] The
figures available on abortion—justified on eugenic grounds—from the
psychiatric ward in Tallinn are modest in number, averaging three cases in seven years.[15] The low number of sterilizations was largely due

to the medical understanding that such practices did not necessarily contribute to achieving eugenic goals.[16]

Yet the law on sterilization should be regarded as a significant episode for characterizing the nature of the state in interwar Estonia. When comparing the legislation on sterilization passed in Nazi Germany and Scandinavian countries with the Estonian case, differences and similarities are detectable. In the German case, biological factors were stressed when targeting patients for sterilization, while in Scandinavian countries sterilization was mainly justified on social grounds.[17] As far as the eugenics underlying the law on sterilization in Estonia is concerned, this tended to follow the German example. In practice, however, eugenic policies followed the Scandinavian model. (This perhaps explains why in Estonia more women were sterilized than men.)[18] The introduction of the law on sterilization should thus be viewed as a by-product of a much larger Estonian eugenics project: pro-natalism.

The Pro-Natalist Character of Estonian Eugenics

The roots of "demographic hysteria" in Estonia can be traced back to the pre-independence era, to a period dominated by fears of social degeneration. In Estonian historiography such anxieties are often linked to the problem of a "small nation's self-perception."[19] The achievement of independence in 1918 meant that concerns about ethnicity were complemented by new concerns for the welfare of the state. The young Estonian nation was faced with the daunting task of establishing membership in the family of "nations with history."[20]

The demographic situation was considered to be catastrophic, and was heightened by fears of national extinction. A low population density posed a serious problem given that it heightened the possibility of external intervention in a Europe intoxicated by ideas of *Lebensraum*.[21] As a result, the majority of eugenic rhetoric related to demographic matters.[22] The eugenicists insisted that political warfare had long ago given way to biological competition between nations, and Estonia, lacking in military might, was also losing the demographic battle.

In 1922, the Estonian population amounted to 1,107,059; by December 1939 this figure had risen to 1,122,440. Not surprisingly, during the 1930s politicians had supported a twofold increase in population numbers.[23] This numerical inferiority led, however, to ironic reflections: Juhan Vilms put it plainly when he stated that "no one would ridicule a nation of three million people," such as the Estonians.[24]

Eugenicists were able to influence public opinion through events organized by the Congress of National Education in 1927 and 1935. There was a clear tendency in the Congress's deliberations towards a greater biologization of the nationalist debate. Issues regarding the fertility of the race, society, and nation were discussed. Women, in particular, were the focal point of discussion. It was declared that the ideal Estonian woman was, in the first instance, a mother.[25] However, despite their allegedly good physical constitution,[26] commentators noted that Estonian women were unwilling to procreate.[27] One possible explanation was that the modernization of society directly affected them.[28] Paraphrasing the Swedish feminist activist, Ellen Key (1849–1926), who once said that the twentieth century would be the "Child's Century," Vilms considered it to be the "Woman's Century." Before the emancipation of women was made infamous by Nazi pro-natalist policies, Vilms viewed it as a means by which one could strengthen Estonian statehood.[29]

The Authoritarian State

The eugenic message was well received in Estonia, a society in which the new constitution of 1938 included the notion of "socially harmful indigent persons."[30] A breakthrough for the eugenicists occurred in 1934, when the Estonian state introduced an authoritarian political regime, in which the propagandist rhetoric of pro-natalism and eugenics flourished. The new system was known as "managed democracy," and the consolidation of "national entirety" (*rahvusterviklikkus*) became the prime goal of the state, an ideal rooted in a biological understanding of society; namely, the nation as a living organism.[31]

The post-1934 era witnessed the banning of political parties, restricted journalistic freedom and so on. On the other hand, a few political decisions were introduced which enlarged the nucleus of power during this period. Between 1935 and 1939, for example, fifteen trade chambers (*kutsekoda*), which dominated the Upper Chamber of the Estonian Parliament (*Riiginõukogu*), were created. Estonia was developing a corporatist system, and this provided a strong argument in favor of the heightened role of professionals in politics. The creation of these chambers was also a means by which to bring women into the autocratic state-building process. The Chamber of Domestic Economy (*Kodumajanduskoda*), established in 1935, which brought together existing women's organizations, was viewed by several politicians as a feminist Trojan horse. Some women activists saw it as an attempt by the state to institutionalize gender stereotyping.[32] One consequence of the creation of the Chamber was that it restricted women from entering into non-domestic employment. Official rhetoric declared that the new infrastructure proved that the state valued women, especially their domestic role.

A prerequisite of the authoritarian state was the complete subordination of all members of society in the name of "national entirety."[33] Eugenicists and eugenic organizations actively supported the paternalist, solidarist and authoritarian system in the Estonian case. The state, in turn, made full use of science and medicine in order to increase its authority and legitimacy.

It is not simply the negative aspects of racial hygiene (the 1937 "Law on Sterilization") that should be stressed in connection with authoritarianism in Estonia. There was also a positive side. The Congress of National Education, held in 1935, led to the state-sponsored creation of the Board of Population Increase and Welfare (*Rahva Juurdekasvu ja Heaolu Komisjon*). Positive eugenics was officially institutionalized from this moment onwards. The Board was responsible for implementing a six-year national plan (1936–1941) aimed at expanding medical institutions and personnel as a way of improving public health in general; and more specifically, combating the demographic problem. This plan encompassed health education, mother–child programs, and tackling the problem of tuberculosis.[34] The state complemented the plan by easing the tax burden on large families, implementing a child support system, and encouraging a "healthy," rural lifestyle.

Conclusions

The eugenic movement was well received by the public in interwar Estonia. This is partly explained by the syndrome of the "small nation's self-perception," and accompanying fears of extinction. Thus, official nationalism, especially during the years of the authoritarian state, became synonymous with the concept of a "national entirety." Estonian society was mobilized to fulfill a biological task—to breed. In this context, the

Estonian Eugenics Society became a leading authority on demographic matters and public health, and was used to encourage the increase of the population. Eugenics became a social mechanism uniting the popular beliefs of Estonian society.

The development of an autocratic regime during the second half of the 1930s is a factor that should be borne in mind when comparing eugenics in Estonia with other countries in interwar Europe. The relatively low number of sterilization cases and the eschewing of racial criteria in Estonia are factors that distinguish eugenics in Estonia from Nazi Germany. The disproportionately high ratio of women among those sterilized also dissuades parallels with the German case. In Estonia, as in some Scandinavian countries, social inclinations determined the nature of eugenic activities. Moreover, the support of pro-natalist attitudes by Estonian eugenicists suggests similarities with the eugenic movements in Italy and France.

Endnotes:

¹ Hendrik Koppel, "Eessõna," *Eesti Üliõpilaste Seltsi Album* 8 (1910), 4.

² Juhan Luiga, "Rahvaste tõus ja langemine," in Juhan Luiga, *Mäss ja meelehaigus* (Tartu: Ilmamaa, 1995), 221–224. This was delivered as a speech in 1909.

³ "The Great Chain of Being" is a metaphor used for a divinely inspired universal hierarchy ranking all forms of higher and lower life. See Arthur O. Lovejoy, *The Great Chain of Being. The History of An Idea* (Cambridge, Mass.: Harvard University Press, 1964).

⁴ See Juhan Vilms, *Suguhaigused—Tegeliku arsti seletused suguhaigustest hoidumisest ja nende arstimisest* (Tartu: Loodus, 1927), 33.

⁵ Juhan Vilms, *Rahvaste edasielamise alustest*. Manuscript dated between 1938 and 1945, found in Estonian State Archive (4006-2-11), 33.

⁶ "ENSV Rahvakomissaride Nõukogu määrus Steriliseerimise seaduse kehtetuks tunnistamise kohta," *Riigi Teataja* 32, art. 368 (1940).

⁷ *Arstniecibas likums. Likumu un Ministru kabineta noteikumu krajums*, art. 219 (31 December 1937).

⁸ "Steriliseerimise seadus," *Riigi Teataja* 98, art. 776 (1936). An English translation is also available in *Tartu University History Museum Annual Report 1998*, vol. 3 (1999), 15–19.

⁹ "Steriliseerimise seadus." Paragraph 1.

¹⁰ Paul J. Weindling, "International Eugenics: Swedish Sterilization in Context," *Scandinavian Journal of History* 24, 2 (1999), 187.

¹¹ *Valitsusasutiste tegevus 1937/1938* (Tallinn: Riigikantselei väljaanne, 1938), 103.

¹² *Valitsusasutiste tegevus 1938/1939* (Tallinn: Riigikantselei väljaanne, 1939), 108.

¹³ Markku Mattila, "The Alegal Eugenic Sterilizations in Finland—an International Perspective," *Tartu University History Museum Annual Report* 3 (1998), 43.

¹⁴ Hugo Valma, *Isikuraamat, Paranduskasvatuslik isiklikkude andmete ja vaatluste kogu* (Tallinn: Eesti Kir.-Ühisus, 1927), 15; and Hans Madissoon, "Eugeenilise mõtte levingust, eriti seadusandluses," *Eesti Arst* XVII, 9 (1938), 204.

¹⁵ Karl Toomingas, "Vaimuhügieeni ülesandeist Eestis," *Eesti Arst* XIV, 7 (1935), 513.

¹⁶ Maximilian Bresowsky, "Vaimuhaigete sterilisatsioonist," *Eesti Arst* XIV, 7 (1935), 536–539; and Robert Silvester, "Sigivõimetustamine geneetilisest seisukohast," *Eesti Arst* XIX, 5 (1940), 392–393.

¹⁷ Nils Roll-Hansen, "Eugenic Practice and Genetic Science in Scandinavia and Germany," *Scandinavian Journal of History* 26, 1 (2001), 78.

¹⁸ Gunnar Broberg and Nils Roll-Hansen, eds., *Eugenics and the Welfare State. Sterilization Policy in Denmark, Sweden, Norway, and Finland* (East Lansing, Mich.: Michigan State University Press, 1996), 264.

[19] Toomas Karjahärm and Väino Sirk, *Vaim ja võim. Eesti haritlaskond 1917–1940* (Tallinn: Argo, 2001), 259–265.

[20] Karjahärm and Sirk, *Vaim ja võim*, 261–262.

[21] *Eesti rahva tulevik* (Tartu: "Tuleviku" kirjastus, 1935).

[22] "Eesti Eugeenika Seltsi 'Tõutervis' juhtlaused," *Tulev Eesti* 1 (1925), 2–6.

[23] "Peaministri teadaanne riigivolinikule ja riiginõukogule," *Päevaleht* (25 May 1938), 4.

[24] Juhan Vilms, *Suurte talude jagamise tarvidusest (väikemaapidamiste tõutervist ja rahvaarvu tõstev mõju)* (Tartu: Sõnavara kirjastus, 1926), 24.

[25] Juhan Vilms, *Poligaamiline abielu Eestis* (Tallinn: Varrak, 1920), 39.

[26] Richard Weinberg, *Esty* (Moscow: n.p. 1901), 30–31.

[27] Peeter Põld, *Eesti riigi tulevik ja karskus* (Tartu: Eesti Karskusseltside kesktoimkond, 1922), 9.

[28] Luiga, "Rahvaste tõus ja langemine," 224.

[29] Vilms, *Poligaamiline abielu Eestis,* 5; See also Robert Proctor, *Racial Hygiene: Medicine under the Nazis* (Cambridge, Mass.: Harvard University Press, 1988), 123–125.

[30] *Eesti Vabariigi Põhiseadus*, Art. 28 (1938).

[31] Hans Kruus, "Rahvusterviklusest," *Akadeemia* 25, 3 (1940), 143–152.

[32] "Tühjade hällide küsimus naiste seisukohalt," *Päevaleht* (6 March 1937), 5.

[33] August Leps, "Väikeriik ja diktatuur," *Akadeemia* 5, 5 (1937), 294.

[34] (Anonymous), "Seletuskiri tervishoiu ala arendamise kava juurde," *Eesti Arst* 14, 11 (1935), 15–23; and (Anonymous), "Detailne eelarveline kava: 'Tervishoiu ala arendamise kava lähema kuue aasta jooksul'," *Eesti Arst* 14, 11–12 (1935), 9–15, especially 20.

Central Europe Confronts German Racial Hygiene: Friedrich Hertz, Hugo Iltis and Ignaz Zollschan as Critics of Racial Hygiene

Paul J. Weindling

The new national states of interwar Europe were fertile seedbeds for the growth of eugenics as science, ideology and medical practice. Sandwiched between the two pariah states of Germany and the Soviet Union, Central European eugenics was astonishingly diverse. In part, there were influences from abroad. The Rockefeller Foundation sought to promote hygiene and welfare in the European successor states; social medicine in Weimar Germany had fertility control as a core interest; and there were socialist endeavors to produce a "new man." Undoubtedly, there were heterogeneous streams in each country, which meant that population and health policies took on distinctive national forms.

The strength of an articulate opposition to German race theory and the Nazification of racial hygiene merits comparison with the fragile political context of interwar Central and Southeast Europe. In this chapter I shall show how radical critiques emerged from within the Central European crucible, and how it moved from an initial concern with anti-Semitism, and the racial mythology of Houston Stewart Chamberlain (1855–1927) and Arthur de Gobineau (1816–1882), to targeting Nordic racial anthropology and the eugenics of the German ultra-Right.

Three figures took the lead in mounting critiques of the scientific pretensions of racial theory: the social scientist Friedrich Otto Hertz (1878–1964); the biologist and geneticist Hugo Iltis (1882–1952); and the radiologist Ignaz Zollschan (1877–1948). Zollschan was also a committed Zionist, while Hertz was nominally a Roman Catholic, although his father was of Jewish descent. Iltis was also nominally Catholic with a father of Jewish descent. Hertz and Iltis were both socialists and secular in outlook. All shared a common background in the Habsburg Monarchy, where, in Hertz's words, "no race could seriously oppress another." Before 1914, their spur was the Aryan anti-Semitic ideology of Houston Stewart Chamberlain. All three thinkers engaged with eugenics during the 1920s: Hertz did so as a social scientist; Zollschan as a Lamarckian anthropologist; while Iltis did so from a firmly Mendelian position, rooted in biological concerns. By 1930, they rose to the challenge of refuting Nazi racial theories.[1]

While it is clear that German racial hygienists, notably Alfred Ploetz (1860–1940) and Fritz Lenz (1887–1976), were Nordic racial idealists, both were cautious in articulating anti-Semitic sentiments until the patriotic fervor of the First World War brought about an intensification of ideas associated with *Lebensraum* and German racial health.[2] The *Gesellschaft für Rassenhygiene* (The Society for Racial Hygiene), founded in 1905, was broadly *Grossdeutsch* in orientation, and had Central European pretensions. Unification of the scattered German people meant looking beyond the limited frontiers of the German Reich established in 1871. Ploetz intended the *Gesellschaft für Rassenhygiene* to extend beyond the boundaries of Germany, to include members from Austria and Switzerland, as well as to cement links with Hungary. The Society also brought together Swiss nationals like Ernst Rüdin (1874–1952), alongside Austrians like the Munich professor of hygiene, Max von Gruber (1853–1927), and the Lamarckian Ignaz Kaup (1870–1944), a follower of the Austrian anti-Semite, Georg von Schönerer (1842–1921).

The Society was proclaimed the *Internationale Gesellschaft für Rassenhygiene* (International Society for Racial Hygiene) in 1907, and only in March 1910 was a national German umbrella organization instituted. During the First World War, racial hygienists rallied to schemes for expanding racial hegemony over Central Europe. Fritz Lenz contributed to Julius F. Lehmann's bellicose journal *Osteuropäische Zukunft* (East European Future) during 1916–1917, recommending the racial value of population settlement in the East.[3] On 28 September 1918 a meeting was planned in Budapest on racial hygiene and population policy, with notable figures in science, medicine and politics from Germany, Austria and Hungary invited to attend.[4]

Friedrich Hertz

In 1902, Hertz published his first critique of "modern racial theories" in the *Sozialistische Monatshefte* (Socialist Monthly).[5] This pre-dated the founding of eugenics societies in Europe, and was a critique of the rising interest in Aryan racial theory on both the political Right and Left, the latter exemplified by the revisionist socialist Ludwig Woltmann (1871–1907). Hertz attacked the writings of the anthropologist Otto Ammon (1842–1916), as well as Houston Stewart Chamberlain's idea of racial breeding. He also refuted the theory that "over-breeding" resulted in a lack of fitness.[6] Hertz saw nationalism as a destructive force; partly as a result of his socialist academic contacts with Otto Weininger (1880–1903), the author of the idiosyncratic *Geschlecht und Charakter* (Sex and Character).[7]

In 1904, Hertz returned to Vienna from studies with Fabian socialists at the London School of Economics. While in Britain, he published a critique of pan-Germanist and pro-Boer propaganda. He then developed his article on race into a book-length critique of racial biology, *Moderne Rassentheorien* (Modern Race Theories). Concerned about scientific forms of anti-Semitism, he also published *Antisemitismus und Wissenschaft* (Anti-Semitism and Science) in 1904.[8] These early attacks against race theories took place at the point when eugenics was emerging as an organized movement with claims of medical and scientific expertise. Advocates of eugenics took Hertz's view seriously. The founding editor of the *Archiv für Rassen- und Gesellschaftsbiologie* (Journal for Racial and Social Biology), Alfred Ploetz, included Hertz in a list of future contributors, and reviewed his critique of race in 1905.[9] Hertz adopted a legal and sociological approach to the problem of race.[10] He advocated the view that: "Race theories are little else but the ideological disguises of the dominators' and exploiters' interests."[11] He pointed out that there was no significant link between physical and mental characteristics, and that race theories lacked a convincing empirical basis. The welfare-oriented eugenicist Wilhelm Schallmayer (1857–1919), for example, regarded Hertz as an ally. They exchanged reprints, and in 1915 discussed Hertz's projected new edition of the "critique of racial theories."[12]

In 1913, Hertz married a physician, Edith, whose medical training with the charismatic Viennese social hygienist, Julius Tandler (1869–1936), later guided her husband's interest in heredity and welfare.[13]

Moreover, the fact that Edith came from Bohemia strengthened Hertz's interest in the partly German-speaking region, with its major industries and political antagonisms. Tomáš G. Masaryk (1850–1937), future president of Czechoslovakia, contributed regularly to Hertz's liberal-left review. During this period, Hertz revised his work in order to take issue with racial hygiene as the basis of social policy. What emerged was the distinction between the biological values of "human economy" and racial eugenics. The Viennese social scientist Rudolf Goldscheid (1870–1931) introduced the term *Menschenökonomie* (human economy) to underline the point that society had a role in facilitating the healthy reproduction of human life. Hertz was cited by the Berlin anatomist Oscar Hertwig (1849–1922) in his seminal critique of Social Darwinism of 1917.[14] In turn, Hertz was criticized by the anthropologist Ludwig Wilser (1850–1923) and the eugenicist Fritz Lenz. Prompted by Ploetz, Hertz responded in the *Archiv für Rassen- und Gesellschaftsbiologie* during 1916–1917.[15]

In the 1920s, Hertz (by now with the title *Hofrat*) worked in the office of Karl Renner (1870–1950), the federal chancellor of the Austrian Republic. The rise of eugenics spurred Hertz to reassess racial hygiene. The result was a book with a substantially new research perspective compared to its predecessor of twenty years before. Hertz published *Rasse und Kultur* in 1925, which appeared three years later in translation in London and New York as *Race and Civilization*.[16] His revised work on racial theories denounced the German anthropologist Eugen Fischer (1874–1967), who, with geneticists Erwin Baur (1875–1933) and Fritz Lenz, published the noted textbook on human heredity and eugenics. Lenz again counterattacked in the *Archiv für Rassen- und Gesellschaftsbiologie*.[17]

Hertz could accept that races existed as physical types, although rarely were races pure. Yet what he could not accept was the view that race and psychology were inextricably linked. He similarly rejected claims that intelligence and mental abilities were wholly due to inheritance, and noted how minor coincidences were posited as elaborate statistical proof.[18] He was scathing about the Nordic racism of Baur, Fischer, Lenz, and their populist ally, the Nordic thinker Hans F. K. Günther (1891–1968). Hertz cited the characterization of Bavaria and Austria as largely composed of inferior "Dinaric" stock, close in type to the Semites of the Near East, especially Armenians and Jews. He argued that the idea of leading intellectuals and social elites being predominately Nordic was untenable. Julius F. Lehmann (1863–1935)—the publisher and promoter of Hans F. K. Günther (1891–1968), and of the Hungarian eugenicist Géza von Hoffmann—presented the volume on heredity written by Baur, Fischer and Lenz to Adolf Hitler when imprisoned at Landsberg in 1924.[19] Hertz presciently warned of the ominous links between racial eugenics and the political Right.

After working in the Austrian chancellery, *Hofrat* Hertz was appointed in 1930 as a political scientist at the University of Halle. He deliberately took this post so as to have an impact on German public opinion. He antagonized Nazi students by publishing a withering critique of Lehmann's protégé, the Nordic racial ideologist Hans Günther, whom the Nazis supported for a chair at Jena in Thuringia. His work *Hans Günther als Rassenforscher* (Hans Günther as Racial Researcher) meant that Hertz became infamous to the political Right as a leading critic of race theory and National Socialism.[20] In June 1933, after Hertz's resignation was turned down, he was summarily dismissed. By this time Hertz—who was in considerable personal danger—returned to Austria.[21]

Ignaz Zollschan

In 1910, Zollschan published his monograph *Das Rassenproblem* (The Race Problem) as a critique of the Aryan racial theories of Houston Stewart Chamberlain. Zollschan was a key figure in the Jewish response to German racial theories of the period. However, by the time he died in 1948 in Britain, Zollschan's ideas on the Jewish race had undergone considerable change.[22]

Zollschan was a physician and radiologist, a specialism which, like dermatology, offered openings for Jews. In 1904, he qualified in medicine from the University of Vienna. His work as a ship's surgeon allowed him to witness racial variations around the world at first hand. After he returned to Vienna for training in radiology, he published *Das Rassenproblem*, which was reviewed at length in Ploetz's *Archiv für Rassen- und Gesellschaftsbiologie*.[23]

Zollschan outlined his definition of the Jewish race in response to Houston Stewart Chamberlain, who advocated Aryan superiority on a theological and biological basis as a new form of German Christianity.[24] Chamberlain's *Foundations of the Nineteenth Century* (1899) conceived racial struggle in religious terms, with Jesus as a Teutonic redeemer, an idea enthusiastically taken up later by Nazi "German Christians." Cosima Wagner (1837–1930) and Kaiser Wilhelm II (1859–1941) were both enthusiasts for this racist gospel of Aryan racial superiority.[25] Zollschan found that Chamberlain, who lived in Vienna between 1889 and 1909 before moving to Wagner's Bayreuth, was primarily motivated by his opposition to the political-anthropological school.[26] This meant that the right-wing Chamberlain attacked the left-wing socialist revisionist and Social Darwinist Ludwig Woltmann.[27] Zollschan warned against using the physical anthropology of hair and skin color as racial markers, and cautioned that the caricatured feature of the hooked Jewish nose was rare. Evolution thus provided a corrective to anti-Semites, and Zollschan's Lamarckism meant that culture and psychological identity became dominant features of the Jewish race.[28]

Zollschan's premise was that the Jewish race was in the throes of dissolution. He defined Jews according to Darwinist precepts, albeit tempered by environmentalism (thus differing markedly from Mendelism). Unlike the Viennese anthropologist Rudolf Pöch (1870–1921), who sought to map the racial characteristics of Austro-Hungarian skull forms, Zollschan did not conduct research into physical anthropology. Nor did he join any eugenic organization. Unlike Jewish race theorists who sought to work alongside advocates of German race regeneration, from the outset Zollschan adopted a combative tone in *Das Rassenproblem*. It served him well in making the transition from an advocate of Jewish race purity in the multiethnic Habsburg Monarchy before the First World War to taking the lead in an international coalition against Nazi racism during the 1930s. By this time, Zollschan had radically to revise his scientific ideas about the Jewish race, and moderate his Lamarckism.

Two years after completing *Das Rassenproblem*, Zollschan moved to the historic spa resort of Karlsbad (Karlovy Vary).[29] Karlsbad was in the Western Sudeten region of Czechoslovakia, an area that was a hotbed not only of anti-Semitism but also of anti-Czech sentiment. At the nearby state radium mines at Joachimsthal (Jáchymov), German miners felt embittered by the high rates of cancer caused by radium emissions.[30] Zollschan lost his left arm in 1939 due to X-ray research, forcing him to retire from medical work but allowing him to focus full-time on setting up an international network of anthropologists to combat the threat of Nazi racism.[31]

Zollschan nonetheless believed that the state was responsible for sustaining the individual qualities of races. Pure races were intellectually more vital than racial mixtures. Moreover, races needed to recover their primal vigor in order to preserve their creative capacities.[32] Lamarckians had exposed gametes to radium, thus providing important evidence for the impact of the environment on hereditary substance.[33] Zollschan approved of the Lamarckian experiments by the zoologists Paul Kammerer (1880–1926) and Hans Przibram (1874–1944) at the Vienna Vivarium (a research institute close to Prater Park). Such Lamarckism allowed for the modification of the psychology of race, so that Jews did not appear as immutable fixed stereotypes.[34] Ernst Haeckel's disciple, Richard W. Semon (1859–1918), offered technical corrections to *Das Rassenproblem* because he considered Zollschan's book important and timely. Zollschan was congratulated by other Jewish academics, like the Viennese medical historian Max Neuburger (1868–1955). British anthropologist A.C. Haddon (1855–1940) found much to praise as well.[35] With its blend of biology, and cultural and religious history, Zollschan's book earned him academic respect.

In March 1919—writing when Austria's borders were closed—Zollschan advocated the purge of nationalist elements from Zionism, and opposed special laws for Jewish minorities. He further argued for ethnic regeneration based on the long-term processes of selection, in addition to the political cultivating of the "mnemic" hereditary composite with respect to cultural and physical attributes. He called this process "homophonie", that is, aligning instinct with history.[36] He thus hoped that Zionism could be blended with internationalism, thereby upholding the ideals of international peace and cooperation. Zollschan's efforts to reach an international public can be seen in an American edition of his lectures on Jewish questions, as well as in English and French editions of his writings.[37] A revised fifth edition of *Das Rassenproblem* appeared in 1925; while Zollschan modified his views on politics, he remained a Lamarckian. Driven by Mendelian concerns with human heredity, anthropology was under pressure to follow a biological path. Meanwhile, German race hygienists were at the forefront of applying Mendelism to racial questions.

Zollschan was alert to the dangers of eugenics. In 1925, he visited the anthropologist and environmentalist Franz Boas (1858–1942) in New York to collaborate on X-ray investigations into various races, having supported the use of X-rays to eradicate favus among East

European Jewish children. Boas, convinced of the need to refute racial prejudice, was in charge of a special committee investigating anatomical and psychological characteristics at Columbia University.[38] In 1926, Zollschan used a memorandum drawn up by Boas as a basis for intensified lobbying in Europe to refute anti-Semitic racism.[39] In 1930, Zollschan published the essay *The Significance of the Racial Factor as a Basis in Cultural Development* with Le Play House, a sociological research institute in London.[40] Here he discussed issues such as the analogies between Aryan and Nordic race theories, and whether Nordic claims of superior culture had any validity. Zollschan now turned to attack Hans F. K. Günther, as he once attacked Chamberlain.

Hugo Iltis

Iltis was decidedly more populist than Zollschan, being active in education in Mendel's hometown of Brünn (Brno). His writings exuded a strict scientific rigour, while weighing the social implications of Mendelism. He worked as a teacher at the German High School (*Gymnasium*), and as *Privatdozent* in botany and genetics at the German Polytechnic in Brünn (Brno); and in 1921 he founded the *Masaryk Volkshochschule* (Masaryk Adult Education Centre), the largest institution for adult education in Czechoslovakia.[41]

Iltis was not only a Mendelian, he was also Mendel's biographer. His biography aroused the interest of geneticists when published (in German) in 1924, and earned him international recognition after its publication in English in 1932.[42] The extent of his preoccupation with social questions is evident from his copious diary entries. He corresponded with the leading German eugenicist Fritz Lenz. While Lenz was a geneticist of some sophistication, he came out in support of National Socialism in 1930 as the best way to realize eugenic measures.[43] Iltis—like Hertz and Zollschan—had first-hand acquaintance with the dangerous progenitors of racial theory.

Although deeply committed to Mendelian genetics, Iltis considered the experimental findings of the Lamarckian experimental biologist Paul Kammerer, conducted at the Vienna *Vivarium*, as significant. The Vivarium was celebrated as an institute for biological experimentation, where animals were studied over generations under differing environmental conditions. Iltis suspected that a right-wing assistant discredited Kammerer by tampering with the experiments. Iltis also criticized anti-Semitism among socialists.[44]

In 1928, Iltis published articles on eugenics in the leftist book review *Bucherwarte* in Berlin.[45] At this stage Iltis dismissed Gobineau and Chamberlain, as well as Günther and Lenz, as racists,[46] but he was prepared seriously to engage with—and partly to accept—the work of the anthropologist Eugen Fischer, although he remained critical of Fischer's linking of psychological with physical types.[47] Iltis accepted certain "objective" eugenic measures like marriage certificates, and endorsed the social hygiene of Alfred Grotjahn (1869–1931). Iltis's substantive work was published by the socialist Urania organization in Jena.[48] In 1930 his *Volkstümliche Rassenkunde* (Popular Racial Studies) was serialized, appearing as a series of political brochures that ensured wide circulation. This was designed to strike at the heart of the groundswell of support for Nazism in Thuringia, and was targeted at the racial theorist Hans F. K. Günther, who was appointed to a chair at Jena. Iltis, Hertz and Zollschan rallied to denounce Günther as unscientific and politically dangerous. Indeed, Iltis foresaw the ominous possibilities of extermination policies, realizing that anthropologists were guilty of supporting genocide under colonialism. As such, he rejected the notion of the degenerative effects of racial interbreeding as a political falsehood.

Opposing Nazi Race Theory

The post-Versailles era witnessed strenuous efforts by German academics, with the support of the Department of Culture at the German Foreign Ministry, to overcome the international boycott of German science. German academics like Eugen Fischer (director of the Kaiser Wilhelm Institute for Anthropology from its inception in 1927) and Fritz Lenz were supportive of ethnic German minorities, who found themselves amalgamated into the new nation-states of East-Central Europe after 1918. Fischer found notable support among Central European racial nationalists. For example, the Hungarian anthropologist Lajos Méhely (1862–1953) contributed the article "Blut und Rasse" (Blood and Race) to Eugen Fischer's 1934 *Festschrift*. Whereas the Hungarian press praised Hertz's work in crushing race theory on 9 December 1933, Méhely countered by condemning Hertz as an environmentalist "fanatic."[49]

By the early 1930s, the broad consensus on eugenics was undergoing political reconfiguration in response to mounting political extremism. In the Soviet Union, Stalin suppressed research into eugenics and genetics. The papal encyclical *Casti Connubii* (On Christian Marriage) came out in opposition to eugenics, as part of the larger condemnation of abortion and birth control. A new solidarity took root among scientific critics of race theory; here Hertz, Iltis and Zollschan, all directly affected by Nazi popularity, took a courageous lead.

From 1933, German anthropologists intensified their efforts to exercise influence in Central Europe. The dissident anthropologist Karl Saller (1902–1969), an opponent of Nordic racism, was purged for questioning Nazi race doctrine, and the Frankfurt-based Jewish anthropologist, Franz Weidenreich (1873–1948), was forced to emigrate. After the Austrian *Anschluss*, the anthropologist Josef Weninger (1886–1959) was required to relinquish his post.[50] The NSDAP and the German Foreign Office competed to exert control over the international profile of German race theory, and key figures like Eugen Fischer and the psychiatrist Ernst Rüdin (1874–1952) readily complied. The *Deutsche Kongress Zentrale* (German Congress Directorate) intensified its controls, briefing racial scientists on how to project Germany's image abroad at major conferences.[51] The Seventh International Congress of Genetics held in Edinburgh in 1939 condemned both Stalinist and Nazi interference with genetics.[52]

Between 1931 and 1942, Central and Southeast Europeans, invited as academic guests, dominated the Kaiser Wilhelm Institute of Anthropology. These included: László Aport (1942), Mihály Malán (1936), Lajos Csík (1940), and Miklós Fehér from Hungary; Nikolaus Ilkow (1942) from Bulgaria; Franjo Ivanicek (1941) from Croatia; and Witold Sylwanowicz (1937) and C. R. Czapnik (1942) from Poland; other visitors included Marius Sulică (1942) from Romania; and Božo Škerlj (1932) from Yugoslavia.[53] These academics represented voices of support for the increasing German interest in racial surveys in Eastern Europe. The scene was set for racial transplantations, deportations, and the genocide of the *Generalplan Ost*.

The menace of Nazism forced liberal geneticists to make common cause with Lamarckians like Zollschan. In 1933, the Prague Academy of Sciences established a commission on race. Zollschan and Boas had mooted the idea of a research institute on racial problems in 1926; and after 1933 they intensified their efforts for an international committee against racial prejudice.[54] In August 1933, Zollschan drew up memoranda on the necessity of taking a position on scientific anti-Semitism. He suggested the convocation of a Society for Anthropology and Sociology of the Jews, which was to survey and monitor academic opinion. A university institute, or an organization like the Rockefeller Foundation, would act as the scientific sponsor.[55]

Iltis's critique of Nazi race theory as unscientific grew out of his commitment to Mendelism, in sharp contrast to Zollschan's commitment to Lamarckism. Between the two there was also a distinction in tactics and religious outlook. In his work against the unscientific basis of racial science, Zollschan established national committees at the Academy of Sciences in Prague and Vienna, in London, under the auspices of the Royal Anthropological Society and Royal Society, and through Franz Boas in the United States. He hoped that an international panel would conduct an inquiry into race and evaluate the basis for racial theories.[56]

Iltis attacked the "poison gas" ideas of Nazi racial purity in *Der Mythus von Blut und Rasse* (The Myth of Blood and Race), published in Vienna in 1936. In his writings, which appealed directly to Viennese public opinion, he called for popular mobilization against racism. He targeted academic anthropologists as a core group of Nazi supporters, not least because German anthropologists believed that Czechs could be Germanized.[57] For example, Iltis represented the Czechoslovakian League against Anti-Semitism at an international congress held in Paris in 1937.[58] Similarly, Zollschan gained the support of Czech political leaders, particularly the philosophically inclined Czechoslovak president Tomáš G. Masaryk; while the foreign minister and later president, Edvard Beneš (1884–1948), lobbied the supine League of Nations. Support grew for the *Ligue Internationale Contre l'Antisémitisme* (LICA), which included Masaryk, Beneš, and Albert Einstein as members of its Committee of Honour.[59]

Both Iltis and Zollschan hoped that the anti-eugenic Catholic Church could also become an ally against Nazi racism. Some German Catholic theologians were in favor of eugenics, such as Hermann Muckermann (1877–1962), the co-founder of the Kaiser Wilhelm Institute for Anthropology in Berlin. The condemnation of eugenics in the papal encyclical *Casti Connubii* gave rise to the hope, largely unrealistic as it turned out, that the church would take a firm stand against Nazi race ideology and policies. Zollschan approached Cardinal Theodor Innitzer (1875–

1955), the Archbishop of Vienna, and Iltis approvingly cited Cardinal Faulhaber's writings against Aryan race theory. At the very least, they hoped to expose the activities of anti-Semitic pastors such as Pater (Wilhelm) Schmidt (1868–1954) in Vienna.[60] In pursuing this goal, Zollschan had two private audiences with Pope Pius XI in 1934 and 1935.[61]

Zollschan's campaign against Nazi racism led him to adopt the view that the Jews constituted a cultural group rather than a race, which caused him substantially to moderate his own scientific position. He increasingly focused on cultural arguments, and found common ground with Hertz's pioneering critique of Social Darwinism. Zollschan's Lamarckism meant that biology and culture were fluid categories. Yet with this *volte-face* he did not completely renounce his earlier ideas about the Jewish race; he updated them in the light of Boas's theories and developments in modern social science.[62]

Hertz and Zollschan were more sympathetic to Lamarckism than to Mendelian notions of population genetics. By way of contrast, Julian Huxley (1887–1975), who represented the public face of British biology in the 1930s, disapproved on scientific grounds of Zollschan's citation of the environmentalist "engramme theory" advocated by the German biologist Richard W. Semon (1859–1918). By the late 1930s, Hertz, Iltis and Zollschan condemned Nazi racism, while simultaneously retaining ideas about population and inheritance as their special preserve.

Zollschan hoped that leading anthropologists would unite under the auspices of the League of Nations to condemn Nazi racism as unscientific. In 1933, Zollschan gained the support of Masaryk and, in 1936, worked alongside Beneš at the League of Nations in Geneva, organizing a petition signed by Freud and Masaryk among many others.[63] The large number of German and Austrian intellectuals in Prague during the 1930s (including Hertz, who moved to Prague in 1938) resulted in a strident opposition to race theory. Heinrich Mann (1871–1950), Arthur Holitscher (1869–1941), Lion Feuchtwanger (1884–1958), Max Brod (1884–1968), and Richard Coudenhove-Kalergi (1894–1972) published the pamphlet *Gegen die Phrase vom jüdischen Schädling* (Against the Phrase 'Jewish Parasite').[64] Zollschan organized expert groups of anthropologists at the Royal Society and the Royal Anthropological Institute in 1934. This was strategically important because of an international anthropological and ethnological conference held in

London in 1934 and at the French Société d'Anthropologie in 1937. By joining forces with the environmentalist anthropologist Boas at Columbia University, Zollschan reinforced an international coalition of notable intellectuals, anthropologists and scientists against the destruction, both intrinsic and imminent, in Nazi race theory.[65] As a result, committees were established in Prague and in Vienna (under the Viennese dental anatomist Harry Sicher) in 1937.[66]

Iltis targeted academic anthropologists as key Nazi supporters. The *International Federation of Eugenics Organizations* planned to meet in 1938 at the Baltic resort of Parnu in Estonia, when the issue of resisting Nazi eugenics arose. Against this, Eugen Fischer and the German biological anthropologist Otmar Freiherr von Verschuer (1896–1969) enjoyed the backing of the Nazi propaganda machinery and exercised rising influence among other rightward-leaning Central European eugenicists.

By the mid-1930s, Czechoslovakia was in the vanguard of opposition to Nazi race theory. This is exemplified by Hertz's settling in 1938 in Prague, where there was already a community of anti-Nazi intellectuals. Partly due to their anti-fascism, Hertz, Iltis and Zollschan experienced an intellectual renaissance from the mid-1930s.[67] Their alertness to the pernicious effects of racial science afforded them status at a time when liberal intellectuals belatedly realized the need for a common front against biological racialism. This episode came at the end of protracted efforts to modernize eugenics, discarding its imperialist racial hierarchies and adapting it to the nascent welfare state, while disassociating it from Nazi racism as a bogus and socially pernicious pseudoscience. For them, Nazi racial science simply had no valid intellectual credentials.[68] This laid the groundwork for a possible rapprochement with Zollschan's idea of an international scientific coalition.

During the war a group of Czech and Allied anti-racial thinkers coalesced around Zollschan. Moreover, the emigration of Iltis to the United States in 1939 meant that he was able to develop contacts with liberal-minded geneticists like L.C. Dunn (1893–1974). A strident critic of the Ku Klux Klan, Iltis once again found that he faced racism on his doorstep.[69] His characteristic response was to plan the founding of an Institute for the Study of the Problems of the Human Races.[70]

Conclusions

Eugenics was a politically contested scientific field in Central Europe, raising issues of ethnic and national identity, political legitimacy, and social justice. Hertz, Zollschan and Iltis were prescient critics of Aryan racial theories and racial hygiene. They correct the views currently dominant in the scholarly literature that there was no substantive opposition to German racial hygiene. Far from racial hygiene coming to Austria, Czechoslovakia, and elsewhere in Central and Southeast Europe only in the wake of the Nazi occupations, these three thinkers engaged with early signs of racial ideologies. They formed networks linking Vienna with Bohemia and Moravia, and these crystallized into an international scientific front against scientific racism. In short, these Central European scientists were early opponents of racism and its stranglehold on German hygiene.

Yet, in recent years, the tendency has been to remember Hertz as a social scientist in exile, Iltis as Mendel's biographer, and Zollschan as a Zionist eugenicist. But the work of all three in organizing international criticism of Nazi race theory merits recognition. As race became increasingly identified with the specter of National Socialism, the alternative aspirations of eugenics among anti-Nazi experts on human biology should not be forgotten.

Endnotes:

[1] Günther Luxbacher, "Zwischen 'Gott erhalte' und republikanischer Propaganda: Der Publizist Friedrich Hertz," in Wolfgang Duchkowitsch et al, eds., *Kreativität aus der Krise* (Vienna: Literas, 1991), 183–200.

[2] Paul J. Weindling, *Health, Race and German Politics between National Unification and Nazism 1870–1945* (Cambridge: Cambridge University Press, 1989).

[3] Paul J. Weindling, "The Medical Publisher J. F. Lehmann and Racial Hygiene," in Sigrid Stöckel, ed., *Die "rechte Nation" und ihr Verleger. Politik und Popularisierung im J.F. Lehmanns Verlag 1890–1979* (Berlin: Lehmanns Media, 2002), 159–170.

[4] Ploetz Papers, the invitation annotated by Ploetz. Private collection Wilfrid Ploetz, Herrsching.

[5] Friedrich Hertz, "Moderne Rassentheorien," *Sozialistische Monatshefte* 6 (1902), 876–883.

[6] Friedrich Hertz, "Die Rassentheorie des H. St. Chamberlain," *Sozialistische Monatshefte* 8 (1904), 10–15.

[7] Otto Weininger to Fr. Hertz, 11 November 1898, Nachlass Friedrich O. Hertz, Archiv für die Geschichte der Soziologie in Österreich (hereafter AGSÖ), Graz.

[8] Reinhard Müller, "Friedrich Hertz. Ein biobibliographischer Beitrag," *Jahrbuch des Dokumentationsarchiv des österreichischen Widerstandes* (1994), 58–74.

[9] See "Liste der Autoren, die ihre Mitarbeit in Aussicht gestellt haben," *Archiv für Rassen- und Gesellschaftsbiologie* 1 (1904) inside cover, not paginated; and *Archiv für Rassen- und Gesellschaftsbiologie* 2 (1905), 860–861.

[10] Friedrich Hertz, *Moderne Rassentheorien. Kritische Aufsätze* (Vienna: C.W. Stern, 1904).

[11] Friedrich Hertz, *Race and Civilization* (London: Kegan Paul, 1928), 311.

[12] W. Schallmayer to Hertz, 11 April 1915, Nachlass Friedrich O. Hertz, AGSÖ. Schallmayer sent reprints to Hertz on 13 March 1905 (*Selektionstheorie, Hygiene und Entartungsfrage*) and 1915 (*Unzeitgemässe Gedanken über Europas Zukunft*).

[13] Hertz assisted his wife Edith in British exile, acting as receptionist. See Bodleian Library, Oxford, Society for the Protection of Science and Learning (hereafter SPSL), 351/1Hertz file. Hertz received patients and accompanied his wife on home visits. The author was a young patient.

[14] See Paul J. Weindling, *Darwinism and Social Darwinism in Imperial Germany: The Contribution of the Cell Biologist Oscar Hertwig (1849–1922)* (Stuttgart: G. Fischer in association with Akademie der Wissenschaften und der Literatur Mainz, 1991).

[15] Fritz Lenz, "Antwort an Hertz," *Archiv für Rassen- und Gesellschaftsbiologie* 12 (1917), 472–475; and Friedrich Hertz, "Rasse und Kultur. Eine Erwiderung und Klarstellung," *Archiv für Rassen- und Gesellschaftsbiologie* 12

(1917), 468–472. Alfred Ploetz to Fr. Hertz, 9 January 1918, Nachlass Friedrich O. Hertz, AGSÖ.

[16] Friedrich Hertz, *Rasse und Kultur. Eine kritische Untersuchung der Rassentheorien* (Leipzig: Alfred Kröner Verlag, 1925).

[17] Fritz Lenz, "Friedrich Hertz. Rasse und Kultur. 3. Aufl. Leipzig 1925," *Archiv für Rassen- und Gesellschaftsbiologie* 18 (1926), 109–114.

[18] Hertz, *Rasse und Kultur*, see chapter 3, "Rasse und Seelenleben."

[19] The edition bearing Lehmann's dedication to Hitler is in the Library of Congress. Erwin Baur, Eugen Fischer, and Fritz Lenz. *Grundriss der menschlichen Erblichkeitslehre und Rassenhygiene*, 2 vols. (Munich: Lehmanns Verlag, 1923).

[20] This was a translation of Friedrich Hertz's *Rasse und Kultur* (1925).

[21] Luxbacher, "Zwischen 'Gott erhalte' und republikanischer Propaganda: Der Publizist Friedrich Hertz," 183–200.

[22] S. Winniger, *Grosse Jüdische National-Biographie*, 7 vols. (Cernăuţi: Tipografia ARTA, 1931–32), 6, 369. See also Mitchell Hart, "Racial Science, Social Science and the Politics of Jewish Assimilation," *Isis* 90 (1999), 268–297; and Mitchell Hart, *Social Science and the Politics of Modern Jewish Identity* (Stanford: Stanford University Press, 2000).

[23] Ignaz Zollschan, *Das Rassenproblem unter Besonderer Berücksichtigung der Theoretischen Grundlagen der Jüdischen Rassenfrage* (Vienna: Wilhelm Braumüller, 1910). See also Paul Kaznelson, "Zollschan, Ignaz. Das Rassenproblem under besonderer Berücksichtigung der theoretischen Grundlagen der jüdischen Rassenfrage," *Archiv für Rassen- und Gesellschaftsbiologie* 10, 6 (1913), 796–802; and Paul J. Weindling, "The Evolution of Jewish Identity: Ignaz Zollschan between Jewish and Aryan Race Theories, 1910–1945," in Cantor Geoffrey and Mark Swetlitz, eds., *Jewish Tradition and the Challenge of Darwinism* (Chicago: Chicago University Press, in press).

[24] Geoffrey C. Field, *Evangelist of Race: The Germanic Vision of Houston Stewart Chamberlain* (New York: Columbia University Press, 1981), 188–189. See also Central Zionist Archives, Jerusalem (hereafter, CZA), A 122/4/1, Zollschan draft letter to Chamberlain, n.d.

[25] Zollschan, *Das Rassenproblem*, 220.

[26] Zollschan, *Das Rassenproblem*, 152–153, and 205.

[27] CZA, A 122/4/1, Zollschan draft letter to Chamberlain, n.d.; Chamberlain card to Zollschan, 17 December 1909.

[28] Zollschan, *Das Rassenproblem*, 62, 96, 99–102, 135, and 155.

[29] Elizabeth Wiskemann, *Czechs and Germans* (London: Macmillan, 1967), 101, 163, 179–180.

[30] Wiskemann, *Czechs and Germans*, 178–179.

[31] CZA, A 122/11/1, Zollschan to Chief Rabbi J. H. Hertz, 5 April 1944.

[32] Zollschan, *Das Rassenproblem*, 420–421.

[33] For early gamete irradiation and "Radiumkrankheit," see Weindling, *Darwinism and Social Darwinism in Imperial Germany*, 252–253.

[34] Weindling, *Darwinism and Social Darwinism in Imperial Germany*, 252–253. See also Arthur Koestler, *The Case of the Midwife Toad* (London: Hutchinson, 1971).

[35] CZA, A 122/3, Semon to Zollschan, 22 October 1910, 12 January 1911; A 122/3, Max Neuburger to Zollschan, 26 October 1910; A 122/11/3, regarding Haddon's review of *Das Rassenproblem* in *The Morning Post*.

[36] See Ignaz Zollschan, *Revision der Jüdischen Nationalismus*, 2nd. rev. ed. (Vienna: Wilhelm Braumüller, 1920), 125–145; and John Efron, *Defenders of the Race: Jewish Doctors and Race Science in Fin-de-Siècle Europe* (New Haven: Yale University Press, 1994), 154.

[37] Ignaz Zollschan, *Jewish Questions: Three Lectures* (New York: Bloch Publishing Company, 1914).

[38] American Philosophical Society (hereafter APS), Boas Papers, Zollschan to Boas, 11 November 1925. Ignaz Zollschan, "Die Favus-Ausrottungsaktion in Osteuropa," *Medizinische Klinik* 3 (1924), 100–102. See also Efron, *Defenders of the Race*, 166, 180.

[39] APS, Boas Papers, Zollschan to Boas, 31 December 1933; Boas to Zollschan, 29 January 1934.

[40] Ignaz Zollschan, *The Significance of the Racial Factor as a Basis in Cultural Development* (London: Le Play House Press, 1934).

[41] See L.C. Dunn, "Hugo Iltis: 1882–1952," *Science* 117, 3027 (1953), 3–4.

[42] Hugo Iltis, *Gregor Mendel: Leben, Werk und Wirkung* (Berlin: J. Springer, 1924); Hugo Iltis, *Life of Gregor Mendel* (New York: W.W. Norton, 1932).

[43] Fritz Lenz, "Die Stellung des Nationalsozialismus zur Rassenhygiene," *Archiv für Rassen- und Gesellschaftsbiologie* 25 (1931), 300–308.

[44] Hugo Iltis papers, San Jose, California.

[45] Hugo Iltis, "Eugenik," *Bücherwarte Zeitschrift für sozialistische Buchkritik* 3, 4 (1928), 97–108.

[46] Hugo Iltis, "Rassenwissenschaft und Rassenwahn," *Die Gesellschaft* (1927), 42–44.

[47] Iltis, "Rassenwissenschaft und Rassenwahn," 44.

[48] Nicholas Hopwood, "Biology between University and Proletariat: The Making of a Red Professor," *History of Science* 35 (1997), 367–424; and Nicholas Hopwood, "Producing a Socialist Popular Science in the Weimar Republic," *History Workshop Journal* 41 (1996), 117–153.

[48] Lajos Méhely, "Blut und Rasse," *Zeitschrift für Morphologie und Anthropologie* 34 (1934), 244–257.

[50] SPSL. Josef Weninger 360/2; Hertz 351/2 and Iltis 200/1 files.

[51] Hoover Institute, Stanford University, Deutsche Kongress Zentrale no. 256 for 1940 criticism of the Italian anthropologists Pende, Bottai, Buffarini (in contrast to Landra) for not signing a racial manifesto. Leo Alexander, "Methods of Influencing International Scientific Meetings as Laid Down by German Scientific Organizations," CIOS File no. 24, item no. 38–8, 1945. See Paul J. Weindling, *Nazi Medicine and the Nuremberg Trials: From Medical War Crimes to Informed Consent* (London: Palgrave Macmillan, 2004); and Madeleine Herren, " 'Outwardly…an Innocuous Conference Authority': National Socialism and the Logistics of Information Management," *German History* 20, 1 (2002), 67–92.

[52] Nikolai Krementsov, *International Science between the Two World Wars: The Case of Genetics* (London: Routledge, 2004), 121–123; and "Men and Mice at Edinburgh," *Journal of Heredity* 30 (1939), 371–374.

[53] Nils Loesch, *Rasse als Konstrukt. Leben und Werk Eugen Fischers* (Frankfurt-am-Main: Peter Lang, 1997), 577.

[54] APS, Boas Papers, Zollschan to Boas, 31 December 1933; Boas to Zollschan, 29 January 1934.

[55] Ignaz Zollschan, *Zwei Denkschriften über die Notwendigkeit der Stellungnahme zum wissenschaftlichen Antisemitismus* (Karlsbad: privately printed, 1933).

[56] CZA, A 122/4/1, I. Zollschan, 'Die Bedeutung einer Rassen-Enquete' [typescript], Amsterdam 1938; A 122/4/10, concerning support from Gowland Hopkins and the Royal Anthropological Institute.

[57] Hugo Iltis, *Der Mythus von Blut und Rasse* (Vienna: Rudolf Harrand, 1936).

[58] APS L.C. Dunn Papers, Iltis Correspondence folder 1: Iltis to Dunn, 9 February 1939; folder 4: Iltis to Mrs Simon Guggenheim, 7 October 1951.

[59] *Cahiers de la Licra* (Paris: Licra, n.d.); CZA, A 122 4/13, Einstein to Zollschan, 12 August 1936.

[60] Hugo Iltis, *Volkstümliche Rassenkunde* (Jena: Urania, 1930); Iltis, *Mythus von Blut und Rasse*. I am grateful to Hugh and Fred Iltis for information on their father, and to Veronika Lipphardt (Humboldt Universität Berlin) for copies from the Boas and Dunn Papers, American Philosophical Society. APS, Boas Papers, Zollschan to Boas, 28 April 1935.

[61] Efron, *Defenders of the Race*, 165.

[62] Ignaz Zollschan, *The Significance of the Racial Factor as a Basis in Cultural Development* (London: Le Play House, n.d.).

[63] "Für das Initiativcomité zur Veranstaltung einer Welt-Enquete über die Rassenfrage, 'An die Vertreter der Wissenschaft!'," Archives of the Presidential Office Prague, (A KPR Praha), file Ignác Zollschan. I am grateful to Michal Šimůnek for locating this document.

[64] Heinrich Mann, Arthur Holitscher, Lion Feuchtwanger, Max Brod and Coudenhove-Kalergi published the pamphlet *Gegen die Phrase vom jüdischen Schädling* (Prague: Amboss-Verlag) in 1933. See Arthur Holitscher's posthumous critique of racial theories published in Vienna by Willy Verkauf in 1948.

[65] CZA A 122 4/6. See also Stefan Kühl, *Die Internationale der Rassisten* (Frankfurt-am-Main: Campus, 1997), 146–151.

[66] CZA A 122/4/1.

[67] SPSL, 432. See Friedrich Hertz file for his London activism. See Marietta Bearman, Charmian Brinson, Richard Dove, Anthony Grenville and Jennifer Taylor, *Das Austrian Centre in London 1939 bis 1947* (Vienna: Czernin, 2004), 16, 24–25.

[68] Geoffrey Cantor, "Charles Singer and the Early Years of the British Society for the History of Science," *British Journal for the History of Science* 30, 1 (1997), 5–24, especially 14.

[69] Recollections of Iltis's sons Fred and Hugh communicated to the author.

[70] APS Dunn Papers. Iltis to Dunn 9 Feb 1939.

Part III

Religion, Public Health and Population Policies

"Moses als Eugeniker"?
The Reception of Eugenic Ideas in Jewish Medical Circles in Interwar Poland

Kamila Uzarczyk

On account of eugenics being associated with Nazi racial policy, until recently there has been little discussion of Jewish eugenics. There is no doubt that eugenics, effectively a value judgment about the worth of human beings, has racist connotations. However, its evaluation solely in the context of the Nazi experience disregards the enormous popularity of eugenic ideas that related to social life, and the diversity of the international eugenic movement in the interwar period. The movement was not an ideological monolith, and the solutions to social problems postulated by the advocates of eugenic ideas were attractive for many societies aiming at the biological regeneration of the population. Historical sources reveal that the concept of a distinctive Jewish race, and the program to eliminate "unfit" elements of Jewish society, did receive backing from Jewish medical circles during the interwar period.

Zionism and the Concept of Jewish Eugenics

In its heyday, eugenics was enthusiastically endorsed by advocates of the Zionist movement, which, as Raphael Falk has noted, evolved from a similar intellectual background, and had analogous goals, to the eugenic movement: "Whereas eugenics aspired to redeem the human species by forcing it to face the realities of its biological nature, Zionism aspired to redeem the the people by forcing it to face the realities of its biological existence."[1] In Zionist circles, it was emphasized that the preservation of the biological and cultural distinctiveness of the Jews living in diaspora would be much more advantageous for humanity than their assimilation into a foreign milieu. According to the Jewish anthropologist Arthur Ruppin (1876–1944):

The Jews have not only preserved their great natural gifts, but through a long process of selection these gifts strengthened. The result is that, in the Jew of today, we have what is in some respects a particularly valuable human type. Other nations may have other points of superiority, but in respect to intellectual gifts the Jews can scarcely be surpassed by any nation. For this fact alone the Jews may well claim their right to a separate existence and resist any attempt to absorb them.[2]

Advocates of eugenics warned that the assimilation process was bringing about the adoption of negative cultural patterns, which, in turn, might affect the wholesome condition of the Jews. Intermarriages and late marriages, birth control as well as alcohol abuse and venereal diseases, previously occurring sporadically among the Jews, were viewed as indicators of the inevitable decline of the Jewish race. One of the zealous supporters of Jewish eugenics, Zewi Parnass, wrote:

During the historical evolution of the Jewish race, capability to maintain racial features found its expression in separatist tendencies, that is, in the whole range of regulations that organized everyday life of single individuals and the nation, and distinguished them from the surrounding people. The Jews, living as a diaspora, maintained symbiotic relations with local people, while keeping enough distance to prevent any changes in racial texture, and to preserve the mechanism of racial selection unchanged.[3]

According to Parnass, this specific form of existence, a state of *vita latens*, was a favorable condition for survival, "just as it can be noticed in some species of plants, which do not get rooted deep in the ground but thrive on what they can find near the surface; hence they can move from place to place."[4] Similarly, argued Parnass, the Jews have always been engaged in trade, an occupation that involved frequent wanderings. This made it more difficult for Jews to settle down and integrate in different cultures, a condition which prevented the loss of traditional values. Thus, in Jewish eugenic circles, cultural assimilation of the Jews was seen as a process endangering national identity, and as a dysgenic factor.

The dysgenic effects of assimilation also attracted the attention of eugenicists from countries with a considerable influx of immigrants and from those with numerous ethnic minorities. However, in interwar Poland, despite the ethnic mix and diversity of national minorities, the question of racial purity, from a eugenic point of view, was an issue of secondary importance. The political preferences of a majority of eugenicists in Poland, recruited from liberal and left-wing circles, were reflected in their attitude towards race. Nevertheless, the dysgenic effect of assimilation troubled the nationalist anthropologist Karol Stoja-nowski (1895–1947). In 1927, Stojanowski noted: "writing about sub-stitution processes, which transform the quality of the nation, we can-not ignore the assimilation processes. Is assimilation, from the eugenic point of view, beneficial for the nation? If positive elements are con-cerned, yes. It would be advisable to Polonize the German element and to consider the possible assimilation of the Russian one, but the Jewish must be seen as undesirable."[5] Stojanowski feared that assimilation would gradually pave the way for the Jewish political and socioeco-nomic domination of state life. He was convinced that:

> maintaining close cultural relationship within the Jewry would com-pletely paralyze Polish national life. Besides, we should keep in mind that the Jewish population is not such a tremendous asset as has been generally assumed. There is no doubt that, as a result of advanced medical care enabling even the weak to survive, Jewish military recruits are the worst. Moreover, it should be noted that their intellectual supe-riority is, to a large extent, a legend built on the self-advertising of this, a more cunning than skilled, nation. To sum up: assimilation of the Jews is eugenically unadvisable. They must either emigrate or limit their birthrate and die out.[6]

For the Jews living in the diaspora, the symbol of national rebirth and the undisturbed development of national life was Palestine, the country that "played an important role in the development of the specific char-acter of the Jewish race and society, and as such should become the cradle of the renaissance of the racial and social life of the Jews."[7] Returning to the Land of the Fathers (as Palestine was referred to in Zionist writings) was perceived as a "huge eugenic revolution in the life of the nation,"[8] provided that future settlers embodied eugenic cri-teria. Parnass proposed that an Office for National Eugenics in Palestine (*Urząd dla Eugeniki Narodowej w Palestynie*) would be in charge of selecting candidates for emigration to Palestine, and only those who, from the eugenic point of view, received positive health certificates would be entitled to cross the borders. Immigration law should exclude "idiots, epileptics, the mentally defective and the men-tally ill, those unable to support themselves and their families, those suffering from infectious diseases and diseases making it impossible to work, prostitutes, anarchists and all others posing a threat to law and order, or who are a burden on society." Parnass proposed that those who do not contribute to national wealth should not be allowed to enter the country. The new social organization in Palestine should be based on hygienic rules and, in order to neutralize the harmful influence of city life, Parnass suggested the creation of "garden cities."[9]

Immigration to Palestine obviously demanded sufficient "human resources," so it is not surprising that Zionist-oriented supporters of Jewish eugenics were equally concerned about what they called the "neo-Malthusian epidemic." "Among those Jews who assimilated into a strange milieu," warned Parnass, "the tendency to limit the number of children causes the Jewish population to die out. While it is true that progress in medicine and decline in death-rates of the newborn make this process slower, it is also undeniable that keeping alive weak individuals will result in a worse biological constitution of the population and bring about an inevitable decline of the quality of the race and an increase in death-rates."[10] Since Jewish assimilation was more advanced in Western Europe, the implementation of eugenic policies with particular emphasis on anti neo-Malthusian campaign was a task of primary importance.[11]

According to contemporary authors, Eastern Jews maintained a relatively high birthrate as compared to their Western counterparts. Moreover, their obedience to the Law made them less prone to dangerous addictions and sexually transmitted diseases. Therefore, the main task of Jewish eugenics in Poland was not the adoption of an anti-neo-Malthusian campaign, but the reform of the health-care system, with particular reference to the health of mother and child, the reservoir of racial forces for future generations. Some physicians even argued that, from a eugenic perspective, it would be advisable to instruct the poor in birth control methods. Not only would it directly improve the living standard of families, but it would also result in the higher "eugenic quality" of future generations.[12]

Although individual eugenic programs for Eastern and Western European Jews differed slightly, the goal was the same: the biological regeneration of the Jewish nation to ensure its survival in the Social Darwinist struggle for existence. To strengthen the "vital forces" of the nation, Parnass insisted on the validity of the selection process:

> Society, through its direction of individual selection, affects the evolution of the race. Stronger societies, stronger numerically or in terms of their social structure, given the same conditions of the environment, will triumph over the less fit. Thus, society has an impact on

the internal and external struggle for existence of the race and the direction of its evolution. (…) The influence of the racial elements in society is even more important. Race provides a biological substratum for society. (…) Social development and the survival of society depend on racial texture, racial species, or the mixture of races that form the society.[13]

In the early 1920s, Darwinian rhetoric was widely accepted within medical circles and a discussion about race also ensued among eugenically inclined members of the Jewish medical community in Poland. Social Darwinism, as Michel Foucault printed out, had become "the way to translate political discourse into biological language, not only the way to disguise political discourse in a scientific cover, but the way to understanding colonialism, wars, criminality, mental disorders, and so on."[14] Besides, Social Darwinism, assuming the existence of a spontaneous generation of races and their capability to adapt under environmental influences, undermined the thesis common in classical racist theories relating to immanent and unchangeable racial features.

At the beginning of the twentieth century, Jewish racial status became a matter of heated discussion among anthropologists.[15] Many scientific authorities argued that Jews, during the course of their history, had "absorbed blood" through intermarriages, and therefore could not be seen as racially pure, but rather as a mixture of races. Yet, for diaspora Jews, the Social Darwinist approach was an argument in favor of their biological, and not only cultural, and religious entity. Visible differences between Jewish and indigenous people in various countries were seen as a result of their biological composition, and as a proof that the Jews constituted a separate racial group. "The Jews," wrote Parnass, "as the group stemming from the same ancestry and distinguished by similar reactions to environmental influences, form a biological entity which continues to prevail through giving birth to offspring, one that bears the same racial features, which, via heredity, have been passed from generation to generation."[16] Consequently, according to Jewish eugenicists, the Jews, as a biologically distinct group, required their own eugenic program. Some sought to prove that, in fact, they had already found such a program in the Bible and the Talmud.

"Moses als Eugeniker"

Due to the lack of eugenic legislation, an irrevocable element of eugenic social instruction was the strict obedience to traditional religious regulations, which, as contemporary authors pointed out, were largely similar to eugenic ideas. Thus Parnass noted:

> Our religious regulations indicate that hygiene, and particularly race hygiene, is what we were aiming for in social life. Let us revive old rules in accordance with the spirit of the past; revive them, and we will get the solution to all problems, solutions which are an ideal for the European eugenicists. They dream of the time when the necessity of race hygiene will be so deeply rooted in social consciousness that it becomes a kind of social religion. We have had this religion for a long time; it arose in the Jewish tradition in Palestine. (...). The whole legislation of Babylonian and Jerusalem Talmud, in the chapters relating to national and racial life, forms the greatest book of eugenic laws.[17]

According to the rabbi Max Reichler (1885–1957), observation of these old instructions, enforced by religious sanction, determined the survival of the Jewish nation against the "onslaughts of time, when others, numerically and politically stronger, succumbed."[18] Therefore the history of the Jewish nation was perceived as "a eugenic experience on a large scale,"[19] and Moses as the first to implement the eugenic ideal. As the dermatologist Hans von Pezold (1870–?) observed in 1932: "Only a hero with extraordinary willpower, (such as Moses) could, during forty years of exile save the nation from degeneration—the result of centuries of subjugation—and train it militarily and educate it morally so that it became so advanced in hygienic standards that, despite all storms and defeats, it was able to preserve its forces, conquer and settle in its own vital space (*Lebensraum*), and build a strong state."[20]

For enthusiastic supporters of eugenics, the social hygiene legislation introduced by Moses—comprising both hygiene instructions in the narrow sense and social legislation, including a ban on work on the seventh day of the week—had explicit eugenic meaning. Yet more than anything else, what made Moses the "father of eugenics" was his concept of sexual hygiene (*Geschlechtshygiene*) and race hygiene (*Rassenhygiene*), both with the goal of producing "valuable" offspring in terms of quality and quantity. Pezold gave an indication as to how this goal was to be achieved: Talmudic instructions relating to the prohibition of marriages with epileptics and lepers, as well as the banning of castration and deviant sexual intercourse, were to be observed. Sexual offences were to be severely punished, including capital punishment. It was stressed that Moses had recommended early marriages, in some cases even polygamy, while opposing marriages where the age difference between spouses was pronounced. Prostitution and adultery were stigmatized to the extent that the person found guilty was stoned to death. Severe limitations on sexual life and circumcision were designed to protect against venereal diseases. According to Pezold, Moses was also the first to enact measures against venereal diseases. And finally, in order to maintain the purity of the race, intermarriages with non-Jewish people were strictly forbidden.[21]

Contemporary advocates of eugenics acknowledged the significance of the eugenic legislation contained in the Bible and the Talmud. "The core of Talmudic hygiene," stated Gerson Lewin (1868–?), "is cleanliness in both its literal physical, and spiritual, meaning. The first was to serve health protection (...), the second eugenic purposes."[22] While the congruity between Jewish tradition and eugenics is a debated issue, it is indisputable that, at the start of the twentieth century, the proponents of Jewish eugenics sought to find evidence of eugenic thinking in traditional teaching.[23] The concept of Jewish endogamy—the intra-Jewish concern with biological lineage and instructions about sexual life—served as a classic example of eugenic consciousness.

Jewish Eugenics in Interwar Poland

In interwar Poland, enthusiasts of Jewish eugenics came together in the Section for Social Hygiene and Eugenics (*Sekcja Higieny Spolecznej i Eugeniki*), formed within the Society for the Health Protection of the Jewish Community in Poland (*Towarzystwo Ochrony Zdrowia Ludnošci Żydowskiej w Polsce*—TOZ). The Society was founded in 1918 in Warsaw as an initiative launched by a group of physicians and social workers; Gerson Lewin was appointed its first president. It was only later, in 1923, that the Polish Ministry of the Interior recognized its statute and the Society was officially registered; yet it received little financial aid from the Polish government and was financed, in the main, by Jewish foundations from abroad as well as local Jewish communi-

ties. The Polish authorities only supported the campaign against tuberculosis and ringworm of the scalp, as well as programs promoting
mother and child health care.[24] While the founders of the Society
admitted that the enforcement of public health was the task of the state,
they also stressed that, in order to improve the health conditions of
the Jews in Poland, it was necessary to raise awareness of dangers to
health and persuade every individual of the necessity to follow eugenic
precepts in private life.

The goal of the TOZ was "the transformation of physical fitness
and the biological value of the Jewish masses in Poland"[25] through a
campaign against sexually transmitted diseases, social diseases (such
as tuberculosis), mother and child health care, job counseling, and a
hygiene campaign. The implementation of preventive measures seems
to have been of the highest importance, which is indicative of the inter-
relation between eugenics and social hygiene in the Jewish interpretation of the eugenic ideal in Poland. Unlike the German racial hygienist
Alfred Ploetz (1860–1940), who stressed that "whereas social hygiene
aims for the substantial development of social institutions, the goal of
race hygiene is to maintain and even strengthen egoism, as it is advantageous for the individuals in their struggle for existence,"[26] Jewish
eugenicists in Poland sought to combine both. Parnass interpreted
Social Darwinism in accordance with the organic concept of society
developed by Ernst Haeckel (1834–1919), whose ideal was the harmonious coexistence of different elements of society, selected according to
eugenic criteria.[27] "Who would doubt," wrote Parnass, "that, although
single individuals of the particular race remain mutually in loose relationship, they depend on each other just as the single cells of the organism do."[28] In consequence, certain social actions could be justified.
However, eugenic selection and the elimination of undesirable individuals remained an option.

The practice of Jewish eugenics in interwar Poland conformed to
the so-called Latin model, which did not entirely abandon the Lamarckian theory of the inheritance of acquired characteristics and therefore limited its purview to an essentially moderate program of preventive and positive eugenics. This did not mean, however, that there were
no proponents of restrictive and radical measures, comprising prohibi-
tive marriage legislation and sterilization laws, but the legal system in
interwar Poland did not allow for the introduction of such measures.
Although the Polish Eugenic Society (*Polskie Towarzystwo Eugeniczne*)

launched a campaign for the implementation of eugenic enactment—in 1935 the draft legislation was submitted to the Health State Council for further scrutiny—the project was rejected and never came into effect.

Negative eugenics was mostly advocated by those physicians who favored Mendelian genetics and genetic determinism. Supporters of Jewish eugenics in interwar Poland sought to implement a program of genetic research that aimed to identify and establish the rules of inheritance as regards positive and negative characteristics; and to evaluate the impact of environmental factors on the biological and psychological constitution of Jews. An anthropological study of the racial features of the "healthy Jew" in Poland was also on the agenda.[29] By comparing Polish Jews with the non-Jewish population, physicians identified distinct biological features in Jews, including lower body mass index; on average lower fertility (particularly in large cities); lower death-rates in the working classes (despite poor living standards); racially conditioned immunity to tuberculosis; and less frequent occurrences of syphilis and dangerous addictions, including, most notably, alcoholism.[30] On the other hand, physicians warned that diabetes and illnesses relating to the nervous system were twice as high among Jews compared to other nationalities.[31] A commonly shared view was that Jews constituted a unique psychological type. According to one author, "Different living standards, religion and jobs performed by the Jews played a role in the process of shaping distinctive national psychological features. Persecution and hard living conditions eliminated weak members of the population. Only stronger individuals (those who did not fear to belong to the group that throughout the centuries played the role of a scapegoat) were accepted through conversion. In terms of psychology, the Jews constitute a unique type and it would be most appropriate to say that they constitute a psychological race."[32]

According to contemporary authors specific racial features of the Jews manifested in characteristic facial and bodily features. Physical appearance was seen as a reflection of inner qualities. Thus, Jewish facial composition was the "expression of the soul," or "an emblem of the ceaseless wanderings and countless agonies."[33] Similarly, corporal structure indicated susceptibility to certain mental illnesses. As one physician argued, the theory of relations between physical type and predilection to specific mental disorders, developed by the German psychiatrist Ernst Kretschmer (1888–1964), provided a tool with which to characterize nations. In order to describe national character scientifically, it was necessary to define the biological composition of individuals and identify mental disorders that occurred most frequently.[34] One view common among Jewish physicians was that the Jews were particularly vulnerable to schizophrenia and manic-depressive psychosis (*psychosis judaica*), which, in turn, suggested the commonality of leptosomatic and physical types; the athletic body build, which, according to Kretschmer, was prone to epilepsy, was considered to be rare among the Jewish population. These findings could potentially be used to prove the "higher" eugenic value of Jews. According to the physician I. Kantner, the typical biological constitution of Jews explained the low crime rate in Jewish communities—yet another phenomenon perceived to be biologically determined. According to the view at the time, patients suffering from schizophrenia and manic-depressive psychosis were less likely to commit crimes (in only 5 per cent and 3.5 per cent of cases, respectively), whereas the epilepsy-suffering athletic type was to be found among criminals relatively frequently.[35] Another indicator of the eugenic value of the Jewish race was the rare occurrence of diseases commonly associated with "racial poisons." According to available medical statistics, the Jews, particularly those living in Eastern Europe, suffered less than non-Jews from psychosis in the course of syphilis (0.17 per cent and 5 per cent respectively), and were also less likely to develop alcoholism (5.4 per cent and 12 per cent, respectively).[36] Because there was no clear scientific explanation for these differences, most authors accepted convenient theories that asserted a distinct biological nature and religious orthodoxy for Jews, both of which reduced the likelihood of deviance.

Furthermore, a relatively high index of illnesses afflicting the nervous system, particularly schizophrenia, raised understandable concern in medical circles. Again, the reasons were unclear and, more often than not, physicians agreed that the condition was the result of environmental and hereditary influences. "The high frequency of mental disorders in Jews," wrote Parnass, "is a result of their constant struggle for existence. (...) A stressful lifestyle, which lacks harmony, causes nervous breakdowns and results in a frightening number of mental illnesses and neuroses. A predilection for diseases of the nervous system is of a hereditary character; therefore, we must focus our efforts on measures that stop the possibility of the transmission of these evils via heredity. Thus, the only solution is sterilization of less valuable individuals."[37] The psychiatrist R. Becker expressed a similar view:

> there is something specific in the living conditions of the Jewish population in Poland, as well as the structure of their nervous system, which makes them particularly prone to highly endogenous diseases. The spread of schizophrenia shows how important it is to fight it. It should be treated as one of the most important issues, from a social medicine and racial-eugenic point of view. Because Jewish people suffer mostly from hereditary psychoses, it would be advisable to eliminate handicapped and sick elements through the control of the reproduction of hereditarily burdened individuals. The chronic mentally ill and the mentally handicapped, as well as sufferers from severe epilepsy, alcoholics and drug addicts should not be married, so as to prevent births of hereditarily burdened offspring. Marriage should also be forbidden for disabled individuals, including the deaf and the blind, as well as sexual deviants such as sadists, masochists and homosexuals.[38]

It was assumed that people suffering from epilepsy posed a particular threat, because they were inclined to criminality and because they were "individuals with a particularly strong sexual drive and, as such, were capable of producing offspring which become a burden on society."[39] Despite the fact that it was already known that epilepsy could be caused by physical injuries, some physicians nevertheless called for measures to control the sexual activity of this group. In 1928, the psychiatrist Zygmunt Bychowski (1865–?) stated: "Unfortunately we still lack eugenic marriage legislation. Some German and American towns have already introduced compulsory health certificates for those who plan to get married. In our legislation this issue has not been tackled yet. But it is beyond question that epileptics should be subject to very strict control and we should not allow them to be married."[40]

A similar attitude was adopted regarding diabetes, a disease which occurred in the better-off classes of Jewish communities and one which caused a death-rate six times higher than in non-Jewish nationalities. Whereas most authors agreed on the importance of environmental etiology, with diet and stressful lifestyle as the primary causes of endocrine system disorders, some also emphasized the role of heredity in the transmission of diabetes. "Therefore all those suspected as hereditarily burdened," noted internist L. Szyfman, "should undergo eugenic counseling to prevent marriages between those who are already ill, and this includes individuals that do not as yet display clinical symptoms but might be hereditarily burdened."[41]

Under the system of law in Poland, the introduction of marriage regulations on health grounds—not to mention negative eugenic measures—was considered *pium desiderium*. Supporters of eugenics hoped, however, that an efficiently directed campaign would gradually transform social attitudes towards eugenic ideas, and that obedience to the eugenic ideals would become something of an inner imperative with an almost religious meaning. "In the course of time," insisted Parnass, "each incurably ill patient will voluntarily undergo sterilization. And those that oppose it will be stigmatized by public opinion as social outcasts who dared to contradict nationally sacred values."[42]

The transformation of the social attitudes of Polish Jews was a pressing issue precisely because state intervention and the implementation of eugenic legislation were unlikely to occur in the near future. Parnass insisted that Europe follow the American example: "I cannot predict when the Europeans will follow the Americans and adopt race hygiene measures. It will take decades to introduce a race hygiene policy in Europe, particularly in Eastern Europe, where the majority of Jewish stock lives. We cannot wait so long."[43] Interestingly, one of the advocates of Jewish eugenics, neurologist Henryk Higier (1866–1942), praised German racial hygiene legislation as late as 1938, particularly the law of 14 July 1933 (*Gesetz zur Verhuetung erbkranken Nachwuchses*), which attempted to prevent the birth of hereditarily burdened offspring. "The Germans," wrote Higier, "pay little attention to environmental influences and stress hereditary factors as the source of diseases. Therefore, according to German law, it is legitimate to sterilize some individuals if medical knowledge and experience justify the prediction that their offspring will suffer from severe physical and psychological disorders."[44]

Conclusions

In interwar Europe the destructive potential of the eugenic ideal was rarely discussed. Eugenics was considered a branch of science that provided explanations for social regress through the principles of genetics. The biological view of man, as a source of national wealth and power, justified eugenic notions about the need to shape social structure. Indeed, as Raphael Falk remarks: "While in the first half of the century Zionist ideology was very much grounded in the anthropological notions of Darwinism which became increasingly eugenically oriented, these sentiments were formally replaced by a withdrawal from any eugenic or biologically racist thought after the experiences of the Nazi era."[45] Only the impending mass extermination of the European Jewry, justified by Nazi ideology on the grounds that Jews were "degenerate" elements of the human species, highlighted the dangers hidden behind demands for the elimination of a section of humanity in order to improve the "genetic stock" of future generations.

Endnotes:

[1] Raphael Falk, "Zionism and the Biology of the Jews," *Science in Context* 11, 3–4 (1998), 587.

[2] Arthur Ruppin, *The Jews of To-day*, trans. by Margery Bentwich (London: G. Bell and sons, 1913), 214–215.

[3] Zewi Parnass, *Kwestia żydowska w świetle nauki* (Lvov: Beth-Israel, 1921), 38.

[4] Parnass, *Kwestia żydowska w świetle nauki*, 38.

[5] Karol Stojanowski, *Rasowe podstawy eugeniki* (Poznań: M. Arcta, 1927), 67.

[6] Stojanowski, *Rasowe podstawy eugeniki*, 68–69.

[7] Parnass, *Kwestia żydowska w świetle nauki*, 65–66.

[8] Yoram Rubin, "The Ingathering of the Exiles from a Eugenic Point of View", *Moznayim* 1, 4 (1934), 89–93.

[9] Parnass, *Kwestia żydowska w świetle nauki*, 80.

[10] Parnass, *Kwestia żydowska w świetle nauki*, 76.

[11] Parnass, *Kwestia żydowska w świetle nauki*, 77.

[12] See H. Rubinraut, "Propaganda świadomego macierzyństwa i jej znaczenie dla ludności żydowskiej," *Medycyna Społeczna* 6, 11–12 (1933), 1; E. Birzowski, "TOZ wobec zagadnień świadomego macierzyństwa," *Medycyna Społeczna* 7, 3–4 (1934), 4; S. Chazan, "Nieodzowność propagandy świadomego macierzyństwa," *Medycyna Społeczna* 7, 3–4 (1934), 5.

[13] Parnass, *Kwestia żydowska w świetle nauki*, 14.

[14] Michel Foucault, *Trzeba bronić społeczeństwa. Wykłady w College de France, 1976*, trans. by Małgorzata Kowalska (Warsaw: KR Publishing House, 1998), 254.

[15] John M. Efron, *Defenders of the Race: Jewish Doctors and Race Science in Fin-de-Siècle Europe* (New Haven: Yale University Press, 1994).

[16] Parnass, *Kwestia żydowska w świetle nauki*, 13.

[17] Parnass, *Kwestia żydowska w świetle nauki*, 78.

[18] Max Reichler, *Jewish Eugenics* (New York: Bloch Publishing Company, 1916), 7. See also Noam J. Zohar, "From Lineage to Sexual Mores: Examining 'Jewish Eugenics'," *Science in Context* 11, 3–4 (1998), 577.

[19] Hans von Pezold, "Moses als Eugeniker", *Deutsche Medizinische Wochenschrift* 58, 35 (1932), 1370; and Gerson Lewin, *Ochrona zdrowia i eugenika w Biblii i Talmudzie* (Warsaw: Brothers Wójcikiewicz Publishing House, 1934).

[20] Pezold, "Moses als Eugeniker," 1370.

[21] Pezold, "Moses als Eugeniker," 1371.

[22] Lewin, *Ochrona zdrowia i eugenika w Biblii i Talmudzie*, 10.

[23] Zohar, "From Lineage to Sexual Mores: Examining 'Jewish Eugenics'," 577.

[24] See "Sprawozdanie z lustracji TOZ-u w dniu 20.04.1934". *Ministerstwo Opieki Społecznej, Archiwum Akt Nowych w Warszawie*, sig. 15 I/665, 2–5.

[25] *Księga Pamiątkowa I Krajowego Zjazdu Lekarskiego "TOZ-U"* (24–25 June 1928), ed. by Towarzystwo Ochrony Zdrowia Ludności Żydowskiej w Polsce – TOZ (Warsaw: TOZ, 1929), 42.

[26] Alfred Ploetz, "Zur Abgrenzung und Einteilung des Begriffs Rassenhygiene," *Archiv für Rassen und Gesellschaftsbiologie* 3, 6 (1906), 865–866.

[27] Daniel Gasman, *The Scientific Origins of National Socialism. Social Darwinism in Ernst Haeckel and the German Monist League* (London: Macdonald and Co., 1971).

[28] Parnass, *Kwestia żydowska w świetle nauki*, 39.

[29] "Uchwały I Krajowego Zjazdu Lekarskiego 'TOZU', sekcja eugeniczna," in *Księga Pamiątkowa I Krajowego Zjazdu Lekarskiego "TOZ-U"*, 335–336.

[30] Henryk Higier, "Zadania pracy eugenicznej w warunkach bytowania mas żydowskich," in *Księga Pamiątkowa I Krajowego Zjazdu Lekarskiego "TOZ-U"*, 37.

[31] R. Becker. "Umysłowo chorzy Żydzi w Polsce i sprawa opieki nad nimi," in *Księga Pamiątkowa I Krajowego Zjazdu Lekarskiego "TOZ-U"*, 112.

[32] B. Brutzkus, "O pochodzeniu narodu żydowskiego", *Medycyna Społeczna* 6, 3–4 (1933), 9.

[33] Salaman N. Redcliffe, "Heredity and the Jew," *Journal of Genetics* 1, 3 (November 1911), 278.

[34] I. Kanter, "Konstytucja a cechy charakteru narodowego," *Medycyna Społeczna* 11, 1–2 (1938), 3.

[35] I. Kanter, "Psychozy a alkoholizm u Żydów. Ich charakterologiczno—kryminologiczne znaczenie," *Medycyna Społeczna* 12, 5–6 (1939), 2–3.

[36] Becker, "Umysłowo chorzy Żydzi w Polsce i sprawa opieki nad nimi," 114.

[37] Parnass, *Kwestia żydowska w świetle nauki*, 77.

[38] Becker, "Umysłowo chorzy Żydzi w Polsce i sprawa opieki nad nimi," 115.

[39] Zygmunt Bychowski, "Padaczka a opieka społeczna" in *Księga Pamiątkowa I Krajowego Zjazdu Lekarskiego "TOZ-U,"* 29.

[40] Bychowski, "Padaczka a opieka społeczna," 29.

[41] L. Szyfman, "Cukrzyca u Żydów (w związku z zadaniami poradni eugenicznych)", in *Księga Pamiątkowa I Krajowego Zjazdu Lekarskiego "TOZ-U,"* 75.

[42] Parnass, *Kwestia żydowska w świetle nauki*, 78.

[43] Parnass, *Kwestia żydowska w świetle nauki*, 77.

[44] Henryk Higier, "Psycho-higiena społeczna a sterylizacja eugeniczna", *Medycyna Społeczna* 11, 1–2 (1938), 3.

[45] Falk, "Zionism and the Biology of the Jews," 590.

Eugenics and Catholicism
in Interwar Austria

Monika Löscher

Since the end of the nineteenth century, various proposals for the genetic betterment of human beings have been posited. These plans were not designed solely in Nazi Germany, but represented a worldwide trend. Eugenics, or "racial hygiene" as it was referred to in German-speaking Europe, was simultaneously a scientific and a political program, and was shaped by the interaction of science, politics and the interest of the general public. It was not only the political Right that elaborated eugenic proposals; all interwar movements across the political spectrum were influenced by this new science, although they developed different approaches. Protestant countries such as the US, some cantons of Switzerland, and Scandinavia, for example, all opted for "surgical solution"[1] (including sterilization laws); however, eugenic movements also existed in Catholic countries, and were developed on the basis of social concerns for the "hereditarily healthy" (*erbgesunden*) and the reversal of social degeneration. The *Fédération Internationale Latine des Societés d'Eugénique*, founded in Mexico in 1935, represented leading eugenicists from the Catholic countries of southern Europe and South America. According to Stefan Kühl, the work of the federation was not simply undertaken against Nazi race policy, but rather against the wider negative impact of Anglo-American eugenics. Kühl also states that the federation was closely connected to the Catholic Church.[2] Moreover, Nancy Leys Stepan suggests that Latin Hispanic eugenicists were in fact liberals and anti-clericals.[3]

The aim of this chapter is to examine the relationship between Catholicism and eugenics in interwar Austria. Unfortunately, a systematic analysis of Catholic eugenics in Austria, focusing on the interaction between science, Catholicism, politics and society, is still lacking. I shall therefore distinguish between the "official" church and its teachings (including encyclicals and pastoral letters) and the "Catholic milieu," which accepted the authority of the Catholic Church, but also expressed independent opinions. One important point is worth remembering: the ideas surveyed were not simply accepted by the Catholic Church; nor did all Catholics in Austria think in the same way. The demands of the German theologians, Joseph Mayer (1886–1967) and Hermann Muckermann (1877–1962), for instance, who argued in favor of eugenic sterilization, were certainly not representative of the majority of views within the Catholic Church, despite the fact that their work was widely read in interwar Austria.

The main contention of this chapter is that the Catholic Church had no genuine interest in eugenics, in part because most eugenic demands conflicted with the ideas about sexual morality and natural justice prescribed by Catholicism. However, the very existence of the traditional Catholic family was seen to be in danger, and therefore eugenics was advocated as a means to return to and retain Catholic values. Although the Catholic Church rejected sterilization, as well as abortion, there was no real criticism of eugenics. Thus eugenics, particularly the criteria of "superiority" and "inferiority" upon which it was based, was not officially condemned.[4] Some eugenic claims were condemned, such as the prescription of conscious "human breeding," but eugenic thinking on a general level was accepted. According to the Catholic understanding of eugenics, therefore, the principle of employability (*Arbeitsfähigkeit*)—that is, to value people by means of their capacity to work—and the diagnosis of degeneration (*Entartung*) were accepted as irrefutable arguments. Moreover, a number of characteristics in the discourse playing a role in *völkisch* eugenics were also important for the "Catholic milieu," including, most notably, the fear of a "dying nation" caused by the declining birthrate, the depiction of sexually transmitted diseases, and alcoholism as the "gravedigger" of society.

In interwar Austria, eugenics was shaped by particular political and social circumstances. Each milieu, and each political camp, formed its own conception about eugenics. Catholic eugenics attempted to refashion sexual morality as a "better form of eugenics" by justifying its claim on the basis of natural science.

The "Catholic Milieu" in Interwar Austria

In the First Austrian Republic, Catholicism constituted a defensive ideology against modernity, reflected in the anti-revolutionary and anti-Enlightenment spirit of the Catholic Church.[5] Although the percentage of Catholics dropped between 1910 and 1934 from 93.7 per cent to 90.5 per cent, Austria was considered a Catholic country. In Vienna, during the same period, the number of Catholics dropped from 87 per cent to 79 per cent.[6] In the capital, only 10 per cent of all Catholics attended Holy Mass; however, in the countryside the figure was between 80 per cent and 90 per cent.[7] According to Ernst Hanisch, being baptized and confirmed, attending Catholic ceremonies, and receiving a Catholic funeral were all typical characteristics of *Milieukatholizismus* (Catholic milieu).[8]

I do not use the expression "Catholic milieu" in the same way as Hanisch, because his definition seems too general in this context. Instead, when discussing eugenics in the "Catholic milieu," I refer to those efforts directed at saving at least sections of society from the troubles of modernity through the construction of a compact religious subculture.[9] This "Catholic milieu" was shaped by its own ways of thinking and living (*Denk- und Lebensreform*), in addition to its organized structures, symbols, images, and language.[10] Similar to socialists, Catholics harbored the same desire to create a system controlling all aspects of life. Moreover, this "Catholic milieu" was forged by a network of parties, clubs and associations, such as the Christian Social Party (*Christlichsoziale Partei*) and Catholic Action (*Katholische Aktion*), the main lay organization of the Catholic Church.[11]

Catholicism and the Natural Sciences

Referring to Germany, Michael Schwartz has suggested that the pro-eugenic engagement in the confessional milieu reflected an open-minded attitude towards modernity and the natural sciences.[12] But what was the precise nature of the relationship between Catholicism and the natural sciences? Tensions resulted from different value systems, as evidenced by the controversy arising from Charles Darwin's theory of descent during the second half of the nineteenth century. Yet what influences shaped discussions inside the Catholic Church about eugenics as an applied science?[13]

Catholic resistance to modernity reached its peak during the pontificate of Pope Pius IX (1792–1878), with the issuing of the circulars *Quanta Cura* and *Syllabus errorum* (both in 1864) that condemned modernism. His successor, Pope Pius X (1835–1914), also fought against modernism, and in the 1907 *Pascendi* encyclical denounced the new rationalistic tendencies in the exegesis of the Bible and the history of dogma. In 1910, Pius X released the *Anti-modernism Oath*, taken by priests and theologians, which was abolished by Pope Paul VI (1897–1978) only in 1967.[14]

Debates over the interactions between modernity and Catholicism focused on the relationship between the church and government on the one hand, and ecclesiastic authorities on the other. Besides, several moral and theological questions, such as "truth and history, experience and reflection, subjectivity and objectivity, ecclesial practice and theological theories" were stressed. This debate was not simply about modern, critical methods of theology, but also about finding a way to combine the new language of natural science with traditional Catholic understandings of the world.[15] A few questions need to be raised: Were there clandestine modernists in the Catholic Church during the 1920s and 1930s? How far can those theologians that supported eugenics be considered modernists? None of these questions has been sufficiently answered to date.

A simple dichotomy between "modern" and "anti-modern," or "progressive" and "un-progressive" with regard to the modern discourses of sexuality in biology, pedagogy, medicine and demography is not sufficient. Thus it is difficult to call these people "progressive," or "liberal," because their demands were in many cases much more repressive than that of the Catholic tradition.[16] How can we explain the "backwardness" of Catholic eugenics, which constantly reacted to a modernist discourse but did not act? Has Catholic eugenics been transmitted as a "better," thus "modern," form of the well-known Catholic sexual doctrine and morality?

The Public Understanding of Science

Eugenicists had interacted with the general public since the beginning of the twentieth century, when eugenics was defined as a political and scientific program. Eugenicists saw their subject as an applied science, and the public as a potential field for eugenic practice. For some time, the popularization of science was perceived as a linear process of providing information in simple language direct to the public; thus the relationship between science and society was understood as one-dimensional.[17] The relationship between science and the public is now understood as various processes of interaction and negotiation, in which value systems and politics influence the understanding of the world. In this sense, the popularization of eugenics is about much more than just the transfer of knowledge. But in the interwar period it also involved an apprehension about the existence, maintenance, and position of the dominant class or elite, alongside a belief in the social regulation regarding the offspring of the underclasses and marginal groups.

In order to understand the implementation of eugenics and the nature of the popularization of science in the "Catholic milieu," it is necessary to explore the forms of communication, the language used, and particularly the role of religious metaphors in elucidating the "new science." Ulrike Felt argued that, at the turn of the twentieth century, journalists in Vienna used quasi-religious language in their articles on science because this was the language people readily understood.[18] If we consider the publications of Catholic eugenicists, it is evident that the authors used language directly appropriated from discourses on morality and ethics. Catholic values were equated with the values of eugenicists.

In the "Catholic milieu," the catalogue of eugenic measures was reduced to a moral appeal to rationality. In this context, abortion, sterilization and family planning were seen as morally unacceptable. Enlightenment and the education of the coming generation according to eugenic principles (*eugenischen Verantwortung für das kommende Geschlecht*) were limited by the Catholic Church's perception of morality (*Sittlichkeit*).

The Encyclical of Pius XI: "On Christian Marriage"

Although individual theologians have commented on eugenics in Germany and Austria since the first decade of the twentieth century, an official statement on eugenics issued by the church hierarchy only came in 1930. The papal encyclical *Casti Connubii* (On Christian Marriage) condemned the control or regulation of human reproduction, but, at the same time, it accepted basic eugenic ideas. "On Christian Marriage"—obligatory reading for all Catholics—signified a turning point in the discourse about eugenics and was received differently in Germany and Austria. In Germany, the papal encyclical constrained the work of Joseph Mayer and Hermann Muckermann, who had argued in favor of sterilization during the 1920s. In Austria, the encyclical marked the beginning of a discussion on eugenics within wider Catholic circles. "On Christian Marriage" can also be seen as an expression of the trend prevalent at the time, in which eugenic ideas were used to propagate ideas about Catholic sexual morality. The Hungarian theologian Tihamér Tóth (1889–1939) declared that positive eugenics constituted "the best apposite apologia for the entire sex education conducted by the church."[19]

The authoritarian doctrine introduced by the papal encyclical indicated a categorical acceptance of eugenic ideas. Pope Pius XI, and particularly his advisers, the Jesuits Franz Hürth and Arthur Vermeersch, explained that the "giving of beneficial advice for the realization of a strong and healthy offspring is not against healthy rationality."[20] They favored some positive eugenic measures with respect to education and marriage counseling. However, eugenic measures were rejected at the point of contradiction with the integrity of the body; therefore, the encyclical totally condemned sterilization and abortion. Even voluntary sterilization was rejected because no one "had the right to dispose of human abilities, organs and body parts."[21] The main argument against "negative eugenic" measures was the doctrine of natural law (*Naturrecht*). Pope Pius XI later declared, during an audience with members of an international medical congress in Rome in 1935, that the prohibition of sterilization resulted from concerns for humanity rather than from religious doctrine. Natural law stressed that human action should not be restricted: the privilege of a happy family could not be enforced by eugenics.[22]

Eugenic Sterilization and the Various Interpretations of the Papal Encyclical

Initially there were two versions of "On Christian Marriage." Both texts had been approved by the pope, but differed slightly in their conception of sterilization as a possible punishment for criminals.[23] The Latin text was originally released in the organ of the Holy See (*Acta Apostolicae Sedis*)[24], while the German version was published with a Latin translation. The original Latin text condemned all forms of sterilization. Indeed, Pope Pius XI did not even agree to sterilization as a form of punishment for criminals. However, this point was not clear in the German version and the question remained unsolved in German-speaking countries.

The German theologian Joseph Mayer had intended to raise the issue of eugenic sterilization prior to the release of the encyclical, "because it has not yet been proved that government-enforced sterilization is bad and against natural law."[25] After the publication of the encyclical he did not give up the argument for the sterilization of criminals, explicitly pointing out the difference between the two versions of the encyclical.[26] Similarly, Hermann Muckermann considered "sterilization as a punishment for crime and as a prevention of further crimes by these individuals." Concerning penal castration, Muckermann believed that "the encyclical does not pass judgement on whether sterilization is a proper punishment for crime; neither does it tackle the issue of a decline in 'sexual appetite'."[27]

The Papal Encyclical "On Christian Marriage" and the "Catholic Milieu"

Although "On Christian Marriage," or at least parts of it, were published and discussed in various Catholic journals and newspapers, particularly its ramifications for contraception, eugenics as such was never properly analyzed. For example, Peter Schmitz (1891–1941), a member of the Order of the Divine Word in St. Gabriel (near Mödling) and author of numerous books on marriage and eugenics, made no mention of eugenics in his 1931 book *Die natürliche und übernatürliche Heiligkeit der Ehe* (The Natural and Supernatural Holiness of Marriage).[28] These publications and commentaries released by the dioceses for the benefit of the general public (including, most notably, the *Kirchenblatt* and the *Wiener Diözesanblatt*) did not discuss eugenics in relation to "On Christian Marriage."[29] Catholic conceptions of morality and the ideal of the family were communicated without using the word "eugenics" or "sexual hygiene."

Before the Papal Encyclical "On Christian Marriage"

With the introduction of the first sterilization laws in America in the first decade of the twentieth century, discussion about the "moral permissiveness" (*sittliche Erlaubtheit*) of sterilization started among moral-theological circles in Austria as well.[30] The moral-theologians, who had been involved in this debate, showed no interest in the science of eugenics. Despite accepting eugenic ideas, they did not focus on basic research and did not develop eugenic measures or communicate eugenic knowledge.[31] The reception of eugenic ideas before the publication of "On Christian Marriage" was visible only in the work of certain individuals. The case of Johann Ude is illustrative in this sense.

The Case of Johann Ude

Eugenics was particularly considered in moral philosophy circles and in the life-reform movement (*Lebensreform*). In 1917, the theologian and philosopher Johann Ude (1874–1965) founded the *Österreichs Völkerwacht* (Austrian People's Watch League) in Graz. As he claimed, "the Austrian Völkerwacht promotes racial hygiene and population policy in the best sense of the word and, therefore, it represents authentic patriotic work."[32] Eugenics was part of a wider program of social reform and moral regeneration. Though Ude did not advocate sterilization, he was in favor of certain forms of "negative eugenics," such as the detention of prostitutes and the "morally degenerate."[33] Central to his program was the fight against alcoholism; Ude advocated the foundation of temperance movements. Furthermore, he propagated social reform and life-reform societies to combat "immorality" (*Unsittlichkeit*).

Yet Ude remained outside the Catholic mainstream. His sermons on the dangers of venereal disease, alcoholism, as well as his political activities, met with criticism. In 1930, for example, the bishop of Graz-Seckau, Ferdinand Pawlikowski (1877–1956), forbade students of theology from attending Ude's university lectures, and the Catholic Church banned him from speaking in public.[34] In 1934, Ude had been dismissed from the University of Graz for his political criticism of the *Ständestaat,* as the Austrian fascist regime between 1934 and 1938 was known. Initially he had sympathized with National Socialism, and was a supporter of the plebiscite (*Volksabstimmung*) after the Anschluss in March 1938. However, after the November pogrom in 1938 he changed his views: in a letter addressed to Gauleiter Siegfried Uiber-reither (1908–1984) he protested against excesses against the Jews.[35] As a result, he received a *Gaulandesverweis* and had to leave Graz. He lived in a small village in Oberdonau as a pastor and teacher of religion until his death in 1965.

Medical Catholic Eugenics after "On Christian Marriage"

The Catholic medical community, which expressed an interest in the "new science" of eugenics, developed their views relatively late. Some time after eugenic demands for the sterilization of the "hereditarily inferior"—and after theological circles warned of the "death of the nation" (*Volkstod*) and constructed the image of an ideal family—Catholics in the fields of natural science and medicine organized themselves into associations devoted to propagating eugenic ideas. This happened after the publication of the papal encyclical in 1930. Within medical circles this form of Catholic eugenics not only indicated the defense and condemnation of eugenics based on the idea of selection, but simultaneously sanctioned a Catholic version of eugenics. As a result, two Catholic medical associations were founded: the *St Lukas Gilde* (St. Lucas Guild) and the *Vereinigung christlich-deutscher Ärzte* (Association of Christian German Doctors), which together endeavored to disseminate eugenic ideas to a broader Catholic audience.[36] The Association of Christian German Doctors had more economic interests, but inserted eugenic demands into their "Guidelines for Cultural-Political Tasks." Conversely, the "Lucasians" understood the transfer of knowledge about eugenics to be one of their primary tasks. These organizations worked together closely, which can be observed

by looking at members of both associations. Even though the membership of the St Lucas Guild was smaller than that of the Association of
Christian German Doctors, their activities were under close surveil-
lance by the National Socialists. This was on account of their interpretation of eugenics, which was in significant contrast to the ideology of
Erb- und Rassenflege (hereditary and racial care) advocated by the
Nazis.

St Lucas Guild

This association of Catholic physicians was founded in 1932 in Vienna.
Its aim was "to defend our [medical] profession against immoral principles, the rise of an un-Godly and wrong eugenics, against un-Catholic
thinking in medical-ethical questions."[37] Members of the St Lucas
Guild understood Catholic eugenics as an educational program, in
which medical doctors should act as mediators and guardians of both
"public health" and "hereditary health" (*Erbgesundheit*). Their health-
care and political agenda was eugenicist in nature and the Guild was
the most determined association within the Austrian Catholic environment propagating and adopting eugenic ideas. Besides scientific lectures,
the St Lucas Guild offered courses on heredity and eugenics for those
training as priests and for those working in the field of social welfare.

Did the members of the *St Lucas Guild* consider themselves to be a
loyal Catholic group, or did they aim to realize their own eugenic program in the face of the church's hostility? To understand the activity of
the St Lucas Guild it is important to mention its members' common
ground and what bound them together. Was it Catholicism, or was it
their middle-class origin and professional background? The St Lucas
Guild was composed of about 80 members, most of whom practiced in
Vienna. Its membership was more heterogeneous than expected: in
addition to Catholic members of Jewish background and conservative
Catholics, there were members of the NSDAP at a time when it was
legally prohibited in Austria (1933–1938).

These Catholic medical eugenicists generally believed in the following principles: 1) the hereditary condition of mental, as well as
venereal diseases and alcoholism; 2) that alcohol and an "excessively
active sexual life" would damage the genetic makeup of the community and result in criminality, mental disease and "moral feeblemindedness" (*moralischen Schwachsinn*); and 3) that the differential birthrate

between social classes (the lower classes were considered to have more children than the racially "worthy") posed the greatest threat to hereditary health (*Erbgesundheit*). Finally, the St Lucas Guild combined a form of hereditary medicalism with an economic emphasis on cost-benefit calculations.

Therefore, the main task of the St Lucas Guild was the propagation of a hereditary (*erbbiologisch*) Catholic sexual morality. Catholic ideas of morality were strongly advocated, along with Catholic values concerning celibacy before marriage and matrimonial fidelity. According to Peter Schmitz: "Real eugenics without high moral values is not possible over time. Modern humanity, and above all sexuality, must be rooted in moral values (*Sittlichkeit*). Nearly all eugenicists preach *Sittlichkeit*."[38] Consequently, one of the most important eugenic activities undertaken by the St Lucas Guild was marriage counseling.

The Marriage Counseling Centre in Vienna

Catholic eugenicists were fascinated by the possibility of a moral reversal in society through individual self-control. Negative eugenic measures would not be necessary in the near future on account of this voluntary control. Sterilizations, as well as any prohibitions on marriage, were anathema to most Catholics, so eugenic marriage counseling seemed to be a modest instrument for the relatively pain-free integration of eugenics into the "Catholic milieu." Furthermore, Catholics intended their marriage counseling centre to detain the secular trend of sexual counseling and birth prevention. Since June 1922 the so-called *Gesundundheitliche Beratungsstelle für Ehewerber* had existed in "Red Vienna," which was organized by the municipal health office.[39] Over the years it had expanded to cover the task of sexual counseling as well. This resulted in keen protests from Catholic middle-class circles and, after the battles of the brief civil war in February 1934, the marriage counseling centre was closed. It reopened in June 1935, although now under different premises: there was a co-operation with the *Mutterschutzwerk der Vaterländischen Front*, which would cover social questions, and with the Josefswerk, which was founded in 1909 as *Hilfsverein für christliche Ehen* (Association for Christian Marriages).[40] This association had as its main focus marriage law and spiritual guidance. The director of the marriage centre was now Albert

Niedermeyer (1886–1957), a gynecologist who had emigrated from Germany because he refused to participate in the implementation of the sterilization law.[41] He was a self-proclaimed "advocate of Christian eugenics" and a member of the St Lucas Guild.[42]

Catholic Eugenics and the *Ständestaat*

The evolution of Catholic eugenics in Austria can be divided into two parts: before and after the promulgation of the papal encyclical "On Christian Marriage." Before the papal encyclical, eugenics was reduced to a moral-theological discussion about the permissiveness of some eugenic measures such as sterilization and birth control. After the publication of "On Christian Marriage," Catholic doctors argued that there was no antithesis between eugenics and Catholicism. This form of Catholic eugenics became visible during the *Ständestaat*, mainly because alternative interpretations of eugenics were repressed by the new regime. Nevertheless, the *Ständestaat* did not support the variant of Catholic eugenics advocated by the St Lucas Guild. As a result, support for the regime was limited both financially and politically. The *Ständestaat* was concerned more with its political program than with eugenics. There was no place for eugenics, or for plans of "genetic betterment." The focus of attention thus turned to the principle of Catholicism, and topics such as sexuality and "breeding" were marginalized.

International Catholic Eugenics

In 1936, the largest gathering of Catholic eugenicists took place in the form of the Second International Congress of Catholic Physicians in Vienna, with over five hundred doctors participating in the meeting. The main themes of the session were "Medical Care in the Missions" and "Eugenics and Sterilization."[43] Vienna's Cardinal Theodor Innitzer (1878–1955), and the Austrian federal minister of social administration Josef Dobretsberger (1903–1970), presided over the meeting.[44] The Congress was opened with the following words: "The main theme of your counseling [eugenics and sterilization] has an outstanding meaning; the entire world has to show interest and has to listen attentively. The blessing or the banes of all nations depend on the solution to the big question that you have put in your program. It is you that are most qualified to work out a solution: Along with the church, you condemn every form of eugenics, which is against natural law and divine, Catholic imperatives."[45]

One of the main speakers, the head of the Catholic University in Milan, Agostino Gemelli (1878–1959), underlined the aim and purpose of this event: "The Pope expects us doctors to show him that the Catholic Church has not acted amiss when she condemned some eugenic trends. We shall propagate the doctrine among people as contained in the encyclical 'On Christian Marriage'."[46] Participants from Belgium, Germany, England, France, Holland, Italy, Poland, Switzerland, Spain, Hungary and Austria all agreed on the following: 1) The medical profession should reject sterilization as a method by which to eradicate the threat of hereditary disease; 2) Catholic physicians were warned of the "slippery slope" from eugenics to euthanasia; 3) Eugenic and penal castration were rejected outright, with the exception of castration in the cases of "psychopathic sex criminals"; 4) Positive eugenic methods should be reaffirmed, including the creation of Catholic counseling centers; and, finally, 5) the participants called for international cooperation by all Catholic medical associations in order to discuss the question of eugenics and genetics.[47]

"...On Slippery Slopes..." From Eugenics to Euthanasia

The question here is whether Catholic eugenicists were generally aware of the "slippery slope"[48] connecting eugenics with euthanasia, and whether one could have understood some of the aspects of this relationship from the dispute between Albert Niedermeyer and Julius Wagner-Jauregg (1857–1944), the latter a psychiatrist and president of the *Österreichischer Bund für Volksaufartung und Erbkunde* (Austrian League for Racial Regeneration and Heredity), who demanded eugenic sterilization.[49] In contrast, Niedermeyer was one of the Catholic eugenicists in Austria warning that negative eugenics would result in euthanasia. He had made his views public in the article "Betrachtungen über die Frage der Vernichtung 'lebensunwerten' Lebens" (Considerations about the Destruction of Life 'Unworthy of Living'), which appeared in the *Allgemeine Deutsche Hebammen Zeitung* in 1928.[50] Niedermeyer saw an intimate connection between sterilization, abortion, and the killing of the so-called "inferiors" (*Minderwertige*). Shortly before Niedermeyer's appointment as head of the communal marriage centre in Vienna, Julius Wagner-Jauregg criticized his conception of eugenics in general, and marriage counseling in particular. In 1935, Wagner-Jauregg published an article on eugenics in the January issue of the *Wiener klinischen Wochenschrift*, in which he demanded steriliza-tion.[51] Niedermeyer countered by invoking the "slippery slope" argu-ment: it would not be possible to prevent compulsory sterilization after the introduction of voluntary sterilization.[52] However, members of the St Lucas Guild did not share his ideas. As Niedermeyer remarked, the majority of them were "fascinated by the ideology of a selective eugenics, as represented by Muckermann. They saw in my resistance against sterilization nothing more than an exaggerated Catholic point of view; it was said that I wanted to be more Catholic than the pope."[53]

Conclusions

The 1930 papal encyclical *Casti Connubii* condemned "negative eugen-ics" but was not opposed to "positive eugenics." Reducing eugenics to a catalogue of measures for eugenic enlightenment and the responsibil-ity of promoting the "hereditary health" of society was therefore char-acteristic of Catholic eugenics. These measures converged with the traditional Christian emphasis on morality and *Sittlichkeit*. Moreover, the idea of self-control was attractive to Catholic eugenicists. In their view, negative eugenics would not be necessary if voluntary self-con-trol were to succeed. Eugenic marriage counseling was seen as a way to avoid sterilization and restrictions on marriage, while simultaneous-ly integrating eugenics into the "Catholic milieu." Positive eugenics, including increasing the birthrate of the biologically "worthy," and social eugenic instruction, was seen as likely to be accepted by the large "Catholic milieu," including both the Catholic hierarchy and the-ologians.

Catholic eugenicists demanded a moral reformation of society in the name of eugenic responsibility, the sacredness (*Gottbezogenheit*) of marriage and sexual life, Christian morality, and the ideal of a Christian family. Catholic eugenics more or less added up to present-ing the well-known Catholic ethics of sexuality as the "better eugenics"—now with the legitimating of natural science. As one contributor to the Catholic journal *L'Osservatore Romano* proclaimed in 1933: "If someone lives an authentic Christian life, legislative measures for eugenic aims are not needed. The support of Catholic morality is enough to eliminate the harshest of hereditary burdens: hereditary syphilis will disappear due to the absolute prohibition of intercourse outside of Christian marriage, and hereditary alcoholism because of the prohibition of drunkenness and immoderateness."[54]

Endnotes:

[1] Géza von Hoffmann, *Die Rassenhygiene in den Vereinigten Staaten von Nordamerika* (Munich: Lehmanns Verlag, 1913); Nils Roll-Hansen, "Geneticists and the Eugenics Movement in Scandinavia," *British Journal for the History of Science* 22, 74 (1989), 335–346; and Philipp R. Reilly, *The Surgical Solution: A History of Involuntary Sterilization in the United States* (Baltimore: John Hopkins Unversity Press, 1991).

[2] Stefan Kühl, *Die Internationale der Rassisten. Aufstieg und Niedergang der internationalen Bewegung für Eugenik und Rassenhygiene im 20. Jahrhundert* (Frankfurt-am-Main: Campus Verlag, 1997), 144.

[3] Nancy L. Stepan, *The Hour of Eugenics: Race, Gender, Nation in Latin America* (Ithaca: Cornell University Press, 1991).

[4] Michael Schwartz stresses that the Catholic Church cannot be viewed any longer as an "anti-eugenic fortress." See Michael Schwartz, "Konfessionelle Milieus und Weimarer Eugenik," *Historische Zeitschrift* 261, 2 (1995), 403–448.

[5] Urs Altermatt, *Katholizismus und Moderne* (Zürich: Benziger, 1989); and Urs Altermatt, "Katholizismus als sozialdisziplinierte Lebenswelt Historicum," *Historicum. Zeitschrift für Geschichte* 15, 2 (1995–1996), 11–14.

[6] *Statistische Nachrichten* 13 (1935), 118. Quoted in Ernst Hanisch, *Die Ideologie des Politischen Katholizismus* (Vienna: Geyer, 1977), 55.

[7] Hanisch, *Politischer Katholizismus*, 56.

[8] Hanisch, *Politischer Katholizismus*, 56–57.

[9] See www.diozese-linz.or.at/pastoralamt/weltanschauungsfragen (Accessed 21 October 2004).

[10] Doris Kaufmann, *Katholisches Milieu in Münster 192- 1933* (Düsseldorf: Schwann, 1984), 8.

[11] See Alfred Diamant, *Die österreichischen Katholiken und die Erste Republik* (Vienna: Verlag der Wiener Volksbuchhandlung, 1965), 73.

[12] Schwartz, "Konfessionelle Milieus und Weimarer Eugenik," 446.

[13] Franz Padinger, "Zum Verständnis des Konfliktes zwischen Gaube und Wissenschaft im Modernismusstreit," in Erika Weinzierl, ed., *Der Modernismus. Beiträge zu seiner Erforschung* (Graz: Styria, 1974), 43–56, 43.

[14] Anton Michelitsch, *Der biblisch-dogmatische 'Syllabus' Pius X. samt der Enzyklika gegen den Modernismus, erklärt von Dr. Anton Michelitsch* (Graz: Styria, 1908); Peter Neuner, "'Modernismus' und kirchliches Lehramt. Bedeutung und Folgen der Modernismus-Enzyklika Pius' X," *Stimmen der Zeit* 97, 190 (1972), 249–262. See also Erika Weinzierl, *Ecclesia Semper Reformanda. Beiträge zur österreichischen Kirchengeschichte im 19. und 20. Jahrhundert* (Vienna: Geyer 1985).

[15] Otto Weiß, "'Sicut mortui. Et ecce vivimus'. Überlegungen zur heutigen Modernismusforschung," in Hubert Wolf, ed., *Antimodernismus und Modernismus in der katholischen Kirche* (Paderborn: Schöningh, 1998), 42–63.

[16] Friedrich Wilhelm. "Moderne Modernisierer, modernitätskritische Traditioanlisten oder reaktionäre Modernisten? Kritische Erwägungen von Deutungs-

mustern der Modernismusforschung," in Wolf, ed., *Antimodernismus und Modernismus in der katholischen Kirche*, 85.

[17] Constantin Goschler, ed., *Wissenschaft und Öffentlichkeit in Berlin, 1870–1930* (Stuttgart: Steiner, 2000), 9.

[18] See Ulrike Felt, "'Öffentliche' Wissenschaft. Zur Beziehung von Naturwissenschaften und Gesellschaft in Wien von der Jahrhundertwende bis zum Ende der Ersten Republik," *Österreichische Zeitschrift für Geschichtswissenschaft* 7, 1 (1996), 61.

[19] Tihamer Tóth, *Die Eugenik vom katholischen Standpunkt* (Vienna: Fürlinger, 1932), 64.

[20] *'Die Eheenzykliken des Hl. Vaters Pius XI, Casti connubii und Divinus illius magistri über Ehe und Erziehung. Authentische deutsche Übersetzung* (Innsbruck: Tyrolia, 1936), 25.

[21] Hermann Pfatschbacher, *Eugenische Ehehindernisse? Eine kirchenrechtliche Studie* (Vienna: Mayer, 1933), 80.

[22] Letter from Hans Harmsen to Ernst Rüdin. Quoted in Jochen-Christoph Kaiser, Kurt Nowak, Michael Schwartz, eds., *Eugenik, Sterilisation, 'Euthanasie'. Politische Biologie in Deutschland 1895–1945. Eine Dokumentation* (Berlin: Buchverlag Union, 1992), xxiii.

[23] Ingrid Richter, *Katholizismus und Eugenik in der Weimarer Republik und im Dritten Reich: zwischen Sittlichkeitsreform und Rassenhygiene* (Paderborn: Schöningh, 2001), 267–279.

[24] (Anonymous), "Casti Connubii," *Acta Apostolicae Sedis* 22 (1930), 539–592.

[25] Joseph Mayer, "Verlangt die Moral eine uferlose Fortzeugung Schwachsinniger?" *Das Neue Reich* 1 (1930), 925–927.

[26] Joseph Mayer, "Die wichtigsten Entscheidungen der Enzyklika 'Casti Connubii'," *Theologie und Glaube* 23 (1931), 572–586.

[27] Hermann Muckermann, *Eugenik und Katholizismus* (Berlin: Metzner, 1933), 34.

[28] Peter P. Schmitz, *Die natürliche und übernatürliche Heiligkeit der Ehe. Zur Ehe-Enzyklika unseres Heiligen Vaters Papst Pius XI* (St. Gabriel: Missionsbuchhandlung, 1931).

[29] Josef Dillersberger, "Ein Brief, eine Zeitung und anderes über die Enzyklika," *Katholische Kirchenzeitung* 71 (1931), 34; and Josef Dillersberger, "Der Papst mit dem Herzen von Stein," *Katholische Kirchenzeitung* 71 (1931), 54.

[30] Albert S. J. Schmitt, "Vasectomia. Eine neue Operation und ihre Erlaubtheit." *Zeitschrift für katholische Theologie* 35 (1911), 66–78.

[31] See Albert Schmitt's review of Joseph Mayer, "Gesetzliche Unfruchtbarmachung Geisteskranker," *Zeitschrift für katholische Theologie* 51 (1927), 421–423; Josef Grosam, "Lehren der Enzyklika 'Casti Connubii' über Sterilisation und Eugenik," *Theologisch Praktische Quartalschrift* 84 (1931), 782–788.

[32] Johann Ude, *Der moralische Schwachsinn. Für Volkssittlichkeit* (Graz: Eigenverlag, 1918), 37.

[33] Johann Ude, *Niedergang oder Aufstieg. Das rassenhygienische Problem* (Graz: Eigenverlag, n.d.).

34 Käthe Moritz, *Sein und Wirken des großen Friedensarbeiters und Lebensreformers Johannes Ude* (Salzburg: Friedens-Verlag, 1960), 15.

35 Christof Karner, *Wir wissen um zu wollen. Die konservativ-katholische Interpretationsvariante der Freiwirtschaftslehre am Beispiel des Lebensreformprogramms von Johannes Ude* (Unpublished Ph.D. thesis, University of Vienna, 2001), 115.

36 Inventory "St. Lukas Gilde," Wiener Stadt- und Landesarchiv R.17/Reg.117, SchA/32/342; Inventory "St. Lukas/Ordinariatsakten/Vereine," Wiener Diözesanarchiv; See also the journals *St. Lukas* 1 (1933) and 5 (1937) and *Mitteilungen der Vereinigung Christlich-Deutscher Ärzte Österreichs* 1 (1935) and 3 (1937).

37 Carla Zawisch, "Was wir sein, was wir tun wollen," *St. Lukas* 1 (1933), 5.

38 Peter Schmitz, *Die modernen eugenischen Bestrebungen in theologisch-soziologischer Bedeutung* (Innsbruck: Tyrolia, 1933), 4.

39 Julius Tandler, *Wohltätigkeit oder Fürsorge?* (Vienna: Verlag der Organisation Wien der sozialdemokratischen Partei, 1925); and Karl Kautsky, "Eheberatung," *Breithscaft* 3, 8 (1923), 14.

40 *Wien im Aufbau. Das Wohlfahrtswesen der Stadt Wien* (Vienna: Magistrat, 1937), 28.

41 Albert Niedermeyer, "Um die rechte Eheberatung. Die Eheberatung als Werkzeug zerstörender Kräfte und im Dienste aufbauender Ideen," *Schönere Zukunft* 10, 3 (1935), 75–79; 756.

42 Franz Sturm, "Familienpolitische Probleme in Österreich," *Schönere Zukunft* 11, 18 (1936), 446.

43 The abstracts of the conference are published in *St. Lukas* 4, 3 (1936), 73–177.

44 (Anonymous), "Ein Blick hinüber," *Volk und Rasse* 1, 6 (1936), 263.

45 (Anonymous), "Aus dem Leben erzählt. Der II. Internationale Kongress Katholischer Ärzte," *Arzt und Christ* 3 (1958), 184.

46 (Anonymous), "Kongreßbericht," *St. Lukas* 4, 3 (1936), 73–76.

47 (Anonymous), "Ergebnisse und Beschlüsse," *St. Lukas* 4, 3 (1936), 173–177.

48 (Anonymous), "Streiflichter vom Erbbiologischen-Rassenhygienischen Schulungskurs für Psychiater in München, Jänner 1934," *Katholische Kirchenzeitung* 74 (1934), 82.

49 Julius Wagner-Jauregg, "Über Eugenik," *Wiener Klinische Wochenschrift* 94, 1 (1931), 1–7.

50 Albert Niedermeyer, "Betrachtungen über die Frage der Vernichtung 'lebensunwerten' Lebens," *Allgemeine Deutsche Hebammenzeitung, Zeitschrift zur Förderung des Hebammenwesens* 43 (1928), 73.

51 Julius Wagner-Jauregg, "Zeitgemäße Eugenik," *Wiener Klinische Wochenschrift* 98 (1935), 1.

52 Albert Niedermeyer, *Wahn, Wissenschaft und Wahrheit. Lebenserinnerungen eines Arztes* (Innsbruck: Tyrolia, 1956), 326.

53 Niedermeyer, *Wahn, Wissenschaft und Wahrheit*, 332.

54 *L'Osservatore Romano* (14 August 1933). Quoted in Schmitz, *Die modernen eugenischen Bestrebungen in theologisch-soziologischer Bedeutung*, 47.

From Welfare to Selection:
Vienna's Public Health Office and the Implementation of Racial Hygiene Policies under the Nazi Regime*

Herwig Czech

Eugenics had been always characterized by a discrepancy between the utopian character of its ambitions and the actual possibilities for the realization of its projects. This was to change when National Socialism came to power in Germany. The phantasm of a "national body" (*Volkskörper*), which would be racially homogeneous (*rassenrein*) and free from hereditary pathologies (*erbgesund*), was one of the key elements of Nazi ideology and politics.

The consequences of this ideology are well known: tens of thousands of inmates in German psychiatric institutions were murdered in the course of several "euthanasia" campaigns; around 400,000 persons were subject to forced sterilization between 1933 and 1945. The whole health-care system was radically restructured in order to impose the systematic discrimination of individuals according to their "worth" for the *Volksgemeinschaft*. Entrusted with the double competence of medical science and public authority, the "public health offices" (*Gesundheitsämter*) of the German Reich assumed a leading role of this major project of population policy. Unlike the health organizations of the NSDAP, only the public health administration possessed the structural requisites to put into practice the politics of racial hygiene with the desired efficiency and on a broad scale. The basis for this was the 1934 "Law for the Standardization of Health Care" (*Gesetz zur Vereinheitlichung des Gesundheitswesens*). Germany obtained thereby, and for the first time, a unified, centrally regulated system of public health offices and specialist doctors (*Amtsärzte*), which were endowed with numerous competencies. Racial hygiene, institutionalized under the label "hereditary and racial care" (*Erb- und Rassenpflege*), became a field of high priority for the health authorities. In this fashion, an effective instrument of bio-politics was created by the Nazi regime, allowing for

the permanent and comprehensive medical policing of the whole population.[1]

In this chapter I shall provide an overview of the implementation of
racial hygiene policies in Vienna, particularly between 1938 and 1940.
In this short period, significant structural changes took place in the
sphere of health care, before the war increasingly limited the scope for
further change. The Department for Hereditary and Racial Care
(*Abteilung Erb- und Rassenpflege*), founded in October 1938, played a
key role in the implementation of racial hygiene policies in Vienna's
health-care system. Furthermore, it also served as the city's central
agency in regard to racial measures relating to Jews, Roma, Sinti and
others.[2]

Following the *Anschluss* in March 1938, Vienna, thereafter the second largest city of the German Reich, attained a special position with
regard to racial policy. One of the largest German-speaking Jewish
communities resided in Vienna, and according to Nazi propaganda, the
city's population contained a high percentage of biologically "inferior
people"—an assumption based on Vienna's close proximity to Eastern
and Southeast Europe and the tradition of immigration from these
regions. Unlike the German case, Vienna experienced the rapid implementation of radical measures in the health-care system during a relatively short period, with the main organizational changes taking place
between 1938 and 1940. In this way, Vienna could soon claim to lead
the way in many areas of Nazi health policies, with public health offices
often playing a decisive role.

Eugenics and Public Health Care before 1938

Upon examination of the rapid, and practically unimpeded, implementation of Nazi racial hygiene after the annexation of Austria, it becomes
evident that the question of the structural preconditions underpinning
this process must also be studied. Did Nazi health reform really constitute a radical change in the practice of Austrian public health policies,
or did it rather continue tendencies that existed before? Time and again
this question leads to controversies in Austria, often focusing on the
person of Julius Tandler (1869–1936), professor of anatomy at the University of Vienna.[3] Until his dismissal by the Austro-fascists in 1934,
Tandler was the Viennese city counselor responsible for the health and

welfare system. He was one of the central figures associated with the socio-political reform movement commonly referred to as "Red Vienna." Tandler employed eugenic arguments in order to justify his ideas of a modern social policy; for example, he defined social policy as the rational management of human capital and differentiated between productive and unproductive costs. Thus, investment in youth welfare and health care were considered productive costs par excellence. Unproductive costs, on the other hand, consisted of public expenditure on people who could no longer be expected to perform productive tasks. On several occasions, Tandler warned of the possibility that "minus variants" or "inferiors" (*Minderwertige*) could numerically become dominant, thus having a negative impact on the genetic heritage of Austrian society. However, he expressed much more reservation when it came to concrete measures, such as forced sterilization.[4]

Taking into account not only Tandler's rhetoric but also the social reforms implemented under his responsibility, the difference between his approach and the racial hygienic project associated with National Socialism becomes clear. The actual implementation of eugenic principles in the field of health care and welfare in Vienna before 1934 was limited to a small number of projects, the most important being the Office of Marriage Counseling (*Eheberatungsstelle*), founded in 1922 and directed by Karl Kautsky, Jr. (1892–1978). The institution was created to advise couples on personal health care and eugenics. Attendance was voluntary. Although the counseling office was celebrated as a pioneering project, attendance rates remained low and its success questionable. It was closed under the clerical Austro-fascist regime, reopening again in 1935, but this time streamlining its message according to Catholic principles.[5]

Certain eugenic principles were widely accepted by academics and health-care professionals. The fact that, after 1924, public health officers were trained in the Department of Hygiene at the University of Vienna under the supervision of Heinrich Reichel (1876–1943), one of the most prominent promoters of racial hygiene policies in Austria, is a prime example of the widespread acceptance of these ideas.[6]

Regarding public health care for alcoholics, the concern over racial hygiene played an important role, as evidenced in the writings of Ernst Gabriel (1899–?), director of Vienna's sanatorium for alcoholics (*Trinkerheilstätte*), founded in 1922 as part of the largest psychiatric institution in Vienna, the *Steinhof*. The sanatorium only admitted men (and a small number of women) capable of reproduction, and only those who had a fair chance of rehabilitation. Hopeless cases remained in standard psychiatric wards. For those, Gabriel demanded permanent detention in labor camps so as to relieve the tax burden on the general public.[7]

A more thorough analysis of the Viennese health-care system between 1918 and 1938 is necessary in order to respond convincingly to the question of continuity during the periods before and after 1938. In any case, numerous organizational changes introduced by the Nazis clearly indicated how limited such continuities were at the institutional level—after all, the *Abteilung Erb- und Rassenpflege* was an entirely new body, with no precedent in public health administration. After its establishment, the principles of racial hygiene could be implemented on a large scale, as never previously attempted.

The Nazification of the Viennese Health Care System

The psychiatrist and neurologist Otto Reisch (1891–1977) played a key role in the Nazi reorganization of the Viennese health-care system after 1938. Reisch was in charge of Vienna's health and welfare administration between March 1938 and March 1940. He graduated from Innsbruck University, where he became a leading member of the then illegal section of the NSDAP in Tyrol during 1934/1935. On 15 March 1938, Reisch embarked on his Nazi career in Vienna, beginning with the removal of Jewish or politically undesirable doctors from their positions, replacing them with NSDAP members.[8] The next day, Hitler ordered all functionaries to take an oath of loyalty to the new regime. Jews (according to definitions laid down by the Nuremberg Laws) were excluded from taking the oath, and therefore lost their jobs automatically.

It is not known precisely how many people lost their jobs in the Viennese health care system because of the Nazi anti-Semitic policies. At the University of Vienna, approximately 51 per cent of staff members from the Faculty of Medicine were dismissed after 1938. Michael Hubenstorf estimates that the total number of Viennese physicians was reduced from 4,900 to 1,700 between March and October 1938, with perhaps 3,000 physicians escaping abroad.[9]

Medicine and the Holocaust

With the progressive exclusion of the Jews from all quarters of public life, especially given the implementation of an ever more rigid policy of segregation, corresponding measures were carried out as regards health care and social welfare. The public health office enforced segregation via a special Public Health Office for Jews (*Amtsärztliche Dienststelle für Juden*), with the purpose of excluding all Jews from public health facilities.[10] Consequently, Jewish recipients of welfare policies became increasingly numerous due to a combination of Aryanization (the confiscation of their wealth and businesses), forced unemployment, and the emigration of economically active family members. As Jews were increasingly excluded from public institutions, the Jewish Community Representation Council was responsible for looking after them. The Council came close to bankruptcy following increasing costs—at one point, Adolf Eichmann (1906–1962) eventually had to contribute funds to keep the council afloat.[11] A large proportion of the Council's budget was devoted to provision for old people; however, the available funds were far from being sufficient to help all those in need. The city of Vienna was directly responsible for overseeing the removal of the aged from communal nursing homes and assembling them in preparation for their deportation—a decrepit hostel for the homeless was used as an "isolation home" (*Absonderungsheim*) prior to deportation.[12]

In July 1940, Max Gundel (1901–1949), the city councilor in charge of health and welfare, wrote to the mayor, Philipp Wilhelm Jung (1884–1965), about how to accelerate the deportation of the Viennese Jews. He had learned that the deportation of 450,000 Jews from the Reich to Poland was planned. According to his calculations, the Jews were straining the Viennese welfare budget by more than a million Reichsmark. Gundel suggested the mayor consult with the *Reichsstatthalter* and *Gauleiter* of Vienna, Baldur von Schirach (1907–1974), in order to speed up the deportation of the Jews to Poland.[13] A few months later, in October 1940, Schirach presented the same request to Adolf Hitler. Schirach justified his request with reference to the housing shortage in Vienna. On 3 December 1940, Schirach received Hitler's approval for the deportation of 60,000 Jews from Vienna to Poland. In February 1941, the first deportation trains since the interruption of the *Nisko-Aktion* of October 1939, headed to Poland. Within less than three years, approximately 65,000 people were deported from Vienna. With a few exceptions, they all died in Nazi ghettos and extermination camps.[14]

This example shows how those responsible in the health and welfare systems embraced the anti-Jewish measures; that is, the persecution and later extermination of the Jewish population. Deportation was seen as an opportunity to increase the benefits for the non-Jewish population, as long as they were classified as "hereditarily healthy" (*erbgesund*) and "worthy of assistance" (*förderungswürdig*).

The Institutional Prerequisites of Nazi Racial Hygienic Policy in Vienna

Nazi reform of the health-care system was based on the "Law for the Standardization of Health Care" (*Gesetz zur Vereinheitlichung des Gesundheitswesens*), effective in Austria (*Ostmark*) as of 1 December 1938. To this day, the law remains the basis for the organization of the Austrian public health service.[15] Since Vienna had a well-developed system of health care in 1938, the city sought to obtain approbation for its public health offices (*Gesundheitsämter*) from the German Ministry of the Interior. This involved extensive organizational changes. Two experts from Germany were brought to Vienna in the autumn of 1938; SS doctors Arend Lang (1909–1981) and Hermann Vellguth (1906–?), both charged with the specific task of setting up the "hereditary and racial care" scheme (*Erb- und Rassenpflege*). During his time in Vienna, between October 1938 and May 1941, Lang held a predominant position in the NSDAP's Office for Racial Policy (*Rassenpolitisches Amt*). In May 1941, he was promoted to a post concerning racial policy at the Ministry of the Interior.[16] Vellguth was in charge of the reorganization of the Viennese public health office as a whole from December 1938, but his work also focused on questions of racial hygiene. From February 1940, Vellguth was the director of Vienna's main public health office, and from 1941 he additionally ran the Office for Racial Policy.[17] In the two and a half years following the annexation of Austria, Vellguth oversaw the extension and centralization of the Viennese public health system, starting with the creation of a public health office in each district, directed by a full-time public health officer representing the interests of the Nazi "health leadership" (*Gesundheitsführung*).

Besides the already mentioned *Erb- und Rassenpflege*, public health offices were also responsible for "health protection" (*Gesundheits-schutz*) and "health care" (*Gesundheitsfürsorge*). "Health protection" corresponded to the classical tasks of the so-called health police (*Gesund-heitspolizei*), whose mission was to protect the health of the population against infectious diseases. However, problems arising during wartime soon dominated the work of these administrative organs. The struggle against venereal diseases and tuberculosis resulted in the establishment of a repressive, police-like apparatus within the framework of the public health services.[18] In this way, personal liberties were sacrificed for the sake of the *Volksgemeinschaft*.

In the field of health care, high priority was accorded to pregnant women, toddlers and infants, in line with the regime's pro-natalist policy. All benefits arising from the system of health care were conditional upon the positive evaluation of "hereditary value," thus reflecting the dominance of the racial hygiene paradigm.

The pervasiveness of Nazi medical ideology—especially through the preoccupation with hereditary and racial care—ultimately determined the direction of the Viennese system of health care, towards systematic discrimination and the persecution of alleged "inferiors" (*Min-derwertige*). A handbook issued specially for the public health service in 1939 (*Der öffentliche Gesundheitsdienst*) reflected this trend: "every measure, undertaken in all areas, must be examined from the point of view of population policy, hereditary and racial care."[19]

The Public Health Office and the "Euthanasia Program" ("T4")

Whereas "infant euthanasia" was carried out in facilities provided by the city, the case of the "T4" program (the deportation and killing of psychiatric patients at special extermination facilities, codenamed after the address of its Berlin headquarters at Tiergarten straße 4) was different. Nevertheless, the deportation of thousands of patients from Vienna's *Steinhof* (around 3,200 persons were transported to be killed in early summer 1940[20]) was made possible by the co-operation of doctors, nurses and administrators. Erwin Jekelius (1905–1952), director of the "death clinic," *Spiegelgrund*, played a key role in the selection of people for the T4 program; he also co-ordinated deportations from Vienna.[21] Jekelius negotiated the financial terms of the extermination campaign; the municipal administration was obliged to pay ten Reichs-marks per person to cover the cost of deportation. The extermination process itself was financed by the secret "T4" organization, in this case acting as the *Reichsarbeitsgemeinschaft Heil- und Pflegeanstalten* (Reich Working Group on Psychiatric Hospitals).[22]

The Project of the "Hereditary Inventory"

The most important practical prerequisite for the implementation of Nazi "hereditary health policy" (*Erbgesundheitspolitik*) was the cata-loguing the population according to racial hygiene criteria (*erbbiolo-gische Bestandsaufnahme*). The collation of medically and socially "incriminating" information meant, in practice, the supervision of the population along bio-political lines. In compiling information for this inventory, records from numerous institutions were accessed. In 120,000 cases, for example, files were drawn from the archive of the Steinhof.[23] Other sources included police records on around 60,000 mentally disabled persons and "psychopaths," 40,000 alcoholics, and 60,000 prostitutes. The Youth Welfare Office (*Jugendamt*) supplied information on approximately 40,000 "difficult and psychopathic chil-dren from anti-social families"; and a dermatological clinic provided information on 8,000 patients with sexually transmitted diseases. The health authorities were involved in anti-Jewish policies as well: the Police Passport Agency (Polizei-Passamt) and the NSDAP relayed information on 6,000 Jews and Jewish *Mischlinge*.[24] This last point indicates that the health authorities also played a role in anti-Semitic policies, namely in the enforcement of the so-called "Blood Protection Law" (*Blutschutzgesetz*), one of the Nuremberg Laws, as well as in the anthropological examination and assessment of so-called *Mischlinge*.

The "hereditary inventory" thus constituted a kind of register of the *Volkskörper*. Physical, psychological and social deviations were for-mulated as medical diagnoses, suggesting a "hereditary burden" (*erb-liche Belastung*) of single individuals and entire families. This served as a basis for systematic discrimination and persecution. Yet hereditary pathology played only a marginal role in the "inventory," as actual hereditary diseases were only rarely recorded. The Nazi authorities were completely aware of this fact. In 1940, the Interior Ministry of the German Reich released guidelines on the evaluation of hereditary health (*Richtlinien zur Beurteilung der Erbgesundheit*), which outlined the socioeconomic components of hereditary and racial health care: "In the selection according to eugenic criteria, evaluation of personal productivity must be of decisive importance."[25] The pathologization of human existence enabled health authorities in the Third Reich to expand their interventionist social role beyond all known limits, reflected most notably in the extensive range of activities undertaken by the Department for Hereditary and Racial Care.

The compilation of the "hereditary inventory" required a large number of personnel. At least seventy people were working on the central register alone. The register was progressively extended through the inclusion of new data from reports submitted by public health offices and other health-care facilities. According to one document, dating from March 1941—that is, two years after the start date of the project—the central register contained data on 500,000 persons, then already constituting one of the most exhaustive hereditary inventories in the Reich.[26] The highest peak came in March 1944, with 767,000 persons registered.[27]

Systematic registration of the population commenced at birth. In May 1939, a new method for the registration of births, designed according to principles of racial hygiene, was introduced.[28] As of 1 July 1939, all children born in Vienna were registered in special files,[29] with the use of birth records in the compilation of the "hereditary inventory" from 1 January 1940.[30] On several occasions officials in the Viennese health administration claimed that they had at their disposal one of the most extensive hereditary inventories in the Reich. The fragmentary knowledge about the implementation of bio-politics in the Third Reich makes it difficult to verify this claim; however, in a letter to Heinrich Himmler, Leonardo Conti (1900–1945) emphasized the achievements of the "hereditary inventory" in the Reich, noting that "ten million individual files have been compiled by the public health offices, which today contain most of the negative and incriminatory facts on every German."[31] This meant that an average of 12 per cent of the population of the German Reich was recorded on the register. Existing research on the "hereditary inventory" in different regions of the Reich shows that, with the registration of approximately 770,000 persons, Vienna was one of the leading cities in this field. Even if similar numbers were recorded in other regions, it was exceptional that this work was completed within such a short time, between 1939 and 1941.[32]

Population Policy, Pro-Natalism and "Positive Eugenics"

Traditionally, the propagators of eugenics made a distinction between "negative" and "positive" measures. Whereas negative eugenics aimed at the prohibition of reproduction of "inferior" individuals, positive eugenics encouraged the proliferation of those deemed superior (*Höherwertige*). In this context, the Nazi regime was primarily interested in measures that would have a tangible, quantifiable impact on the whole population.

Therefore, the Nazis promoted childbirth on a general level, but combined this general policy with methods by which to exclude a minority of "undesirables" from reproduction for reasons of health or race. As has been suggested, the system of health care played a leading role in the implementation of policies according to these precepts.

The extension of health care for pregnant women, toddlers and infants should be viewed against the background of dramatic demographic developments: between 1937 and 1940, the number of births rose by a factor of three. In 1937, a total of 10,032 babies were born in Vienna; in 1938 the figure was 12,645, followed by 28,645 in 1939. The year 1940 witnessed a peak, with the registration of 30,330 newborn children. From 1941 onwards the number dropped due to the intensification of the war.[33]

This rapid increase in the birthrate was mainly due to the economic upswing, for after years of depression many postponed marriages now took place. This did not necessarily reflect an unconditional vote of confidence in the new regime, as those promoting Nazi health policies claimed. Nevertheless, this trend clearly indicates a certain level of public optimism regarding the new political and social conditions of the Nazi era. Yet even the peak birthrate of 1940 was not sufficient in the eyes of Nazi technocrats, who frequently wrote about their fear of racial extinction (*Volkstod*). They claimed that, in order to avoid this fate, an "adjusted birthrate" of 21 births per 1,000 inhabitants was necessary.[34] "Adjusted" meant taking into account measures that prevented the birth of "undesirable offspring" by means of forced sterilization, abortion, the interdiction of certain marriages, and murder.[35]

The Department for Hereditary and Racial Care had obliged to ensure that all those defined as "inferior" were excluded from receiving health care and welfare benefits, which included marital loans, allowances for children, and a number of other benefits. For example, applications for marital loans submitted by approximately 43,000 couples in 1938 were re-evaluated. Of these, 682 were not only refused the loans, but the authorities forbade them from being married. The most frequent reasons for such actions included "feeblemindedness," "schizophrenia," and "infertility" in individual candidates or their families.[36] In the case of adoption or the fostering of children, racial hygienists examined both children and prospective parents in order to determine their "hereditary value." The hereditary inventory was also consulted in assessing applications for communal housing, in order to exclude those deemed "unworthy."

The Supervision of Reproduction: "Marital Health" and the "Protection of Blood"

The "Law for the Protection of the Hereditary Health of the German People" (*Gesetz zum Schutz der Erbgesundheit des Deutschen Volkes*), also known as the "Marital Health Law" (*Ehegesundheitsgesetz*), came into effect in 1935 at the same time as the "Blood Protection Law" (*Blutschutzgesetz*), one of the first Nuremberg Laws. Both pieces of legislation were designed to limit marriages and undesirable sexual relations. Whereas anti-Semitic and racist thinking informed the "Blood Protection Law," the "Marital Health Law" was enforced for eugenic reasons.[37] In the *Ostmark* the latter came into effect on 1 January 1940, together with the "Law for the Prevention of Hereditarily Diseased Offspring" (*Gesetz zur Verhütung erbkranken Nachwuchses,* GzVeN).[38] These laws gave jurisdiction to public health offices, in collaboration with the civil registry offices, over all intended marriages; from this point forth a special certificate of "medical fitness" (*Ehetauglichkeitszeugnis*) was a prerequisite for marriage. All prospective couples were then screened according to the information collected in the hereditary registry. In case of any doubt, couples were required to undergo a medical examination. Certificates were conferred upon prospective couples if the potential offspring was considered genetically desirable. Legal impediments to marriage were seen in contagious illnesses (especially tuberculosis and venereal diseases), mental incapacitation, hereditary diseases and psychological disorders—all of which were supposed to pose a threat to the well-being of the *Volksgemeinschaft*.[39]

The Public Health Office was additionally responsible for enforcing the prohibition of "mixed marriages" between Jews and people of "German blood." In unclear cases, a special anthropological unit within the Department for Hereditary and Racial Care was asked to draft an analysis on the "racial qualities" of those concerned.[40] These laws were also used to target gypsies (Roma and Sinti), blacks and others.[41]

Forced Sterilization and Abortion

On the basis of the "Law for the Prevention of Hereditarily Diseased Offspring," the Reich health authorities commenced a policy of sterilization against persons allegedly suffering from hereditary diseases. The following were posited as indicators justifying sterilization: "congenital feeblemindedness, schizophrenia, manic-depressive disorders, hereditary epilepsy, Huntington's chorea, hereditary blindness or deafness, severe physical deformations and acute alcoholism." Revealingly, the most frequent diagnosis was "feeblemindedness," a term so imprecise that a whole range of "socially undesirable behavior" could be subsumed under it.

On the basis of the new "hereditary health procedure" (*Erbgesundheitsverfahren*), a key role was assumed by public health officials. They tracked down people according to the stipulations of the law, investigated them, and issued reports to the Hereditary Health Courts (*Erbgesundheitsgerichte*). As medical assessors for the special courts, they had direct influence on the hearings. If a legally binding decision for sterilization was made, public health officials were responsible for arranging and overseeing the process.[42] In 1935, an amendment to the sterilization law legalized eugenic abortions in combination with forced sterilization.[43] According to Gisela Bock, between 1934 and 1939 roughly 300,000 individuals were sterilized in Germany. A further 60,000 cases of forced sterilization followed during the war.[44]

When the law was introduced in Austria on 1 January 1940, sterilizations had already been limited to those cases that represented an "extraordinary reproductive danger" (*besonders große Fortpflanzungsgefahr*). Nevertheless, a total of around 1,900 sterilization proceedings were brought before Viennese courts during the war. However, the number of sterilizations that were actually carried out remains unknown.[45] Wolfgang Neugebauer estimates a total of 6,000 cases of sterilization in Austria between 1940 and 1945.[46]

The Extermination of Handicapped Children

Preparations for the systematic registration and extermination of hand-icapped children in the German Reich began in the spring of 1939. The Office of the Führer (*Kanzlei des Führers*) created a special cover organization for this purpose, the Reich Committee for the Scientific Registration of Serious Hereditary Diseases (*Reichsausschuss zur wis-senschaftlichen Erfassung erb- und anlagebedingter schwerer Leiden*). As with forced sterilizations, public health officers assumed a central role in what is commonly referred to as "infant euthanasia." A secret order of the Reich Ministry of the Interior, issued on 18 August 1939, laid the foundations for this policy. Midwives and doctors were instruct-ed to report disabled children to the public health office. Children with the following ailments were targeted: "idiocy, Down's syndrome, micro-cephaly, hydrocephaly and malformations of any kind." These cases were then reported to the Reich Committee in Berlin.[47] At the same time, the children were placed in Special Children's Wards (*Kinderfachabteilun-gen*), which were either installed in existing facilities or as separate institutions. More than thirty of these facilities existed in the Reich.

The Special Children's Ward in Vienna was established in July 1942 on the premises of the *Steinhof*. At first, it constituted part of the Viennese Communal Youth Welfare Institution (*Am Spiegelgrund*), but later became an independent institution. Children were observed and exam-ined and the results were sent to Berlin, where the Reich Committee would reach a final decision on the course of action in individual cas-es. When the authorization for killing was received in Vienna, children were plied with high doses of barbiturates until so weakened that they died of pneumonia or other infectious diseases. Some children were used as guinea pigs in lethal experiments; for instance, Elmar Türk (1907–2000), from the pediatric clinic at the University of Vienna, tested a vaccine against tuberculosis on children from the unit.[48]

In Vienna, the public health office and the "infant euthanasia" proj-ect were closely linked; the first director of the facility, Erwin Jekelius, was also responsible for the public health office's psychiatric care services (*Geisteskrankenfürsorge*).[49] In order to implement the policy of "infant euthanasia," an apparatus for the selection and extermina-tion of "unworthy" (in economic and biological terms) children was created.[50] This policy was intended to become a permanent part of the city's youth care system.

In the years between 1940 and 1945 at least 1,850 infants were admitted to the Special Children's Ward. Of these, 789 died, mostly through poisoning, neglect, hunger and infection.[51] The *Spiegelgrund* clinic thus constituted a death zone in the centre of an extensive and elaborate system of public, private and NSDAP-owned welfare institutions, organized on the basis of racial hygiene. Children were observed, tested, segregated and selected; the "worthy" were separated from the "unworthy," the "fit" from the "unfit." This also remained the case in institutions taken over by the Party's welfare organization, the NSV (*Nationalsozialistische Volkswohlfahrt*). All those unable to comply with the new criteria—on the basis of "willingness to perform" (*Leistungs-bereitschaft*), "hereditary health," and "racial purity"—were exposed to measures of eugenic "eradication" (*Ausmerze*). Disabled or retarded children were killed in the "euthanasia" unit, while troublemakers were brutally disciplined in the "reformatory."[52] Children considered Jewish according to the Nuremberg Laws were caught up in the process of isolation, deportation and, finally, extermination of the Jewish population.[53] This applied to the inmates of public children's homes as well, and to foster children taken away from their families.[54]

Conclusions

What becomes clear from the preceding analysis of Vienna's Public Health Office is that, despite the short time-span between the annexation of Austria and the beginning of the Second World War, the implementation of the principles of racial hygiene within the institutional framework of public health and social welfare was extensive and radical. Although the possibility for new projects was significantly reduced after 1939, certain key initiatives—exemplified by the hereditary inventory, sterilization, and infant euthanasia—continued, and were sometimes even expanded, during wartime. Therefore, it is with good reason that health officials in Vienna claimed that they took a leading role in the practical implementation of bio-politics in the German Reich.

Endnotes:

* This paper was written as part of the research project "Medicine, 'Volk' and Race: National Socialist Health Care and Welfare Policy in Vienna, 1938 to 1945," at the Department of the History of Medicine, Medical University of Vienna, financed by the Viennese Anniversary Fund of the Austrian Academy of Sciences.

[1] See Alfons Labisch and Florian Tennstedt, *Der Weg zum "Gesetz über die Vereinheitlichung des Gesundheitswesens" vom 3. Juli 1934. Entwicklungslinien und -momente des staatlichen und kommunalen Gesundheitswesens in Deutschland* (Düsseldorf: Akademie für Öffentliches Gesundheitswesen, 1985).

[2] For the case of Vienna see Herwig Czech, *Erfassung, Selektion und "Ausmerze." Das Wiener Gesundheitsamt und die Umsetzung der nationalsozialistischen "Erbgesundheitspolitik" 1938 bis 1945* (Vienna: Deuticke, 2003).

[3] See Doris Byer, *Rassenhygiene und Wohlfahrtspflege. Zur Entstehung eines sozialdemokratischen Machtdispositivs in Österreich bis 1934* (Frankfurt-am-Main: Campus, 1988); and Monika Löscher, *Zur Rezeption eugenischen/rassenhygienischen Gedankengutes bis 1934 unter besonderer Berücksichtigung Wiens* (Master's thesis, University of Vienna, 1999).

[4] See, for example, Julius Tandler, "Gefahren der Minderwertigkeit," *Jahrbuch des Wiener Jugendhilfswerks* (1928), 3–22.

[5] For the period 1934 to 1938, see Monika Löscher's contribution in this volume, 299–316.

[6] Alfred Schinzel, "Heinrich Reichel 60 Jahre!," *Forschungen zur Alkoholfrage* 44, 6 (1936), 194–199.

[7] Ernst Gabriel, "Rassenhygiene und Alkoholismus," *Archiv für Rassen- und Gesellschaftsbiologie* 29 (1935), 434.

[8] Municipal and Provincial Archive of Vienna (hereafter WStLA), "Personalakt Otto Reisch."

[9] Michael Hubenstorf, "'Der Wahrheit ins Auge sehen'. Die Wiener Medizin und der Nationalsozialismus – 50 Jahre danach, Teil 1," *Wiener Arzt* 4, 5 (1995), 14–27.

[10] WStLA, M.Abt. 212, A 7/6, 151.51, Rundschreiben Prof. Gundel (29 September 1941).

[11] On the history of the Jewish Community in Vienna, see Doron Rabinovici, *Instanzen der Ohnmacht. Wien 1938–1945. Der Weg zum Judenrat* (Frankfurt-am-Main: Jüdischer Verlag, 2000).

[12] WStLA, NSDAP, Gauamt für Kommunalpolitik, Gundel to Mayor Jung (14 November 1941).

[13] WStLA, M.Abt. 212, A 5/2, Gundel to Mayor Jung (20 July 1940).

[14] Florian Freund and Hans Safrian, "Die Verfolgung der österreichischen Juden. Vertreibung und Deportation," in Emmerich Tálos, Ernst Hanisch and Wolfgang Neugebauer, eds., *NS-Herrschaft in Österreich, 1938–1945* (Vienna: Verlag für Gesellschaftskritik, 1988), 767–794.

[15] J. Dietrich, "Zur Einführung des Gesetzes zur Vereinheitlichung des Gesundheitswesens in der Ostmark," *Mitteilungen der Unterabteilung Gesund-*

heitswesen im Ministerium für Innere und Kulturelle Angelegenheiten 2, 6 (1939), 42.

[16] WStLA, "Personalakt Arend Lang." See also Erich Van Reeken, *Berühmte Ostfriesen* (Leer: Sollermann, 1984), 10–11.

[17] WStLA, "Personalakt Hermann Vellguth."

[18] See Johannes Vossen, "Tuberkulosekrankheit und Rassenhygiene. Die Durchführung des nationalsozialistischen 'Ehegesundheitsgesetzes' im Landkreis Herford 1935 bis 1947," in Andreas Wollasch, ed., *Wohlfahrt und Region. Beiträge zur historischen Rekonstruktion des Wohlfahrtsstaates in westfälischer und vergleichender Perspektive* (Münster: Ardey, 1995), 125–141.

[19] Arthur Gütt, ed., *Der öffentliche Gesundheitsdienst* (Berlin: Carl Heymanns Verlag, 1939), 264.

[20] See Susanne Mende, *Die Wiener Heil- und Pflegeanstalt "Am Steinhof" im Nationalsozialismus* (Frankfurt-am-Main: Lang, 2000), 79–107.

[21] WStLA, M.Abt. 212, A 5/6. See also Ernst Klee, *"Euthanasie" im NS-Staat. Die "Vernichtung lebensunwerten Lebens"* (Frankfurt-am-Main: S. Fischer, 1983), 228, 242, 301.

[22] WStLA, M.Abt. 212, A 5/10, Amtsvermerk Dr. Pamperl (25 November 1940).

[23] During the war, this was known as the *Wagner von Jauregg-Heil- und Pflegeanstalt*; known currently as *Otto-Wagner-Spital*.

[24] WStLA, M.Abt. 212, A 7/7, *Abteilung Erb- und Rassenpflege der Hauptabteilung Gesundheitswesen* (28 July 1939).

[25] WStLA, M.Abt. 121, A 7/7, "Richtlinien für die Beurteilung der Erbgesundheit," *Runderlass des Reichsministers des Inneren* (18 July 1940) - IV b 1446/40-1072c.

[26] WStLA, M.Abt. 212, A 5/4.

[27] WStLA, M.Abt. 212, A 7/7.

[28] WStLA, M.Abt. 212, A 31/1.

[29] WStLA, M.Abt. 212, A 7/1.

[30] Gemeindeverwaltung des Reichsgaues Wien, ed., *Die Gemeindeverwaltung des Reichsgaues Wien im Jahre 1939 vom 1. Jänner 1939 bis zum 31. März 1940* (Vienna: n.p, 1942), 316.

[31] Bundesarchiv Berlin, NS 19, Nr. 2397, Bl.: 16–19. Conti to Himmler, 18 March 1942.

[32] See Czech, *Erfassung, Selektion und "Ausmerze,"* 55–59.

[33] Magistrat der Stadt Wien, ed. *Statistische Jahrbücher der Stadt Wien für die Jahre 1937 bis 1945* (Vienna: Magistrat der Stadt Wien Verlag, 1937–1945).

[34] WStLA, M.Abt. 212, A 7/10, Dr. Wetter, "Geburten und Sterbefälle in Wien, im Deutschen Reich und in den deutschen Großstädten" [1942].

[35] See Gisela Bock, *Zwangssterilisation im Nationalsozialismus. Studien zur Rassenpolitik und Frauenpolitik* (Opladen: Westdeutscher Verlag, 1986), 145.

[36] Anonymous, "Untersuchungsergebnisse der Ehestandsdarlehensbewerber der Ostmark," *Ärzteblatt für die deutsche Ostmark* 3, 7 (1940), 79.

[37] Bock, *Zwangssterilisation im Nationalsozialismus*, 100–104.

[38] Austrian State Archive, AdR, Bürckel-Materie, 2354.

[39] Quoted in Johannes Vossen, *Gesundheitsämter im Nationalsozialismus. Rassenhygiene und offene Gesundheitsfürsorge in Westfalen 1900–1950* (Essen: Klartext, 2001), 327.

[40] WStLA, M.Abt. 212, A 7/7.

[41] WStLA, M.Abt. 212, A 1/31.

[42] Gütt, *Der öffentliche Gesundheitsdienst*, 265–266.

[43] Czarnowski, "Frauen als Mütter der 'Rasse'," 63.

[44] Bock, *Zwangssterilisation im Nationalsozialismus*, 233–234.

[45] I thank Claudia Spring for this information.

[46] See Wolfgang Neugebauer, "Zwangssterilisierung und 'Euthanasie' in Österreich 1940–1945," *Zeitgeschichte* 19, 1–2 (1992), 20.

[47] RdErl. d. RmdI vom 18. 8. 1939, IV b 3088/39-1079/Mi, reprinted in Klee, *'Euthanasie' im NS-Staat*, 80–81.

[48] Matthias Dahl, *Endstation Spiegelgrund. Die Tötung behinderter Kinder während des Nationalsozialismus am Beispiel einer Kinderfachabteilung in Wien* (Vienna: Erasmus, 1998), 110–112.

[49] WStLA, M.Abt. 212, A 7/5.

[50] Götz Aly "Der saubere und der schmutzige Fortschritt," in Götz Aly et al., ed., *Reform und Gewissen. "Euthanasie" im Dienst des Fortschritts* (Berlin: Rotbuch-Verlag, 1985), 9–78, 33.

[51] Herwig Czech, "Selektion und Kontrolle. Der 'Spiegelgrund' als zentrale Institution der Wiener Jugendfürsorge zwischen 1940 und 1945," in Gabriel Eberhard and Wolfgang Neugebauer, eds., *Von der Zwangssterilisierung zur Ermordung. Zur Geschichte der NS-Euthanasie in Wien*, part II (Vienna: Böhlau, 2002), 165–187, especially 186–187.

[52] For an impression of the living conditions in the *Spiegelgrund* institution, see the accounts published by some of the survivors; for example, Johann Gross, *Spiegelgrund. Leben in NS-Erziehungsanstalten* (Vienna: Ueberreuter, 2000); Alois Kaufmann, *Spiegelgrund Pavillon 18. Ein Kind im NS-Erziehungsheim* (Vienna: Verlag für Gesellschaftskritik, 1993); and Oliver Lehmann and Traudl Schmidt, *In den Fängen des Dr. Gross. Das misshandelte Leben des Friedrich Zawrel* (Vienna: Czernin, 2001).

[53] See Rabinovici, *Instanzen der Ohnmacht*, 94.

[54] WStLA, M.Abt. 212, A 5/4, Gundel to Scharizer, 3 April 1941. See Wolf Gruner, *Öffentliche Wohlfahrt und Judenverfolgung. Wechselwirkungen lokaler und zentraler Politik im NS-Staat (1933–1942)* (Munich: Oldenbourg, 2002).

Fallen Women and Necessary Evils: Eugenic Representations of Prostitution in Interwar Romania

Maria Bucur

In the opening sequence of the movie *An Unforgettable Summer*, a group of Hungarian-speaking prostitutes moon several Romanian officers, who bypass their usual stop at the brothel for a more "civilized" night on the town—a ball at the local general's luxurious residence. The movie, based on the 1920s novella *Salata* (The Salad) by Petru Dimitriu (1924–2002), is a bitter critique of both nationalism and the Western civilizing project of the interwar years. The altercation between the officers and the prostitutes offers a stark metaphor for the themes of the movie. As the beautiful women flash the officers in the window of their brothel, the men look on lustfully, while at the same time cursing the prostitutes, calling them Marxists and "Bela Cur," roughly translated as "beautiful ass," a pun on the name of Béla Kun (1886–1939), the leader of the 1919 Communist regime in Hungary. Though intriguing and powerful, the prostitutes fade into the background of the unfolding narrative, relegated to the margins of the story, much as they were in the social landscape of interwar Romania.

Prostitutes figured prominently in the literature of the interwar period as colorful and sometimes powerful characters.[1] They served as vivid props for modernist authors who critiqued bourgeois mores and banal lifestyles, but prostitutes were seldom subject to a great deal of attention in terms of their social and economic position. That is not to say that the existence of prostitutes was simply tolerated at the margins of Romanian society. Between 1918 and 1944, several doctors and social reformers focused on prostitution as a health issue, social problem, and moral question. Debates about these matters materialized into a series of proposals that featured prominently in medical journals and legislative agendas. In interwar Romania, no less than three laws to regulate prostitution were passed in the span of two decades, an unprecedented level of legislative attention to this issue. This chapter investigates the two important positions taken by most participants in the debate about prostitution: regulation versus abolition. These two positions were not particularly novel in the larger context of temperance movements across Europe, but they were certainly new in Romania, and helped define wider agendas for both public health and social reformers after the creation of Greater Romania following the First World War.[2] The analysis will focus on the medicalization of the debate about prostitution, which was a significant shift in relation to previous debates, as it put doctors and the concern with public health at the centre of policy making, displacing the heavy focus on the police and law-enforcement institutions.

There are few processed sociological data about prostitution in Romania during the period 1918 to 1944, despite a wide array of sources ranging from medical to police records. Because, overall, the policy of the Romanian state was to regulate prostitution, this practice was better documented than many other enterprises and professions. However, until a few years ago historians were not particularly interested in this aspect of Romanian social history.[3] Therefore, my discussion will not touch much on prostitution as a phenomenon or on the experiences of prostitutes themselves. Instead, I shall focus on the legal and medical wrangling over the definition of prostitution as a social problem, as well as its proposed solutions. The wealth of publications on this subject, from debates in the press to legislation and different specific regulations for public health institutions dealing with venereal diseases, offers ample evidence that, indeed, prostitution was a matter of great public interest. By focusing on the various representations of prostitution as a "problem," this chapter highlights approaches to the study of social practices and government policies pertaining to gender. As in many European countries, prostitution became a matter of public interest immediately following the First World War. It appeared to both doctors and military leaders that the number of soldiers who contracted sexually transmitted diseases grew significantly during the war, prompting many suddenly to consider prostitution as a matter of grave public-health concern. It is not clear whether the increase in cases was simply a reflection of heightened scrutiny by doctors and social reformers, but there is evidence to suggest that the actual number of individuals who became carriers of venereal diseases grew over the 1914–1920 period.

It was in Transylvania that the interest in controlling sexually transmitted diseases and reforming the legal status of prostitution first developed after the First World War. A group of doctors around Iuliu Moldovan (1882–1966), the interim Minister of Health of this region during the transition from the Hungarian to Romanian administration, elaborated a law on prostitution reform and a series of proposals that underlined the danger posed by prostitution to society, as well as the most appropriate solutions for tackling the issue.[4] The intense degree of interest in venereal diseases expressed by Moldovan is connected not only to the increased incidence of such problems, but also to his medical training and experience. Moldovan had studied in Vienna and concentrated his medical practice on solving the epidemic proportions of sexually transmitted diseases among the Habsburg troops on the Eastern front in Galicia.

At the heart of the proposals formulated by Moldovan and his colleagues was the eugenic notion that venereal diseases constituted a threat to public health and the genetic well-being of society, affecting individuals who came directly into contact with those infected, as well as future generations. Focusing on prostitution was part of a much larger theoretical discussion that suggested—similar to eugenic movements in Germany, the United States, Hungary and France—that the challenges of the modern age, be they economic, political, or related to public health, were directly connected to the "biological capital" of a particular nation, often understood in exclusionary ethnic terms.[5] The movement, led primarily by Moldovan and later at his Institutul de Igienă Socială (Institute for Social Hygiene), affiliated with the medical school in Cluj, aimed to use eugenic arguments to modernize the Romanian state and implement preventive measures in the sphere of public health. This two-pronged approach was at times inconsistent and riddled by paradoxes. For instance, ethnically Romanian peasants were perceived as the cornerstone of the nation's vitality; however, in the eyes of the proponents of eugenics, they also constituted a section of the population plagued by some of the most significant genetic defects such as alcoholism and tuberculosis.[6] Likewise, many leading eugenicists recognized that industrialization and urbanization were necessary as a means of modernization, yet they also sought to curtail the social costs of these processes. In this complicated and often self-contradictory interpretation, the Romanian population appeared as the most valued and simultaneously the most victimized of the ethnic groups inhabiting Romania.[7] And in this often inconsistent view, eugenicists never clarified why a region like Transylvania, generally wealthier and better developed in terms of public health services than the rest of Romania, could at the same time be in great genetic danger at the hands of its non-Romanian elites—who were in fact largely accountable for that superior level of development.[8]

Similar contradictions remained at the heart of the views harbored by eugenicists regarding gender roles. In general, eugenicists viewed gender roles as directly linked to biologically determined factors: femininity and masculinity, for instance, were viewed as innate qualities, rather than environmental or cultural constructions.[9] Healthy men and women were expected to abide by what eugenicists defined as their proper role in the family and society, and as parents in the reproductive chain. Eugenicists defined masculine roles in narrowly normative ways: "healthy" men were to show both physical (and sexual) prowess and certain aesthetic features, such as a muscular body, body hair, and even a certain look of determination in their eyes. Women were generally viewed as weak and unstable, which also made them more dangerous, and thus required the protection and oversight of the state (and indirectly men).[10] In this convoluted and contradictory view of women, what is striking is the obvious sexual double standard, the attempt not only to control women's sexuality but also to mobilize it for the sake of the eugenic ideal: increasing the rate of "healthy births."

Based on these premises, eugenicists articulated a medical argument for regulating prostitution, standing in opposition to both those who advocated its abolition, and those who proposed its regulation on the basis of moral considerations. According to the views of early eugenic reformers, the abolition of prostitution was an ineffective tool, even though it could be viewed as supporting eugenic concerns about eliminating genetically harmful individuals from society. There seems to be a clear contradiction between some of the general statements made by eugenicists regarding public health threats and their position on prostitution. Many considered crime and immorality (among criminals, prostitutes were considered an important group) an inborn quality or a genetic deficiency; it was this deficiency that needed to be eliminated by discouraging such individuals from reproduction.[11] Based on this consideration, it would be unsurprising to see eugenicists embracing an abolitionist view: prostitution would have to be eliminated together with those who practiced and propagated this "genetic illness."

Yet a large number of eugenicists were against the abolition of prostitution during the early 1920s, and it is interesting to speculate why this was so. First of all, many had witnessed the unfolding of the abolitionist-reformist debate during the late nineteenth century, especially in Britain, and observed its results. Romanian eugenicists were convinced that abolitionism had failed miserably, judging by the spread of syphilis and other sexually transmitted diseases following abolition, not to mention the number of clandestine prostitutes caught by the police.[12] After the First World War, many eugenicists thus took a more pragmatic view on this issue: to regulate prostitution rather than abolish it would result in its reduced impact on the rest of the population.

More importantly, such tolerance was connected to the view that male sexuality was genetically predisposed towards greater aggressiveness and activity, which could only be safely satisfied in the realm of regulated prostitution. This was simply an admission of the age-old sexual double standard cloaked in medical language. Several doctors wrote extensively on the innate differences between men and women with regard to sexual aggressiveness, using arguments that tied specific sexual behavior to supposedly inborn characteristics. One doctor, Ovidiu Comşia, stated with certainty that "man's voluntary abstinence is morally superior to woman's voluntary abstinence [because] forced chastity is felt by man as an organic tragedy."[13] Other proponents of eugenics were less ready to accept such conjectural connections between gender, morality and sexuality, and refrained from making such bold statements in their support for the regulation of prostitution. Other advocates of regulation accepted the sexual double standard as a reality too difficult to change. Therefore, prostitution needed to be addressed according to the confines of male sexual promiscuity: "These links [between prostitution and family life] are so deeply rooted in society, that they make [any] radical operation impossible."[14] The medical reference to a "radical operation" is in fact an allusion to the ineffectiveness of abolitionism as a solution to the "deeply rooted" problem of prostitution.

Eugenicists also stood in opposition to abolitionists on moral grounds; this included feminists who viewed the regulation of prostitution as a form of discrimination, one based on the assumption that prostitution was the sole cause of the spread of venereal disease.[15] Eugenicists did not share most views held by liberal feminists, notably that individuals should enjoy equal rights and responsibilities. Yet many did agree that those who had contracted syphilis should be held equally responsible for the eugenic danger they posed to society, including the clients of prostitutes. However, eugenicists used this argument to reframe their views on regulation and increased state control, rather than push for abolition.

Most eugenicists were, in fact, ambivalent towards the view that prostitutes were irremediably "fallen women." They tended to view these women as victims of a cultural, economic, and social predicament that pushed them towards prostitution. Thus, by reversing these predicaments, prostitutes could be brought back into the social fold, as potential future mothers and wives—that is, members of the nation.[16] Such options were open, however, only to ethnic Romanian women, because ethnicity far outweighed gender in terms of dysgenic fears. In other words, eugenicists were generally more concerned with the role women played as members of an ethnic group, than with the way they conformed to their normative gender roles. Therefore only the "fallen women" of one's own ethnic group could be reintegrated into social normalcy. This view was a direct outcome of eugenicists' great interest in reproduction and in limiting all women to their "natural" destiny of mother and wife.

One legislative proposal outlined in 1922 contained some extraordinary claims. Two types of brothel would be recognized as lawful: (1) state institutions led and controlled by specialist doctors, where women could register in order to practice prostitution, much like a hospital; and (2) private boarding houses where prostitutes would be able to practice individually rather than through the mediation of a madam. Each client would have to "go through a medical check-up room where he would be seen by a doctor, and a bath, where he would wash himself, and only after these operations would he be able to enter the prostitute's room. At the exit he would again go through the bathroom and the medical check-up room, where he would receive prophylactic materials before leaving the establishment."[17] This type of intrusive regulation of individual action was unprecedented in Romania and other European countries, with the possible exception of army regulations. What is striking about these proposals, moreover, is the obvious clinicalization of sexual desire. Visiting a prostitute became akin to visiting a doctor. And although their actions were strictly controlled as well, prostitutes were seen to provide a necessary social service.

This was an unprecedented way of thinking about prostitution when considering the antithetical arguments proposed by feminists in favor of abolition. However, this is not to suggest that eugenicists were supporters of prostitution. Rather, they tolerated prostitution, despite describing it as a "social plague" or "gangrene."[18] And some were careful to suggest that prostitution was not "genetically predetermined," that is, an inborn characteristic. Some social reformers went so far as to identify specific social causes underpinning prostitution, placing "social misery" and economic hardship at the top of the list. Of 188 women identified by one study, only 19 had become prostitutes out of "curiosity" or because "they did not like work." The rest gave as the root causes "misery and poverty; family causes; deceit; false pretence; desperation"—all suggesting that these women were victims of both circumstance and social factors.[19]

The ambivalence towards prostitution as a crime versus a necessary social service, as well as a moral evil versus a consequence of social circumstances, led some eugenic reformers to advocate its regulation in conjunction with education and rehabilitation. The same proposal that advocated the clinicalization of client-prostitute relations also supported the move by public health officials to create special hospitals for women, and workshops for sewing and homemaking industries, such as cooking and home cleaning, to encourage rehabilitation. Doctors would send prostitutes diagnosed with venereal diseases to special hospitals and clinics for treatment. Prostitutes would also be encouraged to receive professional training that would enable them to become economically self-sufficient, thereby signaling an end to their career in prostitution and their rehabilitation as healthy and productive members of society.[20] This was, to a large extent, wishful thinking, for the employment offered through government placement agencies did not necessarily guarantee jobs that ensured economic self-sufficiency. Feminist reformers also encouraged rehabilitation, though not because of the same concerns for public health as eugenicists. Feminists saw prostitutes as victims of ignorance, abuse, and poverty, whose only chance to take back their life and become fully integrated in society was to go through a training program followed by employment in a regular job. Various feminist organizations, such as the association The Friends of Young Women (*L'Association des Amies de la Jeune Fille*), ran training workshops and schools, as well as placement offices.[21]

Several reformers proposed more radical measures, such as regulating women's wages to ensure their economic livelihood. This was a position close to the reforms proposed by the most radical feminists in

Romania, such as Calypso Botez (1880–?) and Elena Meissner (1867–1940), though for clearly different reasons. Legislation regarding the minimum wage, especially when it came to women's work, was not an issue that interested most Romanian politicians during the interwar period. Most politicians, following traditional thinking, believed that women belonged in the home. Eugenicists generally stood by this position as well, but in the case of prostitution they took a more realistic view, which recognized the economic needs and social well-being of these women.

Another remarkable feature of this generation of reformers who proposed the regulation of prostitution was the idea that doctors and public health officials, rather than the police, were to be in charge of enforcing the new laws regarding prostitution. In brothels, doctors, not policemen or madams, would be responsible for ensuring that the threat posed to public health by prostitutes and their clients was reduced to a minimum.[22] The police would become involved only in cases of non-compliance by prostitutes and other individuals who did not abide by the regulations set up by the state. It was up to doctors to call on the police to intervene.[23] The police were to oversee the elimination of clandestine prostitution by women not officially registered with the state. By and large, eugenicists demanded the transfer of authority over the practice of prostitution to public health officials.

During the 1920s, prostitution was regulated to some extent by the principles posited by eugenicists. However, in reality the Ministry of Health did not have the resources to set up the "sanitized" brothels mentioned above, nor did it have enough properly trained doctors to check on and treat prostitutes and other potential carriers of venereal diseases. In addition, the projects for training and job placement fell short of initial expectations because substantial funds could not be garnered from the government or private sources. The activities of feminist reformers in Bucharest remained the most important in this respect, while in Cluj, the capital of Transylvania, eugenic initiatives proposed by local public health officials did have an impact.

In 1930, when Iuliu Moldovan was promoted from important public health official in Cluj to minister for public health and social assistance under the National Peasant Party government, a new public health law was promulgated which featured many eugenicist ideas. However, as regards prostitution, the law was clear: "Brothels and any establishments or public places where prostitution is practiced by women living there or frequenting such places are absolutely forbidden. All such establishments currently in existence are to be closed. Fines for disobeying these decisions, as well as for those who exploit women and those that practice prostitution, will be 5,000 to 100,000 lei, and in case of repeat offences, a prison sentence of up to six months."[24]

It appeared that eugenicists had shifted towards an abolitionist stance. There is not enough published evidence to explain this dramatic volteface by a group of reformers who, only eight years earlier, were adamant supporters of regulation. The law represented a compromise between, on the one hand, the eugenicists' concerns for public health and the spread of genetically debilitating diseases, and, on the other, the morally conservative outlook of the National Peasant Party government, and especially of its leader, Iuliu Maniu (1873–1953). Maniu was an austere person of impeccable moral standing, whose decisive stance on ethical principles often went against the realpolitik of pragmatic solutions. It could be argued that Maniu simply could not ally himself with a policy that made legal the immoral practice of prostitution, even if strictly regulated and for the protection of the wider community.

What happened during the 1930s was a combination of abolitionism and the regulation of the trade, regarding prostitutes as individuals free to practice their profession within the confines of regular medical check-ups by specialists in venereal diseases.[25] Yet this policy must have appeared unsatisfactory to public health officials for, by 1943, a new wave of discussions regarding the regulation of prostitution re-emerged and a new law was passed for the eradication of venereal diseases and the control of prostitution. It is in some ways remarkable that Romanian legislators found the time and mustered the interest to focus on this issue in the midst of the debacle of the German-Romanian campaign in the Soviet Union during the Second World War. The unease about venereal disease was expressed just as unequivocally by military leaders during the war in other European countries and the United States of America, due to the perceived threat to the troops by prostitutes.

The 1943 legislation returned firmly to a regulatory position. Not only did it define prostitution properly for the first time, as "procuring material gains from sexual liaisons with different men, like a regular profession," but it also shifted the responsibility for regulating prostitution from the institutional to the individual.[26] Registering could only be done "at the request of the woman who declares herself a prosti-

tute." Conversely, clandestine prostitutes were to be "forcefully" registered, that is, made accountable.[27] Brothels were made legal again. They
were obliged to abide by certain rules, and they could only accept registered prostitutes. Therefore, prostitutes themselves were ultimately
responsible for becoming acquainted with the demands of the public
health authorities; however, they could also freely leave the brothel to
practice alone.

This law did not emphasize the notion that prostitutes were "fallen
women" and morally reprehensible. In fact, neither the language of the
law itself, nor the preamble which outlined the necessity of the new
law, contained the kind of derogatory remarks about prostitutes found
in many other such documents, with two exceptions. First of all, the
preamble defined the threat of venereal diseases as a problem that
affected "not only the individual person, but also the very future of the
race through depopulation and degeneration."[28] Which "race" in particular remains unclear, but viewed in the wider context of the debates
of the time, this is most likely a eugenic argument concerning the need
to increase the powers of the state in matters of public health.

A second similar reference comes much later, with the description
as to who is able to open a brothel and apply for a license to practice
prostitution. Article 51 of the law stipulated that only persons of Romanian nationality could do so.[29] This statement can most likely be interpreted as meaning persons of Romanian ethnicity, rather than Hungarians, Germans, Jews, or other ethnic minorities. This limitation
reveals the eugenicist preoccupation with ethnic hierarchy and the
classification of the population into "acceptable" and "unacceptable"
categories. Only Romanian prostitutes were acceptable and could be
rehabilitated and forgiven by the rest of society. For prostitutes from
other ethnic groups, rehabilitation and trustworthiness remained questionable.

The new law stipulated that the enforcement of these regulations
remained in the hands of public health officials, with the police as a
secondary law enforcement agency. By and large, however, prostitutes
were dependent on doctors for official paperwork certifying that they
could practice their profession legally. Another important aspect of this
law was the punitive measures, which included both hefty fines of up
to half a million lei and a prison sentence of up to five years. Even
harsher punishments were reserved for doctors and police officers who
knowingly misinformed public authorities about the health of prosti-

tutes and thus allowed the spread of venereal diseases through their practices.

Yet this law said nothing about the clients of prostitutes and their responsibility to protect themselves and their families against venereal diseases. Nevertheless, two provisions in the law established the possibility of extending punishment to these men. According to Article 13, two categories of people would serve from three months to two years in prison, with a corresponding fine of 5,000 to 20,000 lei: "a) any person that had not been seen by a doctor as soon as s/he found out that s/he had contracted a venereal disease; and b) members of the same family who live together, if they do not all undertake medical control, after finding out that one of them had syphilis."[30] Both categories included all men who frequented brothels and contracted venereal diseases. The jail sentences were serious and so were the fines, yet evidence is not available on how this law was applied to men, rather than affecting, predominantly, prostitutes.

Another remarkable and entirely new aspect of this law was the provision which stipulated that managers or owners of brothels were to deposit 10 per cent of the earnings of each prostitute into a savings account, to be later withdrawn by the prostitute herself with the approval of the public health authorities.[31] It is not clear whether this 10 per cent constituted the total amount of money the prostitute was obliged to save from her earnings, or whether this amount formed part of an insurance or pension fund in the eventuality of unfair treatment at the hands of brothel owners. The inclusion of such a provision suggests greater concern by the state for the livelihood of prostitutes than before, and also a desire to protect their financial status. In these respects, too, the law was unprecedented.

Although it is difficult to establish lines of continuity between the interwar eugenic preoccupation with controlling prostitution and the attitudes prevalent among policy makers after 1945 and later the collapse of Communism, a few interesting remarks can be made in this respect. After the Second World War prostitution continued to be tolerated, but by 1948, when the Communist Party finally gained control over the state, prostitution was made illegal, as in all other Communist bloc countries.[32] For the remainder of the Communist period, prostitution was considered a "bourgeois atavism" and a symbol of the decay manifest in capitalist countries. Nevertheless, prostitution was practiced, albeit illegally, both against the explicit wishes of the government and, paradoxically, with the tacit approval of the Communists, since prostitutes could be useful tools for blackmail and the gathering of secret intelligence. Rumors abounded about the influence exercised by a hierarchy of prostitutes that frequented luxurious hotels—like the Bucharest Intercontinental—under the control of the Securitate, the Romanian secret police.

After 1989, prostitution continued to be tolerated under the guise of abolitionism. However, under the new free-market system, the economic and social status of prostitutes deteriorated rapidly. The story of the deplorable sex traffic industry is widely known;[33] suffice it to say that by the mid-1990s the practice had become so widespread that it generated interest in the legalization of prostitution. In 1999, the Romanian Parliament debated a legal initiative that would make prostitution lawful, albeit strictly regulated. The law was not passed, but the ensuing debates have proved insightful, if embittering, exposing broader and more fundamental questions about morality, gender roles, sexuality, and cultural mores. Feminists, priests, social workers, psychologists, politicians from the left to the right of the political spectrum, parents, journalists, and sometimes prostitutes themselves, participated in the debates about the social causes that led to the practice of prostitution, the moral grounds for its (de)criminalization, as well as the possibility of eliminating the practice altogether.[34]

By and large, abolitionists continue to view prostitutes as a "moral plague" that must be eradicated. This type of abolitionism is unremarkable when compared to the attitudes of other Europeans; however, in the context of the history of the debate about prostitution in Romania it is a more short-sighted position than that promoted by many eugenicists in the interwar period. Ironically, the position assumed today is closer to that held by the Communists.

Proponents of regulation have been more tolerant but have shown little concern for the prostitutes themselves. They talk about prostitution from a moral high ground, similar to that of the abolitionists. But they view prostitution as an endeavor that created important revenues for the state, and which should be taxed in the same way as other services are. In addition, they want to reinforce the marginality of these women by segregating them physically in "red light" districts away from schools and churches. Although apparently more tolerant, this justification for regulating prostitution is both narrow (in practice, there would be no place where a brothel could be situated, if one were to abide by the law) and callous, in terms of seeing prostitutes as individuals that should contribute, via taxes, to the greater financial good of society, without much in the way of protection for their personal and professional safety.

The way in which eugenicist reformers articulated their ideas about prostitution reform at the beginning of the interwar period has proven the most progressive view of prostitution to date in Romania. This is not to say that eugenicists were progressive reformers.[35] However, it was the eugenicists who directly engaged in interwar debates concerning the regulation as opposed to the abolition of prostitution and other specific legislation aimed at its regulation. The eugenicists had, therefore, a more encompassing view of the subjects to be affected by their reforms (including potential victims of venereal diseases, besides the prostitutes themselves), as well as the socioeconomic and cultural dilemmas associated with prostitution, as discussed in this chapter. Today's social reformers might benefit from a closer reading of the proposals made by their predecessors.

Endnotes:

[1] See some of the descriptions in Adrian Majuru, "Bucureştii—arena a hetairelor," *Adevărul literar şi artistic* 12, 672 (1 July 2003), available at http://adevarul.kappa.ro (Accessed 16 October 2003). Some of these colourful characters can be found in the novels of Ionel Teodoreanu (1897–1954) and Camil Petrescu (1894–1957).

[2] Some of the important works on prostitution include Judith Walkowitz, *Prostitution and Victorian Society. Women, Class, and the State* (New York: Cambridge University Press, 1980); Charles Bernheimer, *Figures of Ill Repute. Representing Prostitution in Nineteenth-Century France* (Durham, NC: Duke University Press, 1983); Jann Matlock, *Scenes of Seduction. Prostitution, Hysteria, and Reading Difference in Nineteenth-Century France* (New York: Columbia University Press, 1994); Alain Corbin, *Women for Hire. Prostitution and Sexuality in France after 1850* (Cambridge, Mass: Harvard University Press, 1990); Laura Engelstein, *The Keys to Happiness* (Ithaca: Cornell University Press, 1992); Laurie Bernstein, *Sonia's Daughters. Prostitutes and Their Regulation in Imperial Russia* (Berkeley: University of California Press, 1995); and Francisca de Haan and Annemieke van Drenth, *The Rise of Caring Power. Elizabeth Fry and Josephine Butler in Britain and the Netherlands* (Amsterdam: Amsterdam University Press, 1999).

[3] One significant departure is a study published recently. See Ghizela Cosma, "Prostituţia în România între războaie," in Ghizela Cosma, Enikö Magyari-Vincze, and Ovidiu Pecican, eds., *Prezenţe feminine. Studii despre femei în România* (Cluj: Ed. Fundaţiei Desire, 2002), 415–437. Another young historian working on the topic of prostitution in connection with "marginality" in urban settings is Adrian Majuru. See Adrian Majuru, *Bucureştiul subteran. Cerşetorie, delicvenţă, vagabondaj* (Bucharest: Ed. Paralela 45, 2005).

[4] Maria Bucur, *Eugenics and Modernization in Interwar Romania* (Pittsburgh: University of Pittsburgh Press, 2002), 26–31.

[5] See Mark Adams, ed., *The Wellborn Science. Eugenics in Germany, France, Brazil, and Russia* (New York: Oxford University Press, 1990); Atina Grossman, *Reforming Sex. The German Movement for Birth Control and Abortion Reform, 1920–1950* (New York: Oxford University Press, 1995); Daniel P. Kevles, *In the Name of Eugenics. Genetics and the Uses of Human Heredity* (New York: Knopf, 1985); Frank Dikotter, *Imperfect Conceptions. Medical Knowledge, Birth Defects, and Eugenics in China* (New York: Columbia University Press, 1998); Frank Dikotter, "Race Culture: Recent Perspective on the History of Eugenics," *American Historical Review* 103, 2 (April 1998), 467–478; and Edward Larson, *Sex, Race, and Science. Eugenics in the Deep South* (Baltimore: Johns Hopkins University Press, 1995).

[6] Bucur, *Eugenics and Modernization in Interwar Romania*, 136–139.

[7] Bucur, *Eugenics and Modernization in Interwar Romania*, 145–148.

[8] For a more in-depth discussion of these inconsistencies, see Bucur, *Eugenics and Modernization in Interwar Romania*.

[9] Bucur, *Eugenics and Modernization in Interwar Romania*, 166.

10 Bucur, *Eugenics and Modernization in Interwar Romania*, 139–144.

11 In the Romanian case, sterilization was not the favoured tool. Instead, most eugenicists insisted on the need to use education to dissuade alcoholics, carriers of tuberculosis and syphilis, as well as social deviants from reproducing.

12 These reformers clearly knew the history of these movements quite well, as they quoted many details from the history of prostitution reform in Western Europe, especially Great Britain. See D[ominic] Stanca, "Rostul ambulatorului policlinic în combaterea boalelor venerice," *Sănătatea publică* 2, 11–12 (1922), 43.

13 Ovidiu Comşia, "Biologia familiei. IV. Biologia sexelor," *Buletin Eugenic şi Biopolitic* 3, 11–12 (November–December 1929), 375–378. Here Comşia states that, as far as men are concerned, sexual desire is "irresistible."

14 Stanca, "Biologia familiei," 41.

15 For the feminist abolitionist position see Elena Meissner, "Lupta contra imoralităţii," *Buletin Eugenic şi Biopolitic* 2, 11–12 (1928), 349–353; and Alice Voinescu, "Primul congres de morală socială," *Asistenţa socială* 5, 1 (1936), 8–13.

16 For instance, Hungarian prostitutes in Transylvania were viewed, by definition, as a eugenic threat to Romanian men, while Romanian prostitutes were considered only potentially dysgenic, if they actually transmitted venereal diseases to their clients.

17 Stanca, "Biologia familiei," 47.

18 Stanca, "Biologia familiei," 41.

19 Stanca, "Biologia familiei," 44. Yet other eugenicists identified prostitution as more closely connected to genetic predisposition. See Ovidiu Comşia, "Consideraţii generale asupra cauzelor prostituţiei," *Buletin Eugenic şi Biopolitic* 4, 1–2 (1930), 47–53.

20 Stanca, "Biologia familiei," 45–46.

21 See, for instance, "Rapport de l'Association des amies de la jeune fille," *Bulletin du Conseil National des Femmes Roumaines, 1921–1938* (Bucharest: Imprimeries "Curentul" S.A., n.d.), 5. The association The Friends of Young Women coordinated the educational, and especially placement, offices. It had offices in the main train stations in Bucharest, where most young women from other cities and rural areas, in search of employment and easily prey to sexual abuse and deceit, could be found.

22 Stanca, "Biologia familiei," 47.

23 Stanca, "Biologia familiei," 48–49.

24 (Anonymous), "Legea sanitară şi de ocrotire," *Monitorul Oficial* 154 (14 July 1930), 5369.

25 Cosma, "Prostituţia în România între războaie," 436.

26 (Anonymous), "Legislaţia pentru combaterea boalelor venerice," *Buletinul Sănătăţii şi Ocrotirii Sociale* 48, 8–9 (August–September 1943), 685.

27 (Anonymous), "Legislaţia pentru combaterea boalelor venerice," 685.

28 (Anonymous), "Legislaţia pentru combaterea boalelor venerice," 657.

29 (Anonymous), "Legislaţia pentru combaterea boalelor venerice," 687.

[30] (Anonymous), "Legislația pentru combaterea boalelor venerice," 662.

[31] (Anonymous), "Legislația pentru combaterea boalelor venerice," 689.

[32] In the archives of the Ministry of Health I found prostitutes' health cards dating from as late as 1947.

[33] See http://www.sens.org.ro/despre_trafic.html (Accessed 16 October 2003).

[34] See http://antiprostitutie/go.ro, (Accessed 16 October 2003); and "Zilnic 'Deschide ochii și iei milionul," *Adevărul* 4038 (23 June 2003) available at http://www.adevarulonline.ro/arhiva/2003/Iunie (Accessed 26 July 2005), and http://humanism.ro/forum/viewtopic (Accessed 26 July 2005).

[35] See my overall critical view in Bucur, *Eugenics and Modernization in Interwar Romania.*

Part IV

Anti-Semitism, Nationalism and Biopolitics

Culturalist Nationalism and Anti-Semitism in Fin-de-Siècle Romania

Răzvan Pârâianu

During the nineteenth century, European thought was deeply influenced by a new heuristic notion used in most of the human sciences at the time. It was the idea of race, which, benefiting from the scientific prestige offered by natural sciences, achieved pre-eminence in many theories about human nature, society, history, and eventually culture. The idea that a group of people may share common physical and psychological features seemed a particularly powerful explanation for many scholars. This way of reasoning was extremely seductive because it established a biological foundation for the "science of man" and created the prerequisites for a positivist approach in understanding human nature. Such a racial interpretation of groups and cultures survived until the middle of the twentieth century, when these ideas were discredited by Nazi racial ideology.

In a completely different political context, intellectuals in Central and Southeast Europe, emulating their Western counterparts, adapted various theories of race to their own national and political programs. This was the case for a young generation of Romanian intellectuals, who, at the end of the nineteenth century, having the opportunity to study at German or French universities, was mesmerized by theories of race. For this generation, race represented something profoundly atemporal and trans-individual, capable of shaping the culture and civilization of a specific *Volk*. Moreover, modern theories about race offered a good opportunity to present and legitimize political claims "scientifically." It was an opportunity to affirm that the true nature of the *Volk* was not altered and, under propitious circumstances, may in fact flourish on fertile soil. The unity of language was the irrefutable proof for the great expectations these young and enthusiastic intellectuals had for their nation and state.

According to contemporary theories of culture, language was the indubitable manifestation of the collective mind of the *Volk*, the result of a primitive intuition revealed to each race, and the expression of the civilizing potential of ethnic groups.[1] This linguistic argument attracted many adherents because it offered a powerful political justification for the foundation of the nation-state. As far as language was concerned, it was by far the most important element uniting people.[2] The unity of language corresponded to the "unity of the collective soul," which in real terms represented the capacity of people to work together and eventually to create a state of their own.[3] Thus, centuries of "historical disunion" were pushed into the background because the "primordial union" of the *Volk* was far more important than accidental history, contingent upon the will of the great neighboring empires.[4]

This chapter explores three perspectives in the way in which the idea of race was adopted and used for political goals by a young generation of Romanian intellectuals at the end of the nineteenth century. The first perspective is the relationship between the idea of race and the idea of culture. The characteristic elements of race were viewed as ingredients of culture, even when race, as a concept, either did not play a central role or was absent in intellectual debates. Often, language had an intermediary position between race and culture, and therefore literature and literary criticism constituted the playground of many theories and disputes. Many authors perceived race and culture interchangeably, the only difference between them was the context in which they were developed and conceptualized. Gradually, these authors refined initial notions of race into more sophisticated theories of cultural specificity and ethnic originality that eventually served as the intellectual foundations of right-wing ideologies during the 1930s.

The second aspect discussed here is the "revolutionary ethos" that the idea of race revealed in its revolt against modern institutions and their "liberal" foundations. Commonly seen as an expression of racial difference, the idea of "culture" played a major role in shaping the new political discourse. It was the peculiar use of the notion of culture, as a specific racial trait of a *Volk*, which became popular in the last decade of the nineteenth century, influencing not only conventional political perspectives, but also literary currents and national literature. A culturalist movement was thus formed as a reaction against the political establishment. This movement was based on the premise that all material or spiritual manifestations of each nation, or *Volk*, were culturally specific, and this "specificity" corresponded to a psychological heritage determined by race. A number of political associations and leagues were extremely active in disseminating or "sowing" these new ideals in literature, culture and—last but not least—politics.[5] In Romania, the most important of these associations was *Liga Culturală a tuturor Românilor* (The League for the Cultural Union of all Romanians), with which an entire generation of Romanian nationalists became associated.

Initially, this nationalist movement mobilized and organized popular awareness about the difficult circumstances under which Romanians from neighboring empires lived. Under these circumstances, the revolutionary ethos assumed an emancipating ideal to which all other principles were subordinated. Yet after the First World War, when Transylvania, the Banat, Bukovina, Maramureş and Bessarabia were incorporated into the Kingdom of Romania, and the envisioned political union of all Romanians was fulfilled, the "revolutionary ethos" gained a new impetus. In these new political conditions, the question of ethnic minorities was the most troublesome for the Romanian state; ethnic minorities were frequently depicted as "a foreign stake thrust into the body of nation."[6] However, for many commentators, ethnic minorities were not the real problem within the Romanian state, so much as the Romanian political class, which was unable to strengthen and centralize state institutions. The same political elite was heavily criticized before the First World War for not adequately supporting the "national ideal:" now they were condemned for "selling" the country to foreigners. Romanian politicians were negatively depicted in political literature as being "estranged by foreign education," "uprooted by modern civilization," "alienated by mass media," "degenerated by urban life," thus becoming the lightening rod for popular discontent exacerbated by political and economic crises.

The third aspect addressed here concerns the way in which the idea of race was used to depict perceived enemies of the Romanian nation. In political pamphlets and literary texts, the national struggle was narrated as a clash between (Romanian) national culture and the Semitic civilization. The syllogism was as follows: modernity is foreign to national culture; the foreigner is the Semite par excellence, therefore anything modern was likely to be Semitic. Romanian national culture was thus constructed as the anti-Semitic reaction to a foreign civilization. The sole purpose of national culture was the preservation of racial traits and the capacity to create a new national modernity. This anti-

Semitic activism was directed against contemporary politicians and journalists and was styled as an insult rather than a general reaction against the Jewish race as a whole. For Romanian nationalists of the fin-de-siècle period, the negative influence of the Jews in corrupting modern man and, most importantly, those that had lost their roots, culture and sense of community, was perceived as the greatest threat. The logic of this argument was centered on "a life and death struggle" between the nation and "foreigners" (Jews). It was another version of Herbert Spencer's "survival of the fittest."[7] Accordingly, if Semitic modernity prospered, the Romanian nation and race would be extinguished; therefore, the Semites should be annihilated altogether. Romanian national culture was consequently deemed the instinct of natural self-preservation, which took the form of an anti-Semitic reaction.

Nationalism and Semitism at the End of the Nineteenth Century

Both the idea of race and the idea of culture originated during the nineteenth century, when many scholars struggled with the following question: How can the level of development reached by Western civilization be explained? Yet during the twentieth century, the destiny of these two ideas was to be radically different. Today, the idea of race and subsequent related terms—racial or racist—have a biological meaning devoid of any cultural content.[8] This was not the case during the late nineteenth century, when the symbiosis between the two terms was still dominant, surfacing in many pamphlets and books. For example, the theory of language developed by Ernest Renan (1823–1892) and the philosophy of art envisioned by Hippolyte Taine (1828–1893), were two of the most successful studies at the end of the nineteenth century illustrative in this respect: What explains the cultural level of development of Western civilization?

As with many other heuristic ideas, the idea of race had more than one source of inspiration and precise moment of birth. Indeed, many disciplines enthusiastically looked towards the new science of race. During the nineteenth century, the entire spectrum of the humanities engaged in an interdisciplinary dialogue that contributed to the creation of an impressive corpus of writings on race. The natural sciences, biology, anthropology, ethnology, anthropo-sociology, sociology, philosophy, the philosophy of history and of culture, literary criticism, philology, Oriental studies, and even biblical exegeses contributed to a new understanding of human life based on the manifestation of race. This is why there was no definitive theory regarding the human races, but a number of interpretations according to various disciplinary canons.

However, there were some common features that gave a certain conceptual profile to the idea of race within a Romanian movement of young intellectuals. Firstly, they considered culture to be the supreme national principle, one to which all other ideals were subordinated. According to them, culture was not the result of education (*Bildung*), but rather the inborn characteristics of the *Volk*. Culture was meant to preserve the traditions, customs and racial individuality of the nation.[9] Moreover, education, science, and civilization were gradually considered as factors of estrangement and "uprootedness" by these young intellectuals, who, on losing their national culture, were consequently lost in an urban alien world. Very few could resist the pressure of schooling in the universities of Western Europe and the "foreign civilization" these universities represented; and even fewer were able to return to their national culture after being "contaminated" by "foreign modernity."

Called the "young steeled," this handful of people claimed to embody the "power of national instincts" and managed to transform the idea of culture into a political weapon. Thus, at the beginning of the twentieth century, traditional politics was rejected while national culture became "true politics." As the university professor and mentor of the interwar generation, Nae Ionescu (1890–1940), remarked later, "the generation of 1906 relized that the nation is not a political instrument but a cultural one."[10] This was a new type of politics which, unlike conventional politics, would unite rather than divide the nation. The unity of the nation was guaranteed only by the unity of culture. According to this generation of Romanian nationalists, national culture united the nation because it was shared by "true Romanians" and not by the foreign, "improvised," and *parvenus* educated abroad. As the novelist Ioan Slavici (1848–1925) affirmed, "the *people* is the guarantee of national unity because the social elite have fatally foreign customs and habits." He went on to explain:

> [Let's take] a lawyer from Bucharest, one from Czernowitz, another from Cluj, one from Şimleu and one from Maramureş, and let us imagine them discussing the general principles regarding family life. The

same with administrative bureaucrats, professors, judges, and even priests (...) nobody could say that we have the same tradition. These different elites are pushing Romanian society in different directions. [On the contrary] the Romanian people has the same habits every-where, the same temper, the same customs, the same way of seeing and feeling, the same sense of moral being, and as long as we do not distance ourselves from the origins of our folk we will be soulfully one once again.[11]

All social and political phenomena were ascribed to national culture, which became the expression of the national spirit. This spirit of the people (*Volksgeist*) radically altered conventional political arguments because a new meaning was ascribed to the notion of "nation" on the basis of new ideological convictions. The political arguments regard-ing culture proved decisive in designing various normative projects of the nation because—unlike the inclusive liberal idea of the political nation—this culturalist idea of the nation was exclusive. As David Carroll has noted in *French Literary Fascism* (1995), culture became a source of authenticity and it was used against those exempted from the national collective. Moreover, by having a "Self," or its own individu-ality, this national collective was envisioned in racial terms—not only as a natural or biological entity but as a cultural ideal created and pro-tected through permanent struggle.[12] Therefore, the national struggle defined the national community according to the new standards estab-lished by the culturalist movement. Whoever was unaware of the necessity of this struggle was lacking in national consciousness and was susceptible of being alienated from the *Volk*. From this point of view, being a "foreigner" was not necessarily a sign of biological or ethnic difference, but the result of cultural influences "foreign" to the national culture.

Moreover, the nationalism of the liberal tradition (connected to the theory of historical rights) gradually became obsolete and was sur-passed by another form of nationalism: *lyrical* nationalism. In an epoch of the social expansion of civil rights and the democratization of national symbols, this new type of nationalism was capable of organiz-ing collective emotions. There was a transformation not only in the rhetoric of public discourse but in the social composition of public opinion and the political class. A new generation of political thinkers, litterateurs in the main, became prominent in politics. Therefore poets, those that harbored an appreciation of the sensibility of the psychological profundity of the *Volk*, were suitable candidates for leading the way forward. This "culture of grace," which tended to replace the "culture of law,"[13] was first conceived by the Romanian poet Mihai Eminescu (1850–1889).[14] However, this culturalist movement went a step further than just creating a symbolic national position for poets.[15] A number of writers entered the political arena simply because they were considered to represent the nation. Naturally, their main political argument was that they represented the true "Romanian Culture."[16]

Finally, this nationalist and culturalist movement encapsulated a "revolutionary ethos" that was anti-bourgeois. The promoters of this ethos were mostly intellectuals who had not found their place within the state apparatus and eventually became embittered. For this neo-Romantic youth, German student radicalism represented an appealing model. As early as the 1860s, student movements in Germany became a vanguard of the political Right, precursors of the revolutionary movements of the 1930s.[17] Beginning with the last decades of the ninetieth century, these generations of intellectuals harbored a profound aversion to modern society, in which they had failed to find their place. They expressed their contagious distrust of the "progressiveness" of humanity, democratic values, and anything connected to Enlightenment thought.

The modern world came under constant fire from works of art to pamphlets and articles. Modern civilization was depicted in terms such as urban, fragmented, mercantile, materialist, capitalist, liberal, rationalist, individualist, selfish, atheist (faithless), cynical, cosmopolitan, internationalist, Bolshevik, estranged, uprooted, improvised, sterile, prosaic, artificial, ignoble, sinful, illegitimate, disloyal, sick, and ugly. For these nationalist rebels, modern civilization was the antithesis of national culture. National culture was thus plied with terms like rural, communitarian, unitary, autarchic, idealist, agrarian, conservative, intuitive, collectivist, altruist, profoundly Christian, traditionalist, rooted in country soil, creative, poetic, noble, virtuous, brave, loyal, healthy, beautiful; and more importantly, it was perceived as seriously endangered by the expansion of modern society. For the "young steeled", there was no other solution than the "national offensive" or "revolution."[18]

By and large these three categories describe the culturalist, anti-political, anti-liberal, conservative, and profoundly anti-modern currents in the nationalism of the early twentieth century. This new nationalism

created a political sphere, based on the aesthetic and emotional funda-
ments of racial characteristics.[19] The battle cry was for the salvation of
the nation, in peril due to the unprecedented expansion of a decadent,
corrupt and foreign modernity. This "ideology of culture,"[20] in which
both culture and politics changed their meanings in order to describe
the more profound, unperceivable, ancestral or ahistorical traits of the
nation, was ultimately based on a racial template. For late-nineteenth-
century Romanian nationalists, race was the sum of all characteristic
features of the nation.[21]

Moreover, for the first time a new form of propaganda revealed the
beginning of anti-Semitism as a political current, independent of other
parties and factions.[22] This new anti-Semitism caused an unprecedented wave of publications, from booklets to newspapers, and from short
stories to novels. This anti-Semitism can appropriately be described as
"modern literary anti-Semitism."[23] Most of these texts are based on
figurative language that invaded the political sphere. This contributed
to a world-view in which the modern world and bourgeois society, with
all the attributes mentioned above, was identified with Semitism.[24]

Literary Anti-Semitism in Fin-de-Siècle Romania

The present section outlines the circumstances in which modern literary anti-Semitism coevolved with the narrative of a "national idea"
centered on the principle of racial difference. This found its literary
expression in the fight of the nation against the Jews, and its political
expression in the elimination of the Jews as the only alternative to
national destruction. This raises the following questions: Why were
Jews the central figure of this narrative? And why did Jews gradually
become the veritable alter ego (the constitutive "other") of the "healthy"
Romanians? In his 2004 *Imaginea evreului în cultura română* (The
Image of the Jew in Romanian Culture), Andrei Oişteanu explored
the stereotypes used by nationalist propaganda, differing from those
found in popular culture.[25] Though the images of Jews were similar in
national and popular culture, their anti-Semitic function was essentially different. Focusing on the "syntax of anti-Semitism," which can be
read as a complementary perspective to the "morphology of anti-Semitism" offered by Oişteanu, will help in responding to the questions
posed above.

Many scholars and commentators have noted that anti-Semitism is meaningless in so far as it is opposed to something that does not exist—namely, Semitism. Yet, many Romanian authors around 1900, perceived Semitism as a real threat to their nations, a "plague" that needed to be eradicated. Primarily, it was not a matter of ethnicity or race, and often Semitism was simply used as a metaphor for the predicament of modern society.[26] Modern society was simultaneously capitalist and socialist, uncultivated and decadent.[27] In other words, Semitism represented an assault, a universal conspiracy against everything local, traditional, autarchic and rural. It was in defense of this society that young intellectuals united their revolutionary efforts. For them, anti-Semitism was a sign of mutual recognition, "a cockade," as Shulamit Volkov has remarked.[28]

This particular "culture" was fundamentally anti-Semitic because it was built on the dichotomy between culture and civilization, a dichotomy literarily translated in terms of the opposition between nationalism and Semitism. This simple equation meant that a "real" nationalist was an anti-Semite, while those deemed Semitic were anti-national. If the victory of Semitism meant the death of the nation, the reverse was also true: the triumph of the nation implied the elimination of the Semites.

Why were Jews painted in such a negative light? The first answer takes into consideration the existence of a significant amount of anti-Semitic attitudes in modern Romanian history. There is no doubt that various pre-modern superstitions played an important role in a country like Romania, where the vast majority lived in pre-modern, if not archaic, conditions. The discussion regarding the political emancipation of the Jews, an issue that resulted in the publication of an impressive number of books and pamphlets, did not eliminate the anti-Semitic prejudices that survived even after the completion of Romanian statehood in 1918. However, it should also be noted that anti-assimilatory prejudices had liberal origins. According to liberal arguments, the Jews were incapable of living in a civilized world. As the writer Constantin Stere (1865–1936) explained in 1907: "If the Jews constitute, as I mentioned before, a *nation*, a distinct *cultural type*, their cultural type is not based on a common language which could serve as a starting point for a philosophical, scientific, literary creation of their own and thus contribute, as a *nation*, to the enrichment of human culture in general (...) *Religion* is what binds the Jews together, what makes them a people and gives them their specific cultural character."[29]

For those embracing this point of view, the only acceptable solution to the Jewish question was the emancipation of individuals, which was conditional nonetheless upon the "civilizational value" of the person in question.[30] In this case, civilization was not diabolized, and the generic image of Jews was considered to have nothing in common with modern civilization; therefore, Jews were not granted civil and political rights in Romania. On the contrary, this generation of young nationalists inverted these arguments, portraying Jews as a danger to national culture because of their identification with a modern, cosmopolitan civilization. The function played by Jews in this rhetoric was radically new, although many of the previous stereotypes were adopted and perpetuated. The Jew was no longer a bigot but an atheist who posed a threat to traditional spirituality; he was no longer a small trader but a socialist; and he no longer received instruction in Talmudic schools in the countryside but was educated at foreign universities. In other words, Jews were no longer despised but feared.

It was not only the anti-Semites who associated Jews with modernity; Jewish intellectuals also characterized Romanian Jewry as deeply committed to progressive values. Those who opposed the political and civil emancipation of the Jews considered them unworthy to join the civilization of the new Romanian state. As in France, Germany, and Austria–Hungary, the Jews in Romania organized themselves into associations that published innumerable periodicals and leaflets in an attempt to defend their rights and their image. A case in point is *Societatea de cultură israelită* (Society for Israelite Culture), founded in 1862 under the direction of Iuliu Barash (1815–1863). The aim of the society was to protect the political and symbolic integrity of the Romanian Jews.

However, the principles and values in the name of which the Romanian Jews pleaded their emancipation—namely modernity, civilization, progress, equality, and so on—came under fire. Young Romanian intellectuals were frustrated by the outcomes of modernity, considering its advancement to be overly fast. They argued that a period of "reaction" was necessary to temper the negative results intrinsic to rapidly adopting Western institutions, customs and, last but not least, culture. The historical irony was that the rhetoric of emancipation and assimilation of the Jewish community continued after the emergence of this culturalist movement, which proved to be against modernization, civilization and progress. In other words, the Jews made themselves conspicuous targets for the anti-modernist fury of the young generation of national-ist revolutionaries. When public opinion turned against Western notions of progress and civilization, modernity was vehemently attacked from both the Left and Right of the political spectrum. Jews thereafter con-stituted the last defenders of the Enlightenment values according to which all people were equal and rational, and a free society was the guarantee of progress and civilization.

A third response to the question as to why radical nationalists in fin-de-siècle Romania focused on the Jews is offered by a quasi-reli-gious interpretation of the national idea. During the revolutions of 1848–1849, a liturgical vocabulary played an important role in the rhetoric of revolutionary propaganda. Yet after the First World War, the national idea assumed the role of a "political religion."[31] The national revolution promised the salvation of the nation within historical time and under certain political circumstances. Thus, prehistoric (and ahis-toric) times, when the Romanians lived undivided in social classes, were equivalent to a veritable Garden of Eden, from which the diabol-ic Fall came in the form of barbaric invasions and the corruption of state institutions. Since that heavenly time, the nation had been perpet-ually corrupted by foreign invasions or dominations (migratory tribes, Turks, Hungarians, Greeks, Jews, Russians, and so on). Following this, a series of national heroes managed to embody both national instincts and the fight against decadent and faithless oppressors. These heroes were the martyrs and prophets of the nation, who carried the "good message" (gospel) of national salvation. This was also the message of those animated by the national spirit, writers and journalists "from around the hearth of national culture," as the historian and politician Nicolae Iorga (1871–1940) noted.[32] This was a messianic message. The undying body of the nation would be cleansed, after which the nation would be ready to fulfil its historical destiny: the creation of a purely Romanian nation-state. A Judgment Day was announced for all sinful people, foreigners and uprooted individuals, who endangered the existence of the nation. The national idea implied the destruction of the modern world and the foundation of another modernity which could better capture the ethnic essence of the nation.

In the Romanian literature of the fin-de-siècle period, the Jew un-equivocally personified this modern condition. Besides literature, the Christian tradition also constituted an important factor in the building up of a national messianism. However, anti-Semitic nationalists were less concerned about theological issues than political plans. In other words, they cared far less about who crucified Jesus than about the fact that those who did not recognize the Messiah in the first place might seek to crucify him again.[33] The poet Octavian Goga (1881–1938), for example, offered himself as a kind of national "Jesus born in a cowshed, among the animals."[34] Other nationalist leaders made similar statements. Besides the "Providential leader," another aspect of the messianic message was the national offensive launched against enemies of the national idea. Once again, the Jew was the figure most frequently depicted as the enemy. If the pseudo-theological argument of Romanian nationalists in the late nineteenth century is followed through, one can interpret their anti-Semitism as the destruction of modernity, or as the modernicide necessary for the foundation of the kingdom of their national ideal on Earth.

Culturalism and the New Nationalism

Towards the end of the nineteenth century, the public spirit in Romania underwent a number of important mutations. Under pressure from the crises of modernity (the formation of the nation-state) and the influence of Western ideas (German idealism and the French Right in particular), a new narrative about the nation was created by the Romanian intelligentsia. The change was gradual; initially the movement grew unnoticed. Yet in less than a decade a new rhetoric pervaded Romanian public space and dominated political discourse to such an extent that few gauged the source of this change. Most commentators viewed the trend as a sign of progress and emancipation. These changes induced a new understanding of the political nation and of the importance of the national writer to politics.

This was not particular to Romania, but was instead a Europe-wide phenomenon. It was against this background that many Romanian intellectuals became acquainted with forms of *völkisch* ideology, Social Darwinism, Wagnerian mythology, Nietzschean philosophy, Wundtian ethno-psychology, and so on. The new slogan of this generation was "togetherness," as opposed to the "loneliness" of the modern world. Their idealist *Weltanschauung* was radically opposed to the liberal vision of the world and challenged the conventional understanding of politics as it was developed in the first part of the nineteenth century.

Even the term "politician" became detested as a symbol of unscrupulousness and moral corruption; therefore politicians were obliged to endorse "men of culture" or "men of letters," writers who were in a seemingly better position to understand the spirit of their nation (*Volksgeist*). In their literary and political writings, modernity was depicted as the rule of civilization, discord and cynicism, trade and personal interests, cosmopolitanism and internationalism. In contrast, the veritable culture was national, *völkisch*, and autochthonous. For the sake of the Romanian nation, culture was considered the source of inspiration for politics. National politics was therefore depicted as the struggle between national culture and a foreign, invading modernity.

Correspondingly, there were a number of political events triggering the emergence of a culturalist movement in Romania.[35] The first was a secret political treaty ratified in 1883 (and renewed in 1892) between Romania, Austria–Hungary, and Germany. This treaty remained secret, but its consequences—namely the adoption of a favorable attitude by the Romanian government towards the Austro-Hungarian Empire— were apparent. But the position of non-interference in the home affairs of neighboring countries by Romanian conservative elites provoked criticism, even among its traditional supporters. However, a tacit consent was needed to keep this alliance secret.

In 1892, Dimitrie Sturdza (1833–1914) became the leader of the Romanian Liberal Party, which was in opposition at that time. Sturdza decided to use the "national question" in Transylvania as a political weapon against the Conservative Party in order to force the hand of the king, Carol I (1893–1953), in demanding that the Liberals form a new Cabinet. As the conservative politician Titu Maiorescu (1840– 1917) stated, it was the first time that the "national question" was raised in the Romanian Parliament.[36] Sturdza's political campaign convinced Romanians in Hungary that it was an appropriate moment for the drawing up of a Memorandum in which their national grievances could be made known to both the Habsburg Emperor and European public opinion.

The Romanian National Party in Hungary (Partidul Naţional Român, PNR) could not reach an agreement about this issue. Led by Alexandru Mocsionyi (1841–1909), there was a sizeable faction in the party opposing the Memorandum, but this group eventually accepted the will of the majority.[37] Another group was similarly undecided about the form of such a memorandum. Ioan Raţiu (1828–1902), the leader of the PNR, led this group. The so-called "Tribunists" (after the journal *Tribuna*) constituted the other group in the party. Although the "Tribunists" sought to ameliorate the relationship between the Hungarian government and Romanian politicians, they were the most radical force supporting the publication of the Romanian Memorandum. It is not surprising, therefore, that when Ioan Slavici wrote the first draft of the Memorandum it was rejected by the Romanian National Committee as too radical. Eventually Slavici found support for his version of the Memorandum in Bucharest. On 24 January 1891, two Transylvanian Romanians, Gheorghe Bogdan-Duică (1865–1934) and Ioan Russu-Şirianu (1864–1909), convened the League of Cultural Unity of all Romanians (*Liga Unităţii Culturale a tuturor Românilor*), also known as the Cultural League (*Liga Culturală*). After the foundation of the Cultural League, the course of events changed noticeably. On the one hand, the memorandum movement continued its course in Transylvania, where the petitioners were trialled (in 1894) and sentenced to prison for several years.[38] On the other hand, when the Memorandum was published it provoked responses from Hungarian and Romanian students in Austria– Hungary. This polemical exchange of pamphlets signified the beginning of a Europe-wide campaign in support of Romanians in Hungary.[39] Nevertheless, on 13 October 1895, a few days after he was appointed prime minister, Sturdza refuted his previous opinions on the "national question" in Austria–Hungary, even as he offered Romanian nationalists in Transylvania the alternative of immigrating to Romania and continuing their work under the banner of the Cultural League. This change of attitude alienated the Liberals from those segments of the public in Romania concerned about the fate of their "Romanian brothers abroad."

With the foundation of the Cultural League, nationalism became the foremost political doctrine in Romania, eclipsing conservatism and liberalism. The name of the league suggested that the solution to the "national question" resided in a doctrine that transcended the conventional political spectrum, seeking the "cultural union" of all Romanians.

Under these circumstances two figures rose to prominence: Nicolae Iorga and A.C. Cuza (1857–1946). Between 1891 and 1910, the movement for the "cultural union" of all Romanians was gradually absorbed into a nationalist doctrine under the direction of these two friends. On the one hand, Iorga brought to this partnership his academic reputation, an admirable work ethic, and a stubborn character. On the other,

Cuza was an agile spirit, and possessed a remarkably witty and speculative mind. Both contributed to the journals *România Junâ* (Young Romania) and *Sămănătorul* (The Sower), with Iorga later becoming the editor of *Sămănătorul* (1905–1906). Eventually, he was appointed secretary of the Cultural League in 1914, and later its president in 1919. Both intellectuals made vital contributions in transforming the Romanian nationalist movement, which in 1910 came to form the basis of a nationalist political party, the National Democratic Party. The publication *Neamul Românesc* (The Romanian People), under the direction of Iorga, became a link between the new party and the Cultural League. Cuza also published numerous anti-Semitic articles in this cultural review.

Iorga expressed his views on "national policy" and its connection to national culture in a political speech delivered in 1908.[40] The speech is significant in so far as it shows the transformation of public discourse at the time. The "national question" was deemed politically paramount, with the Cultural League representing the interests of the entire Romanian nation, which extended beyond the narrow political fragmentation of parties and coteries. The Cultural League was entitled to this prominent position because it advocated "the union of Romanians from all social strata; of those who know and feel Romanian (...) and who, with the help of God and our humble assistance, will one day really know Romania."[41]

For Iorga, the notion of national policy was related to "national culture" in a way that had never been properly explained before. According to him, conventional politics was founded on the interests espoused by particular groups of people. Politics therefore differentiated and separated segments of society, reflecting the interests of individuals rather than the nation as a whole. A new political system could be founded only on the basis of national culture. The proponents of the culturalist argument stressed the cultural unity of the nation. "Culture" was defined as follows:

> Culture is the root of all things; culture is the soul, and everything begins with the soul. Material things never determine the life of a people: a people's life springs from its soul, and its soul is its culture. A people with culture has a soul and is (nationally conscious. It is like in the Scriptures: what was at the beginning? What? The World or the Word? At the beginning, there was only the Word! We who represent the Word, who serve the Word, not only by being above different polit-

ical parties but also [by] ignoring them, we claim: nobody can speak
about a more clean and honest Romanian politics than those around
the fireplace of culture, where the burning fire illuminates the life of
the entire people.[42]

In the 1923 *Doctrina naționalistă* (Nationalist Doctrine), Iorga referred
to the collective wisdom of the people as a foundational source of
national policy.[43] Arguing that there was something particularly valuable in the tradition and popular customs of the nation, he attempted to
provide a stronger foundation for national culture. He thus defined
national culture as something inherited, not acquired. The "soul of the
nation" was a trans-individual reality encompassing the ancestral
experiences of the Romanian people. This was not only a eulogy for
the ancestral qualities of his nation; for Iorga, the state was the creation
of the nation, with the latter originally comprised of only one social
class, the peasantry. The state, then, should be led by its national basis,
in this case the peasantry, and not by the aristocracy as elsewhere in
Europe. At the beginning of the twentieth century, Iorga's idea of class
homogenization remained unpopular in the Romanian Kingdom, but it
found supporters among Transylvanian Romanians. The Tribunists
adopted most of the ideas presented by Iorga in *Semănătorul* and, after
1906, in *Neamul românesc*. Their respect for Iorga grew not only
because of his academic achievements but also due to his persistent
interest in the Romanian cause in the Austro-Hungarian Empire.

Conclusions

The culturalist movement that developed in fin-de-siècle Romania rev-
olutionized political thought and rhetoric. It promoted a new narrative
about the nation and the state, based on a national idea rooted in the
primacy of culture. This new narrative did not represent a logical discourse about the Romanian nation—it was a web of significances
aimed at capturing the emotions of the masses. The main characteristic
of this nationalist discourse, promoted by a young generation of intellectuals, was a figurative language, whose substance was derived from
metaphors not from precise concepts. The development of this nationalist discourse had immediate consequences for the depiction of the
image of the Jew in Romanian culture. A Semite was not necessarily a

Jew. On the one hand, the Semite, the "Yid" (*jidanul, jidovul*), was not always used pejoratively against the Jews, but it was often used with the clear intention to offend certain ethnic sensibilities of political enemies. Just as the German anti-Semite Wilhelm Marr (1819–1904) accused Germany of being "Judaized," Ioan Slavici and, later, Octavian Goga accused Hungary of being "Semitized." For Goga, modernity itself was Semitic and destructive to national interests. These young rebels were not against modernization but the type of modernity promoted throughout the entire nineteenth century—namely, a cosmopolitan, urban and liberal modernity: the heritage of the Enlightenment.

Three aspects of this culturalist movement proved vitally important for its future development during the interwar period. The first was a "revolutionary ethos." This was a by-product of the political challenges to liberal traditions and modern institutions. The second was a racial understanding of culture. "Culture" was less an artificial construct and increasingly natural and innate, gradually becoming a matter concerning the national instinct for self-preservation. Within this culture, the "foreigner" could not be accepted because assimilation was impossible. The "foreigner" represented another culture that should not be assimilated into the national community. The only solution was the radical isolation of national culture from "foreign influences," and eventually the elimination of these sources of cultural estrangement. Finally, the cultural understanding of race became progressively prominent in Romania. It provided the opportunity to interpret the fabric of society as a combination of two distinct layers: one genuinely national; the other a "super-imposed stratum"[44] which was artificially "imported." This new language for interpreting Romanian society was the essential element in establishing the frontline of the "national revolution" and in defining the true enemy of the nation: Semitism. Moreover, it was a way of personifying social evil and offering an object of collective loathing. During the economic crises of the late 1920s, this racial interpretation of culture was interpreted literally by many nationalists: getting rid of the Jews would overturn the primary cause of national crisis.

The narrative of the national idea as an anti-Semitic plot against national culture, and as a temporal manifestation of race, presupposed not only the views of the literary tropes of radical nationalism during the interwar period and the identification of its intellectual sources, but, at the same time, the understanding of a current of thought that survived into the Communist era.

Endnotes:

[1] See Ernest Renan, *De l'origine du langage* (Paris: Michel Levy, 1858), 20.

[2] Dimitrie A. Xenopol, *Politique de la race* (Rome: Forzani E. C. Tip. Del Senato, 1903), 8.

[3] See Ioan Slavici, "Barbaria modernă," *Apărarea Naţională* 3, 79 (1902), 1.

[4] Dimitrie A.Xenopol, "Unitatea sufletului românesc," *Arhiva* 12, 5 (1911), 193.

[5] *Sămănătorul* (1902–1910) was one of the most important cultural journals ascribed to serving the national movement at the beginning of the twentieth century, to which many eminent intellectuals contributed articles and feuilletons. Nicolae Iorga was the editor of *Sămănătorul* between 1905 and 1906, before founding his own journal, *Neamul Românesc* (1906–1940), which for many years was the quasi-official organ of the Cultural League. In 1936, Iorga attempted to resuscitate *Sămănătorul* under a new name, *Cuget Clar (Noul Sămănător)* (1936–1940).

[6] Octavian Goga, "Politica etnică a României," *Ţara noastră* 8, 12 (1927), 292; and Octavian Goga, "Conferinţa dela Teatrul Naţional, rostită cu ocazia sfinţirii Catedralei din Cluj în Noemvrie 1933," in Octavian Goga, *Discursuri* (Bucureşti: Tiparul Cartea Românească, 1939), 53.

[7] This famous expression, usually attributed to Charles Darwin, was actually coined by Herbert Spencer in "A Theory of Population," published in the *Westminster Review* 57 (1852), 250–268.

[8] See, for example, the entry "Sociology of Race" by Nicholas Abercrombie and Stephen Hill, in Bryan S. Turner, ed., *Dictionary of Sociology* (London: Penguin Books, 1994), 406–407.

[9] In his study on French literary fascism, Carroll noted: "in fact *tradition* performs exactly the same function as race in racist theories, enabling the French to be *pre*formed and providing cultural rather than racial typologies of what it is to be French." See David Carroll, *French Literary Fascism: Nationalism, Anti-Semitism, and the Ideology of Culture* (Princeton: Princeton University Press, 1995), 30.

[10] This fragment, originally published by Nae Ionescu in 1928 in *Cuvântul*, is quoted by Mircea Eliade in "Cultură sau politică?" *Vremea* 7, 377 (1935), 3.

[11] Ioan Slavici, "Unitatea Noastră Culturală (part V)," *Tribuna* 11, 11 (1907), 2; reprinted in Ioan Slavici, *Amintiri* (Bucureşti, Editura pentru literatură, 1967), 506–507.

[12] Carroll, *French Literary Fascism*, 20, 26.

[13] Carl Schorske refers to the two cultures coexisting in Vienna: the "culture of grace," which derived from the Counter-Reformation and was manifested at the level of religious life but survived in art; and a "culture of law," which related to Enlightenment values on moral, political and scientific order. Within Viennese modernity, Schorske identifies the interaction of two cultures, Baroque and Enlightenment, which simultaneously complement and compete against each

other. See Carl Schorske, *Thinking with History: Explorations in the Passage to Modernism* (Princeton: Princeton University Press, 1998). Schorske's interpretation is particularly relevant for the case of fin-de-siècle Transylvania, where an anti-liberal ethos animated the young nationalist generation. The scandal caused by the journal *Tribuna* (1910–1912) created two opposition camps which identified themselves with writers and poets on the one side, and lawyers on the other. In this debate, the "culture of grace," in open conflict with the "culture of law," established the parameters of Romanian messianic nationalism.

[14] Mihai Eminescu is considered to be the national poet par excellence, the highest embodiment of the "Romanian soul." His political articles were rediscovered after 1891 as a source of inspiration for the new culturalist movement. A series of editions of his political articles were published by individuals connected to the Cultural League (Gr. Păucescu, 1891, 1894; Ion Scurtu, 1905; G. Teodorescu Kirileanu, 1909; and A. C. Cuza, 1914).

[15] As Carlyle stated in 1840: "Yes, truly, it is a great thing for a Nation that it get an articulate voice; that it produce a man who will speak forth melodiously what the heart of it means! Italy, for example, poor Italy lies dismembered, scattered asunder, not appearing in any protocol or treaty as a unity at all; yet the noble Italy is actually *one:* Italy produced its Dante; Italy can speak! The Tsar of all the Russians, he is strong, with so many bayonets, Cossacks and cannons; and does a great feat in keeping such a tract of Earth politically together; but he cannot yet speak. Something is great in him, but it is a dumb greatness. He has had no voice of genius, to be heard of all men and times." See Thomas Carlyle, *On Heroes, Hero-Worship, and the Heroic in History* (Pennsylvania: Pennsylvania State University Electronic Classic Series, 2001), 98.

[16] Culture, as manifestation of the "genius of race," became an essential feature of the nation. As far as society was regarded as having a (collective) Self, Culture was personified.

[17] See William McGrath, *Dionysian Art and Populist Politics in Austria* (New Haven: Yale University Press, 1974). The case of Romanian intellectuals who emulated the great German models and aestheticized politics by creating "a poetry of politics" (the term used by McGrath) has never been explored.

[18] See Simion Mehedinți, *Ofensiva națională* (București: Tipografia Cooperativă, 1912). Goga used the term "revolution" extensively after 1933.

[19] See Lucian Blaga, "Despre rasa ca stil," *Gândirea* 14, 2 (1935), 69–73; see also Ilarie Dobridor (pseud. Constantin Iliescu Cioroianu), "Sărăciei spirituale a evreilor," *Gândirea* 16, 10 (1937), 502–513.

[20] See Carroll, *French Literary Fascism*, 247.

[21] See Shulamit Volklov, "Antisemitism as a Cultural Code: Reflections on the History and Historiography of Antisemitism in Imperial Germany," *Leo Baeck Institute Year Book* 23 (1978), 25–46.

[22] Seemingly, in 1879, Wilhelm Marr first employed the term "anti-Semitism." Whether or not he coined the term is less important than the resonance of his brochure *Der Sieg das Judentum über das Germantum* (Berne: Rudolph Costenoble, 1879), which synthesized a series of ideas and suggestions already

present at the time in works of various contemporary scholars, one of which was that the Germans were a race conquered by Jews who wanted to "Judaize" Germany. In the same year, Marr founded the Anti-Semitic League, in which many Romanians had a significant role. See Bernard Lazare, *L'Antisémitisme. Son Histoire et ses causes* (Paris: Déon Chailley, 1894), 241–242; and also Moshe Zimmermann, *Wilhelm Marr: The Patriarch of Anti-Semitism* (New York: Oxford University Press, 1986). For the Anti-Semitic Alliance, founded in Bucharest in 1895, see Bernard Lazare, *Les Juifs en Roumanie* (Paris: Cahiers de la Quinzaine 1902), 80–85, where several paragraphs from the Alliance's statutes are quoted. Constantin Istrati, who became the minister of education in Romania in 1900, was one prominent member of the Alliance. *The Antisemitic League* was founded in 1900 in Bucharest and was connected to *Antisemitul*, a journal with a short existence (1899–1901). Between 1905 and 1916, another journal, *Vocea Dreptăţii* (in 1906 renamed *Liga Antisemită*) represented this anti-Semitic trend in Romanian public sphere.

23 It is literary because it employs numerous literary tools, and modern because it was different from any previous anti-Semitic work. In this respect the term is closer to "literary anti-Semitism" as David Carroll understands the term, than the way in which Mark Gelber addresses this issue. See Carroll, *French Literary Fascism*, 170–195; and Mark H.Gelber, "What is Literary Antisemitism?" *Jewish Social Studies* 46, 1 (1985), 1–20.

24 The term was used by Ioan Slavici in his "Semitismul (parts I–V)," *Tribuna* 12, 131–134 (14–18 June 1908), 1–2.

25 Andrei Oişteanu, ed., *Imaginea evreului în cultura română. Studiu de imagologie în context est-central European* (Bucharest: Humanitas, 2004).

26 See Răzvan Pârâianu, "Semitism as a Metaphor for Modernity", in *Studia Hebraica*, vol. 5 (2005): 23–68.

27 See Brian Chayette, *Construction of the Jew in English Literature and Society: Racial Representations, 1875–1954* (Cambridge: University Press, 1993), 9.

28 See Volklov, "Antisemitism as a Cultural Code," 25–46.

29 Constantin Stere, "Problema naţională şi social-democratismul. Antisemitismul," *Viaţa Românească* 2, 11 (November 1907), 190–191; italics in original.

30 See Dimitrie A. Xenopol, "Naţionalism şi anti-Semitism," *Noua Revistă Română* 5, 18 (1908), 276–278.

31 See Emilio Gentile, *The Sacralization of Politics in Fascist Italy* (Cambridge, Mass.: Harvard University Press, 1994)

32 Nicolae Iorga, *Cultura naţională şi politica naţională. Discurs ţinut de N. Iorga în seara de 14 Octombrie 1908 la Bucureşti (Sala Dales)* (Vălenii-de-Munte: Neamul Românesc, 1908).

33 Octavian Goga, "O tempora... tempora!" *Tribuna literară* 239 (1901), 208.

34 Octavian Goga, "Oameni...," *Luceafărul* 11 (11 March 1912), 219.

35 For a move detailed account see Răzvan Pârâianu, "National Culture as a Plot against Modernity", *Studia Hebraica,* 4 (2004), 103–116.

[36] Titu Maiorescu, *Istoria politică a României sub domnia lui Carol I* (București: Humanitas, 1994), 195–204.

[37] Sever Bocu, *Drumuri și răscruci* (Timișoara: n.p., 1939), 43.

[38] On 18 September 1895, King Carol I signed the protocol of a new alliance treaty with Austria and Germany. The next day, Emperor Franz Joseph I pardoned the Memorandists.

[39] See Eugen Brote, *The Romanian Question in Transylvania and Hungary* (Bucharest: Voința Națională, 1895).

[40] Nicolae Iorga, *Cultura națională și politica națională* (Vălenii-de-Munte: Neamul Românesc, 1908).

[41] Iorga, *Cultura națională și politica națională*, 4.

[42] Iorga, *Cultura națională și politica națională*, 7.

[43] Nicolae Iorga, *Doctrina națională* (București: Cultura națională, 1923).

[44] Mentioned, for example, by Mihai Eminescu in his "'Pseudo-Românul' ca să esplice," *Timpul* VI, 185 (1881), 3; included in Mihai Eminescu, *Opere* XII (Bucharest: Editura Academiei, 1985), 312.

The Politics of Hatred:
Scapegoating in Interwar Hungary*

Attila Pók

The loss, between 1918 and 1920, of two-thirds of pre-war Hungarian territory after the Treaty of Trianon (4 June 1920) caused trauma and repercussions, still felt in the present day. Not surprisingly, the "Trianon syndrome" is a standard point of reference when dealing with any aspect of twentieth-century Hungarian history. The argument of this chapter is that the conceptual framework of scapegoating is useful in explicating one of the key problems of twentieth-century Hungarian history—namely the relationship between anti-Semitism and Hungarian involvement in the implementation of the Holocaust in Hungary in 1944.[1] If there had been a Hungarian *Historikerstreit*, this issue could well have been one of its focal points.[2] The same question may also be asked differently: Does the Holocaust in Hungary represent the apogee of a long-term evolution in Hungarian anti-Semitism, one rooted in early modern and contemporary Hungarian economic, social and cultural history? Or, instead, was it the result of short-term antecedents rooted in interwar Hungarian society; can it, in fact, be traced to what we might call the "Trianon syndrome"? This chapter shall discuss the "complexity of complicity" in reference to the socio-psychological tool of scapegoating.

In terms of method and sources much here is collaborative: throughout, the works of many colleagues, including historians, social psychologists, anthropologists and philosophers, are employed. There are five authors, however, whose ideas and insights were of particular importance in my conceptualization of scapegoating: Ferenc Pataki, whose 1993 article effectively described the idea of scapegoating;[3] György Hunyadi, who, for decades, has been trying to build bridges between history and social psychology, and whose 1998 book on stereotypes during the decline and fall of Communism deserves more attention;[4] Randolph Braham, whose studies on the Holocaust in Hungary combined scholarly expertise with humanitarian and democratic commitment;[5] Omer Bartov, whose article, "Defining Enemies, Making Victims: Germans, Jews and the Holocaust," fittingly contextualizes the Hungarian case;[6] and a Ph.D. thesis on the "Trianon syndrome," "Pursuing the Familiar Foreigner: Resurgence of Anti-Semitism and Nationalism in Hungary since 1989," by Jeffrey S. Murer, which emphasizes the importance of critical theory for the study of anti-Semitism.[7] Finally, the choice of this subject was motivated by recent historical and political debates about the variant forms of terror and dictatorship that characterized twentieth-century Hungarian history.

Historical Sources: The Scapegoat in the Old Testament

The original meaning of the scapegoat is first explained in Leviticus, the third book of the Pentateuch. In the course of a ritual, Aaron lays both his hands upon the head of a goat and confesses all the iniquities and the transgressions of the people of Israel, thereby transferring the sins of his kin onto the goat that is, in turn, expelled into the wilderness. The message of the ritual is clear: the scapegoater is fully aware of his guilt and is most consciously trying to get rid of it. The nature of guilt is also determined by context: it infers admittance of a crime, having broken the law, which may arise from bad character or sinful, irresponsible behaviour. According to this ritual, however, the background to guilt is unimportant, for the scapegoater is tortured by remorse. Scapegoating is a comfortable way out of a troubling situation; it provides relief. The scapegoaters of the Old Testament are not guilty; however, they are carefully identified as carriers of a burden. Mediaeval and early modern European cultural history provides numerous examples of this type of scapegoat. Many English sources, for example, write about the "whipping boy," the young or low-ranking person forced to accept punishment for crimes committed by his superiors. English and French sources also describe "sin-eaters" that "ate up" the sins of the dead.[8] In its original conception the ritual is in no way to be confused with a sacrifice, an offering to God that can also be a goat (a bull or a ram).

Methodology: Social Psychology and Scapegoating

The scapegoat of modern social psychology is quite different. Modern twentieth-century scapegoaters are convinced of the guilt of their victims and consider themselves innocent. In Hungary, the loss of territory is the most important source of the greatest national traumas, and therefore scapegoating, on a national level, is connected to this issue. It is at this point that the analytical tools of social psychology may be applied.[9] In particular, the concept of *enforced attribution* has proved the most useful. It describes a simple and common phenomenon neglected by a number of historians. Human nature dictates that both individuals and groups require clear-cut, mono-causal explanations for all—especially negative—events, including complex historical and political phenomena. Questions about how to explain military defeat, deep economic crisis, cultural decline, and so on, cannot be substantially or satisfactorily answered by clearly defining or identifying a perpetrator. In order to preserve or re-establish a group's (be it a society's or national community's) self-esteem and self-respect, the perpetrator (be it an individual, a smaller or larger group of people, or an abstract force) has to be named. In this way, the cohesion of a given group (society) can be strengthened, which is essential for the survival of the group or, in this case, the national community. Instead of opening up internal cleavages, focusing hatred on a scapegoat can be indispensable in preserving the integrity and strength of the group. This directly relates to the mobilizing function of scapegoating, a feature which is a decisive characteristic of emerging totalitarian and authoritarian regimes, as well as the movements behind them. It includes what Eric Erikson has called "pseudo-speciation,"[10] meaning the scapegoat is considered to belong to a species different from that of the scapegoater. Exemplifying this idea is Heinrich Himmler's speech on the destruction of the Jews, delivered to German generals in Sonthofen on 5 May 1944:

> You can understand how difficult it was for me to carry out this military order which I was given and which I implemented out of a sense of obedience and conviction. If you say: "We understand as far as men are concerned but not in the case of the children," then I must remind you of what I said at the beginning. In this confrontation with Asia we must get used to condemning to oblivion those rules and customs of past wars which we have become used to and, indeed, prefer.

In my view, we as Germans, however deeply we may feel in our hearts, are not entitled to allow a generation of avengers filled with hatred to grow up, with whom our children and grandchildren will have to deal because we, too weak and cowardly, left it to them.[11]

This widespread interpretation of the aim of Germany's struggle had a great impact on many German soldiers and civilians alike. They drew the conclusion that the otherwise "normal," "regular" prohibition of aggression against one's "own species" did not apply to Jews and other "inferior" people.

The Ideal Scapegoat

Works on the subject agree that the ideal scapegoat is markedly different from the majority of people in a given society, either in the negative or in the positive sense of the word. Both exceeding and falling short of the norms of the collective may stir up the hostile, exclusionist instincts within society. National minorities invariably serve this purpose well. René Girard has pointed out that the concept of scapegoating is a potentially violent notion.[12] Following Girard, it is not only the deviation from the group's norms that incites hatred and violence but also if a religious or national minority is ready to accept accommodation and complete assimilation. Recent scholarship shows that, for example, anti-Semitic agitation was aimed at those Jews displaying a tendency towards assimilation rather than their co-religionists who openly declared their Jewish identity.[13]

The "Trianon Syndrome" in Hungary

This theoretical overview is directly relevant to the "Trianon syndrome" in shaping political, social and cultural attitudes towards Jews in Hungary. A careful examination of scapegoats of all kinds shows that the scapegoat-victim paradigm incorporates the most despised elements of one's own identity (group or individual). The members of the group project internal tensions, worries, and fears onto the scapegoat. For example, in smaller groups of adolescent children, those with a mature physique and attractive appearance frequently become scapegoats:

the group punishes the member who deviates from customary norms. They punish the scapegoat for what they secretly long to be themselves, and this way, they disown their own desires.[14] The idea of "blame allocation" means not only the evasion of responsibility but also obscuring the original problem. However, the scapegoater is almost always fully aware of the nature of a problem and its causes. The scapegoater is also prisoner of enforced attribution, part of a simplistic, mono-causal explanation of a complex occurrence.

Why was the ire of Miklós Horthy's closest associates (members of the "national government" and the "national army") primarily aimed at the Hungarian Jewry? First of all, the dimensions of the national disaster in 1918 were beyond comprehension. Who or what can bring about such a fundamental change in the life of a nation and a state—especially one with a thousand-year history—to the brink of complete destruction?[15] That force must be of extreme and barely conceivable strength. Resurrection is hardly possible without self-examination and atonement. If an individual or a small group experiences disaster, the first step towards recovery is the ritual of mourning. Mourning, and its expression in the form of various rituals, comprises the acceptance and acknowledgement of tragic loss. This "adaptive mourning" relieves the individual or community from an obsession with the past, opening up the possibility of contemplating a vision of the future.[16] "Adaptive mourning" was not a feasible alternative for Hungarian society after the First World War; no nation would have willingly accepted the loss of two-thirds of its homeland and more than one-third of its population. The lack of "adaptive mourning" has made it difficult to determine the causes of the tragedy, let alone the culprits. To blame the victorious Entente Powers, or Hungary's new neighbors, was not a viable option: some neighbors were in a position to impose further demands on the country. There remained only one serious option: the national community should identify "some part of itself that it can cut off or remove and then project the guilt onto the amputated part, onto the abject."[17] That part of the Hungarian self, the "familiar foreigner," was the Hungarian Jewry. The Hungarian Jewry was sufficiently assimilated to identify with the national self, and yet sufficiently foreign for its exclusion from the new conception of "Hungarianness." This amputation, unfortunately, proved literal: shortly after the Red Terror carried out by the Hungarian Soviet Republic under Béla Kun (1886–1939) in 1919, hundreds of Jews were killed by the White Terror.[18] Politically motivated pogroms with a high death toll constituted a new phenomenon and were certainly not part of previous Jewish–gentile relations in Hungary.[19] The anti-Semitic brutality of the White Terror (even more than the frequently cited "Numerus clausus Law" of 1920)[20] signified a qualitative turning point in the history of anti-Semitism in Hungary, and a critical step forward on the road to the Holocaust. The Holocaust is thus directly connected to the radicalization of nationalism, rather than to historically rooted anti-Jewish sentiment or modern anti-Semitism. The anti-Semitic arguments associated with the period of the Second World War trace their roots to the socioeconomic and political realities of the First World War and the immediate post-war period.

As illustrative examples of this point, I refer to five interpretations of the "Jewish question" in explaining the Hungarian tragedy of Trianon. All personalities discussed here are outstanding figures of twentieth-century Hungarian history; they have all shaped Hungarian political thought, and they have each been the subject of significant historical controversy.

Pál Teleki (1879–1941) came from an aristocratic family in Transylvania. He was a renowned geographer and during two crucial periods (1920–1921 and 1939–1941) was the prime minister of Hungary. From the very beginning of his career, Teleki suggested using biology in the interpretation of social phenomena. The impact of eugenics, emerging as a discipline in fin-de-siècle Europe, was decisive in shaping his views on the social, cultural and political role of Jews in Hungary. In this vein, Teleki viewed Hungary's collapse after the First World War as a direct consequence of the victory of Jewish influence over Christianity in all national spheres. In his view, the regeneration of the nation and the state demanded positive, constructive measures that would, on the one hand, strengthen the patriotic Christian social groups and, on the other, curb and limit Jewish influence in the economy, education and culture.[21]

Lajos Méhely (1862–1953), whose father worked as a caretaker on the estates of aristocratic families in the northeast of Hungary, had a successful career as a biologist prior to the First World War. Méhely interpreted the war as the most advanced form of the Darwinist struggle for existence and survival among and between races. Well read in the eugenic literature of the time, Méhely pointed out that in this struggle the Hungarians, on the basis of their racial characteristics, were doomed to failure. The cause of Hungary's tragedy was the mixing of

Hungarians with other races in the Carpathian Basin. Consequently, in order to overcome this national trauma, the Hungarian race needed to be cleansed of Jewish and German influence.[22]

Dezső Szabó (1879–1945), the son of a Transylvanian radical Calvinist, was one of the most influential Hungarian writers during the 1920s. Szabó blamed a conspiracy of "Syrian bandits" and "immoral, wild Jewish imperialism" for Hungary's ill-fated circumstances between 1918 and 1920. His novels and essays reveal two nationalistic motifs. First, Szabó assumed that the negligent and liberal attitude towards Jews during the nineteenth and early twentieth century displayed by the Hungarian political elite resulted in increased Jewish assimilation; and second, that the Germans were as threatening, if not more so, than the Jews. Both factors supposedly explained Hungary's demise by 1918.[23]

Gyula Szekfű (1883–1955), from a Roman Catholic family, published the most powerful interpretation of Hungary's national tragedy as *Három nemzedék* (Three Generations) in 1920. According to Szekfű, three successive generations of the Hungarian political elite erroneously pursued, and were ultimately misled by, the mirage of Western liberalism. As a consequence, the Jews, in the main, permeated and weakened Hungarian society. It was Hungary's internal weakness, rather than its military defeat, that caused the 1918–1919 revolutions and territorial losses. During the 1920s, Hungarian public opinion focused on the anti-Semitic implications of Szekfű's analysis. Careful reading of Szekfű's book, however, reveals that Szekfű's criticism targeted those negligent liberal politicians who advocated the assimilation of the Jews into Hungarian society.[24]

Oszkár Jászi (1875–1975), the son of a Transylvanian doctor and radical critic of Hungary's political establishment, participated in the 1918 revolution and emigrated in 1919, after its failure. In 1920, in Vienna, he published his views on the causes of the dismemberment of Hungary. Like Szekfű, he also focused on internal factors. However, Jászi did not blame liberalism as such, but rather the lack of liberalism, as the root of national disaster: "All serious liberal-minded intellectuals were silenced during the last quarter century; (...) all liberal, cultural and political aspirations were trampled upon by plundering gangs of leaders intoxicated by nationalism." Jászi's conclusion, however, comes surprisingly close to that of Szekfű: "The Hungarian soul turned out to be sterile and the thinning ranks of the army of culture were increasingly filled by aliens, first of all Jews, which, in turn, led to a disgusting mixture of feudalism and usury."[25]

Different as these personalities—and their approaches to the role of the Jews in the collapse of the Hungarian state at the end of the First World War—might have been, one view was shared by all during the 1920s and 1930s: the victorious powers in Paris had struck the final blow on an already fatally sick Hungary, and the prerequisite for national rejuvenation was finding a decisive solution to the "Jewish question."

The Classical Trope: Jews–Freemasons–Communists

Frequent references to the Jewish–Freemasonic–Bolshevik conspiracy played a decisive role in the "intellectual stimulation" of anti-Jewish rhetoric. Most anti-Semitic treatises refer to Freemasonry as a particularly destructive force working against national interests. This view is presented most forcefully by the Viennese occult writer, Friedrich Wichtl (1877–1922), in *Weltfreimaurerei, Weltrevolution, Weltrepublik* (World Freemasonry, World Revolution, World Republic), first published in 1919.[26] This book, together with the writings of the Swiss theosophist Karl Heise, clearly defined and identified targets of common hatred that could be blamed for the sufferings of the First World War and the following revolutionary anarchy.[27] Connecting Jews to Freemasons as allies in the struggle for world hegemony was common in early-twentieth-century Europe (the "Protocols of the Elders of Zion" being the most notable example).[28] The recovery and resurrection of the nation thus rested on clearly definable perpetrators. Countless publications and political statements in interwar Hungary echoed the identification of Jews and Freemasons with their assumed close allies, the Communists. The Catholic archbishop of Székesfehérvár, Ottokár Prohászka (1858–1927), perhaps the most influential public figure of the early 1920s in Hungary, explained: "The eyes of numerous people were blindfolded and they did not see the true face of Freemasonry. They were told that they [the Freemasons] were an innocent, philanthropic association. Now we see that they are an internationalist, defeatist gang that hates the church. [The Freemasons] opened the gates to Jewish infiltration and trampled upon Christian national traditions."[29] This assertion is practically identical to the view expressed in February 1939 by the Nazi propaganda minister Joseph Goebbels (1897–1945).

Goebbels explained that the driving force behind the campaign against "peace-loving" Germany was "international Jewry, international Free-masonry and international Marxism."[30] These views reflect the socio-psychological phenomenon of enforced attribution. Critical and con-flict-loaded situations—like the aftermath of the First World War, the period of the Great Depression and the international crisis of 1938–1939—were fertile hotbeds for the targeting of scapegoats. Again, anti-Semitic traditions provided the basis for this scapegoating.

Leszek Kołakowski supports this contention in arguing that the seemingly harmless and dispersed elements of anti-Semitism could be easily and quickly blended into an explosive mixture.[31] In 1944, Hun-gary found itself in a political and ideological vacuum, and this again created the prime conditions for enforced attribution. Without German occupation, large-scale deportations would never have taken place. However, once the Germans were in charge they were not short of helpers. The distortion of Hungarian nationalism is well documented by Gábor Kádár's recent study on the economic exploitation of Hun-garian Jews.[32] One example from the summer of 1944 illustrates this point: gendarmes at the Kolozsvár (today Cluj, Romania) railway sta-tion, preparing the deportation of Jews, confiscated some of their lug-gage, explaining that "you should not take everything to the Germans, something should also be left for the Hungarians."[33]

Anti-Semitic inclination does not have a national character, but there are crisis situations in which the identification of a scapegoat is manifested in the expression of anti-Semitism. Such an anti-Semitic climate led to anti-Jewish action as a result of numerous factors in sit-uations when the tragedy and the sins are most visible but the causes are most complex and hard to define. In the case of the greatest tragedy of modern Hungarian history it was the Nazi regime's pressure, fol-lowed by direct German intervention that laid planks over the wide gulf between anti-Semitism and the Holocaust. Elements of nationalism and anti-Semitism were thus blended into a most dangerous mixture.

Conclusions

The call for self-reflection and the re-examination of the national self by István Bibó (1911–1979), and other outstanding Hungarian intellec-tuals during the period 1919 to 1945, was never answered. Decades of

Communist rule swept the issue under the carpet. However, when the lid was temporarily unscrewed during the revolution of 1956, the Jewish question did not surface. Why? If the answer is formulated with the help of scapegoating theory, one response is that, in 1956, the frontlines were clear: Soviet-imposed dictatorship vs. the call for a sovereign Hungary with a democratic political system. The situation in 1956 seemed less complex than the situation following the First World War, or the period 1944–1945. Hungarian society directed blame at the Soviets: they were the obvious subjects of common hatred despite the fact that everyone was aware of the high percentage of Jews in the Communist leadership. The lack of anti-Semitism in 1956 is not explained by the fact that there were quite a number of Jews among Communist reformers, but that the Soviets were perceived as the obvious and major current enemy of national interest. This, however, does not mean that the potential for an outburst of anti-Semitism was non-existent; indeed, its consequences were clearly evidenced in the emigration of Jews in 1956 and in the behavior of those Jews who believed that the best protection from anti-Semitism was to join János Kádár's group that reestablished the Communist system in Hungary.[34]

Why did anti-Semitism surface following the next major turning point in Hungarian history, following the democratic transition of 1989–1990? We can once more utilize the scapegoat theory in searching for an answer. Vast amounts of literature on the subject show that the short-lived euphoria of the summer of 1989 and the spring of 1990 was followed by disappointment arising from the harsh reality of the market economy, declining living standards and so on.[35] Both Hungarian society and the social sciences were faced with the task of making sense of the collapse of the Communist system and the roots of the socioeconomic problems following the transition. In spite of the departure of Soviet troops, no miracle occurred. The security of a protected cage gave way to the dangers of a free jungle. This is a perfect environment for the proliferation of more scapegoats: the problems are visible and powerful, and some of them (unemployment, new forms of violent criminality, extreme poverty and so on) are unfamiliar. With Soviet domination gone, an ideal time for enforced attribution and the mobilization of Hungarian society once again arose. As early as the beginning of 1990, an article in a new newspaper explained that the anti-Semites of "old Hungary" did not hate the capitalist businessman but the "bespectacled Marxist Freemasonic intellectual who sold Transylvania and brought the Communists into power."[36]

At some point, towards the end of the Second World War, István Bibó wrote: "Collective hysteria is a state of the whole community and it is useless to separate or remove the visible carriers of hysteria, if, in the meantime, the preconditions and basic situations conducive to hysteria survive, the traumas experienced at the beginning of the hysteria do not dissipate, the phoney situation at the core of the hysteria is not resolved. Even if we destruct all 'evil' people, the community within one generation will again reproduce the madmen of hysteria, its beneficiaries, its hangmen."[37] It thus seems appropriate to conclude this chapter with words of warning about the capacity for evil in all of us, captured by the Hungarian poet, Gyula Illyés (1902–1983): "Where seek tyranny, think again/Everyone is a link in the chain/Of tyranny's stench you are not free/You yourself are tyranny."[38]

Endnotes:

* An initial version of this paper was published as "Scapegoating and Anti-semitism after World War I," in András Kovács and Eszter Andor, eds., *Jewish Studies Yearbook*, vol. 3 (Budapest: Central European University, 2003), 125–134.

[1] For a short introduction to these concepts see Attila Pók, "Atonement and Sacrifice: Scapegoats in Modern Eastern and Central Europe," *East European Quarterly* 32, 4 (1999), 533–536.

[2] On the *Historikerstreit* in Germany see Christian Meier, *Vierzig Jahre nach Auschwitz. Deutsche Geschichtserinnerung heute* (Munich: Beck, 1987). See also the March/April (1996) issue of the French journal *Le débat*, and A.O. (Andreas Oplatka), "Politisierter Historikerstreit in Ungarn," *Neue Zürcher Zeitung* (24 November 1999), 3; and Anne Applebaum, "Hungary's 'House of Terror'," *The Wall Street Journal Online*, available at http://www.WSJ.com (Accessed on 1 March 2002).

[3] Ferenc Pataki, "Bűnbakképzési folyamatok a társadalomban," in Ferenc Pataki, *Rendszerváltás után: társadalomlélektani terepszemle* (Budapest: Scientia Humana, 1993), 94–106.

[4] György Hunyadi, *Stereotypes during the Decline and Fall of Communism* (London: Routledge, 1998).

[5] See, for example, Randolph L. Braham, *The Destruction of Hungarian Jewry. A Documentary Account*, 2 vols. (New York: Pro Arte for the World Federation of Hungarian Jews, 1963); Randolph L. Braham, *The Hungarian Jewish Catastrophe: A Selected and Annotated Bibliography* (Boulder, Col.: Social Science Monographs, 1984); and Randolph L. Braham, ed., *Perspectives on the Holocaust* (London: Kluwer-Nijhoff, 1983).

[6] Omer Bartov, "Defining Enemies, Making Victims: Germans, Jews, and the Holocaust," *The American Historical Review* 103, 3 (1998), 771–816.

[7] Jeffrey S. Murer, *Pursuing the Familiar Foreigner: Resurgence of Anti-Semitism and Nationalism in Hungary since 1989* (Unpublished Ph.D. thesis, University of Illinois, Chicago, 1999).

[8] See Thomas Douglas, *Scapegoats: Transferring Blame* (London: Routledge, 1995), 3–12.

[9] See Pataki, "Bűnbakképzési folyamatok a társadalomban," 94–106; Douglas, *Scapegoats: Transferring Blame*, 107–148. For a more detailed discussion of "scapegoating," see René Girard, *The Scapegoat* (Baltimore: John Hopkins University Press, 1989). On East-Central European aspects see Vladimir Tismăneanu, *Fantasies of Salvation* (Princeton: Princeton University Press, 1998).

[10] Erik Erikson, *Insight and Responsibility* (New York: W.W. Norton, 1964).

[11] Cited by Bartov, "Defining Enemies, Making Victims: Germans, Jews, and the Holocaust," 785.

[12] Girard, *The Scapegoat*, 14–23.

[13] Rolf Fischer, *Entwicklungsstufen des Antisemitismus in Ungarn 1867–1939* (Munich: R. Oldenbourg Verlag, 1988), 61.

[14] Douglas, *Scapegoats: Transferring Blame*, 135–148.

[15] On Trianon, see Ignác Romsics, *The Dismantling of Historic Hungary: The Peace Treaty of Trianon, 1920* (Boulder, Co.: Columbia University Press, 2003).

[16] The concept was introduced into the "Trianon discourse" by Murer, *Pursuing the Familiar Foreigner*, especially 148–266.

[17] See Murer, *Pursuing the Familiar Foreigner*, 176.

[18] See László Karsai, *Holokauszt* (Budapest: Pannonica, 2001), 215; and Gábor Kádár, Zoltán Vági, *Hullarablás. A magyar zsidók gazdasági megsemmisítése* (Budapest: Hannah Arendt Egyesület –Jaffa Kiadó, 2005), 35–43.

[19] Murer, *Pursuing the Familiar Foreigner*, 179; and György Ránki, ed., *Magyarország története* (Budapest: Akadémiai Kiadó, 1976), 310–311.

[20] The word Jew or Israelite was not used in the legislation that limited the number of Jewish students in Hungarian universities. See Viktor Karády, "A numerus clausus és a zsidó értelmiség," in Viktor Karády, ed., *Iskolarendszer és felekezeti egyenlőtlenségek Magyarországon* (Budapest: Replika, 1997), 235–245.

[21] The most recent work on Teleki is by Balázs Ablonczy, *Teleki Pál* (Budapest: Osiris, 2005).

[22] See János Gyurgyák, *A zsidókérdés Magyarországon* (Budapest: Osiris, 2001), 387–397.

[23] See Gyurgyák, *A zsidókérdés Magyarországon*, 554–559.

[24] See Pók, "Atonement and Sacrifice: Scapegoats in Modern Eastern and Central Europe," 542.

[25] Quoted in Pók, "Atonement and Sacrifice: Scapegoats in Modern Eastern and Central Europe," 542–543.

[26] Friedrich Wichtl, *Weltfreimaurerei, Weltrevolution, Weltrepublik. Eine Untersuchung über Ursprung und Endziele des Weltkrieges* (Munich: J. F. Lehmanns Verlag, 1919).

[27] See, for example, Karl Heise, *Die Entente-Freimaurerei und de Weltkrieg. Ein Beitrag zur Historie des Weltkrieges und zum Verständnis der wahren Freimaurerei* (Basel: Ernst Fluckh Verlag, 1919). For an analysis of works focusing on "conspiracy theories" see Johannes Rogalla Bieberstein, *Die These von der Verschwörung 1776–1945. Philosophen, Freimaurer, Juden, Liberale und Sozialisten als Verschwörer gegen die Sozialordnung*, 2nd. ed. (Frankfurt-am-Main: Peter Lang, 1978), 63.

[28] Jacob Katz, *Jews and Freemasons in Europe 1723–1939* (Cambridge, Mass.: Harvard University Press, 1970); Jacob Katz, *From Prejudice to Destruction* (Cambridge, Mass.: Harvard University Press, 1980), especially 139–144.

[29] Quoted in István Somogyi, *A szabadkőművesség igazi arca* (Budapest: Apostol Nyomda Könyvkiadó Vállalat, 1929), 181.

[30] Quoted in Bieberstein, *Die These von der jüdisch-freimaurerischen Weltverschwörung*, 30–46.

[31] Quoted in Péter Kende, *Röpirat a zsidó-kérdésről* (Budapest: Magvető, 1989), 146.

[32] Gábor Kádár, *A magyarországi Vészkorszak gazdasági vetületei* (Unpublished Ph. D. thesis, University of Debrecen, 2002).

[33] Kádár, *A magyarországi Vészkorszak gazdasági vetületei*, 57.

[34] See György Litván, "Jewish Role in Hungarian Communism, Anti-Stalinism and 1956," in Jónás Pál, Peter Pastor, and Péter Tóth Pál, eds., *Király Béla emlékkönyv* (Budapest: Századvég Kiadó, 1992), 241.

[35] See, for example, György Csepeli and Tibor Závecz, "Várakozások, remények, félelmek: az Európai Unió képe a magyar közvéleményben," in Sándor Kurtán, Péter Sándor and László Vass, eds., *Magyarország politikai évkönyve* (Budapest: Demokrácia Kutatások Magyar Központja Alapítvány, 1997), 650–668. See also Rudolf L. Tőkés, *Hungary's Negotiated Revolution: Economic Reform, Social Change and Political Succession, 1957–1990* (Cambridge: Cambridge University Press, 1996); Georges Mink and Jean-Charles Szurek, *La grande conversion. Le destin des communistes en Europe de l'Est* (Paris: Éditions du Seuil, 1999); and John S. Micgiel, ed., *Perspectives on Political and Economic Transitions after Communism* (New York: Columbia University, 1997).

[36] See *Szent Korona* (21 February 1990), 6–7. Quoted in László Karsai, ed., *Kirekesztők* (Budapest: Aura, 1992), 150–151.

[37] István Bibó, "Az európai egyensúlyról és békéről," in István Bibó, *Válogatott tanulmányok*, vol. 1 (Budapest: Magvető, 1986), 376.

[38] Gyula Illyés, "One Sentence on Tyranny," in Ádám Makkai, ed., *In Quest of the Miraculous Stag: The Poetry of Hungary* (Chicago: Atlantis–Centaur, 1997), 649.

Racial Politics and Biomedical Totalitarianism in Interwar Europe

Aristotle A. Kallis

There is no more pertinent evidence of the totalitarian nature of the National Socialist regime in Germany than its uncompromising ambition to exercise full authority over every aspect of individual and collective life. Firstly through a series of legislative initiatives (including most notably the 1933 "Sterilization Law" and the 1935 "Citizenship and Marriage Laws"), and from 1939 onwards through the torrent of murderous policies (for example the T-4 "Euthanasia Program" and the "Final Solution of the Jewish Question"), the National Socialist state became the primary arbiter of human value, survival and elimination.[1] This was a bio-political project of the most extreme kind, a radical counter-utopia. In their fanatical pursuit of the "ideal *Vaterland*" the Nazis received crucial support from the German biomedical community—support that was verbal and logistic as well as technocratic and political. More than half of German medical practitioners became members of the NSDAP, a quarter joined the SA, and almost one in ten felt that either their professional or scientific interests would be best advanced through the SS.[2] The apparent willingness with which the biomedical community bowed to National Socialist demands for co-operation can be described as "anticipatory" co-ordination (*Gleichschaltung*). This involved the voluntary and pre-emptive implementation of measures aimed to placate the new Nazi authorities and thus to achieve the best possible arrangement with them. Even this interpretation, however, runs the risk of becoming reductionist, assuming that there was a pre-conceived Nazi norm to which the medical profession subscribed, through intimidation, peer pressure, opportunism, or even enthusiastic endorsement. In truth we are dealing with the conjunction, collusion and synthesis of two separate modern visions with totalitarian scope and implications, each with its own distinctive history, values and teleology. The National Socialist vision of a "racially pure" society remained far more nebulous and uncertain of its prescribed form than its biomedical counterpart. The "structuralist" literature on National Socialist policies against various non-normative groups, both in terms of racial hygiene (the mentally disabled) and racial anthropology (Jews, Sinti, Roma), underlined a culture of experiment of different "solutions" and radical prescriptions.[3] By contrast, specific branches of anthropological, biological and medical research that appeared in the late nineteenth and early twentieth centuries developed self-sustaining blueprints of both the desirable goal and the feasibility of the required methodology. Even if there were still fundamental disagreements between sub-groups about what were the best (most appropriate, effective and acceptable) means and objectives,[4] in its ideal-typical form this new scientific counter-paradigm was holistic and "total"; pointing in the direction of what, for the purposes of this chapter, I term biomedical totalitarianism.

The vision of biomedical totalitarianism rested on the idea that biomedical science alone could promote and guarantee the ideal of infinite individual and social perfectibility through eliminating all forms of perceived pathology; and that its practitioners could make the most authoritative and effective decisions about all matters relating to the life and "health" of the individual and society alike. Perhaps more importantly, it entailed a major process of jettisoning deep-seated cultural and moral convictions as well as overcoming the strength of common wisdom about the role and scope of scientific intervention in life. But it also presupposed a new type of relationship between scientific elites and sovereign state power, one providing the legislative, political and bureaucratic framework for the implementation of a far-reaching bio-political vision.

Biomedical totalitarianism shared with traditional representations of medicine the belief that the doctor was a potent agent of "healing," whose main responsibility was to prolong life and battle against the "threat" (real or potential) of "disease." However, it also depicted the existing paradigm of medical and social "health" as deeply misguided and unsustainable, and articulated radically new definitions of all these notions in a way that suggested a genuine "scientific revolution."[5] A series of new foci (such as the overriding concern with the collective national body, the biological origins of racial membership, the role of heredity in explaining human conditions, the application of natural laws in society, and new conceptions of social and racial "threats") marked a dramatic departure from conventional biomedical practices and alluded to an open-ended extension of jurisdiction over the individual and society. The ideal-typical vision of biomedical totalitarianism pointed to a future organic society, where distinctions between the social and political, the individual and the collective, scientific expertise and political power, would be reconciled and incorporated into a holistic paradigm of "health" administered by biomedical experts and practitioners on behalf of the "national community" (*Volksgemeinschaft*).

Was this an allusion to the open-ended capabilities of an allegedly "pure," value-free modern science? Interwar Europe was a laboratory of different solutions for future society, rooted in different structures of tradition and power. While the specific model of liberal modernity appeared to be making decisive inroads in the aftermath of the First World War, this did not mean that either anti-modern or alternative modern prescriptions had been stifled—quite the opposite. In fact, the project of liberal modernity was predicated on a set of checks and guarantees of individual freedom that essentially produced new ethical caveats about the most desirable course of future action. When Karl Binding (1841–1920) and Alfred Hoche (1865–1943) published their radical treatise *Die Freigabe der Vernichtung lebensunwerten Lebens*, on "life unworthy of living," in 1920, they bemoaned the existing force of religious, cultural and legal barriers to the "elimination of physically and mentally defective" individuals for the greater benefit of society.[6] This request for the removal of "non-scientific" caveats alluded to a "total" sphere of biomedical jurisdiction, but it was by no means unaffected by cultural conditioning, albeit of a very different source and character. The scientific counter-paradigm of "negative" eugenics, of aggressive selective breeding, prioritizing notions of collective "health" over individual freedom, reflected a decisive "totalitarian" rationale that also implied the authority to make decisions about life and death. Yet it too was rooted in cultural perceptions of "deviance" and notions of how best to promote health. Such perceptions bore a generic-historical and national dimension. The authority of the scientific paradigm and its potential scope reflected the dynamics of modern deployment in certain societies. The strength of the impediments that Binding and Hoche criticized derived from the persistence of different value systems, whether rooted in liberalism or in traditional ethical codes (such as religion).

This chapter places the development and radicalization of a totalitarian biomedical paradigm into the context of wider developments from the late nineteenth century across Europe. These shifts were evident on four levels: first, in scientifically identifying, defining and attempting to explain forms of "deviance" on an individual and collective basis; second, in extending the biomedical domain of jurisdiction from fighting disease on an individual basis to promoting "health" at a collective level, a process that has been described as the "medicalization" of social space; third, in empowering this new "science" to provide alleged radical "solutions" deemed far more superior and effective than the previous agnosticism of traditional and liberal societies; and, fourth, in promoting an exclusive alliance between state and science with a view to maximizing the scope of scientific intervention in modern society. In this context, National Socialist Germany should be considered by far as the most radical biomedical experiment in interwar Europe. This is the case because it alone witnessed the confluence of two eliminationist visions: a) one rooted in the belief that the medicalization of perceived pathologies opened the way for extreme action predicated on the basis of excluding and eliminating all forms of hereditary biological and social "deviance;" b) a specific National Socialist vision of generating an "ideal national and racial community" through the identification of non-normative behaviour (social, cultural, biological) as detrimental to the health of the *Volk*, and its physical eradication. The fusion between the two visions was made possible by a consensus about what constituted "deviance," an overriding concern for the health of the national body in organic terms, and the benefits of action geared towards exclusion rather than inclusion. This consensus suited the interests of both the medical profession and the functionaries of the National Socialist *völkisch* state. The former could achieve an elevated social status, corporate empowerment, and the removal of conventional moral, political and legislative obstacles to their desired aim of social intervention. The latter could capitalize on the legitimacy provided through the involvement of the scientific and academic communities in an open-ended "gardening" project. As a result, national, social and individual health combined in a holistic, total vision that advocated the "elimination" of allegedly harmful elements. This is the point where totalitarian models of ethno-exclusive racial nationalism and racial hygiene met.[7]

This inextricability of the two distinct brands of totalitarian thinking raises questions about the agency and authorship of what subsequently became a pandemonium of "medicalized killing" on the most horrifying scale.[8] What happened between 1933 and 1945 in Germany and elsewhere under the auspices of the National Socialist "new order" points to the notion that there was something unique in the German interwar context. Although the distinctiveness of National Socialism as a "racial revolution" has been widely documented in literature, its debt to generic cultural and scientific currents by no means restricted to Germany has not been properly acknowledged to date.[9] The persistence of such principles acted in most cases as a disincentive—both as a form of self-censorship of the medical profession and as an obstacle to the adoption of radical prescriptions by state authorities. Cultural and moral barriers to the formulation and implementation of this extreme vision of scientific (technocratic) modernity had varying degrees of potency in each country. However, rather than focusing on the alleged uniqueness of National Socialist Germany, emphasis should be placed instead on why such obstacles were eliminated or sidestepped in some cases while maintaining their paradigmatic potency in others. To put it differently, biomedical totalitarianism had been articulated as a utopian counter-paradigm to social reformism and conventional beliefs regarding the inviolability of individual life long before the rise of National Socialism. The leap, however, from marginal idea to mainstream theory, and from there to action, required further crucial shifts and concessions that, with the exception of National Socialist Germany, proved hard to accept and even harder to promote in interwar Europe. Therefore, the debate concerns both the generic (historical) and specific (national) trends that, when combined, in some cases facilitated the transformation of a totalitarian potential for radical biological engineering into official state policy, and impeded it in others.

A Scientific Revolution: Medicine, Social Intervention and Elimination

The extent to which scientific visions of human science had been undergoing dramatic transformation from the second half of the nineteenth century onwards cannot be disputed. Debates concerning the relationship between nature and nurture, heredity, healing, the sanctity of human life, and the neutrality of the doctor were turned upside down in the light of new "unorthodox" intellectual currents. This was not the first time that radical marginal ideas challenged mainstream and deeply embedded views about individual and social life. The success, however, that this particular set of non-conformist views had in saturating existing discourses in certain societies raises complex questions about the nature of the suggested revision and the circumstances that facilitated it.

In his study of the relationship between modernity and extermination in National Socialist Germany, Zygmunt Bauman underlined the significance of modernity in the conceptualization of new opportunities, functions and ethical codes for society. In his opinion, this made the excesses of the 1939–1945 period not just conceivable but also possible and, more important, desirable.[10] Whatever criticism the historiography of the Holocaust has put forward with regard to the accentuation of the role of modernity in institutionalized killing, it should be stressed that any form of totalitarianism presupposes possibilities inherent in a distorted and decidedly modern framework.[11] It is not just that modernity supplies weapons of unfathomable functionality and efficiency; more importantly, it provides a new conception of social life that employs these weapons in the context of a radical utopia based on "total" intervention, control, and a morally justified direction of widespread engineering. While not being morally neutral, modernity emerges as a form of empowerment without a definitive moral subtext.[12]

Modernity thus supplied the (by then empirically founded) belief in the possibility of remedial action vis-à-vis instances of perceived pathology, both individual and collective. The reclassification of conditions and phenomena previously considered de facto deviations from "normality" generated a further debate about what could be "healed" and what should remain beyond the reach of scientific intervention. While progressive welfare visions extended the benefit of corrective action to an expanding social domain, a parallel enquiry into the link between heredity and pathology also gathered momentum. Unlike social welfare, heredity was exclusionary and erected new, impenetrable walls between "normality" and "deviance."[13] Once a condition had been defined as hereditary, remedial action became redundant, unrewarding, and potentially dangerous.

The idea that any form of "deviance" could refer to empirically ascertainable biological origins was an intoxicating proposition—for the professional groups that claimed to possess the knowledge and the scientific apparatus to perform this function; for society as a whole; and for the state authorities responsible for the welfare of the individual and the community. However, this mindset went much further, implying a biomedical monopoly of jurisdiction over the whole spectrum of life. In this struggle, the advance of modernity was instrumental in mitigating the de facto authority of traditional institutions (church, family, and so on) to rule over such matters. What had been conceptualized as unacceptable moral or social behavior in the past was now articulated as deviation from "health," or as detrimental "sickness" originating from biological pathology. The process of re-labeling forms of social deviance as "disease," moving away from individual concern to a more global, social perspective, and finally of establishing this medicalized "deviance" as a matter of collective concern, brought forth an authoritative diagnosis and a wholesale prescription. By fostering a trend towards closer empirical enquiry into the causes of "non-normative" behavior in all its forms (physical, social, mental, and sexual), and by drawing attention to the alleged biological origins of a wide array of "deviant" conditions, it also enforced a far more rigid paradigm of normality and "health." Any departure from this paradigm, whether seemingly innocuous or overtly unacceptable, constituted a form of "sickness" and, as such, carried detrimental implications not just for the affected individual or group, but for the whole of the social body.[14]

This particular reading of deviance carried a lethal, totalitarian implication. On the one hand, the emerging sense of modern empowerment generated an illusion of ongoing perfectibility and nurtured an illusory belief in the possibility of eradicating forms of perceived pathology. On the other hand, the paradigm of biological heredity pointed to immutable forms of deviance while still propagating the importance of eradicating them. The difference was fundamental: social welfare envisioned the social elimination of offending conditions over time. Negative eugenics could only conceptualize elimination in physical terms, or otherwise allow the condition to perpetuate itself without any possibility for remedial action. This sort of dualistic thinking did not necessarily lead to either a vision of medical totalitarianism or the medicalized killing witnessed in National Socialist Germany. In other words, it was not automatically "eliminationist." Coupled, however, with the belief that hereditary "deviance" was fatal for the health of the community and an impediment to the realization of its allegedly superior "mission," physical exclusion acquired a sense of biological and historic urgency that rendered elimination a less indefensible prospect. Once defended on the basis of collective interest, there was a further step to be taken: presenting elimination as a feasible and desirable option.

Biomedical Totalitarianism in the Nineteenth Century

The scientific trend that favored the emergence of biomedical totalitarianism should be placed into the historical and cultural context of nineteenth- and twentieth-century Europe. The advance of various types of modernity, proceeding in tandem with secularization, opened up new areas of enquiry for science and empowered empirical research to seek wider, more absolute solutions for human problems.[15] Grand scientific theories, such as Charles Darwin's (1809–1882) theory of evolution; Thomas Robert Malthus's (1766–1834) population predictions; or Ernst Haeckel's (1835–1919) Monism, offered universal prescriptions; for a better human order, making increasing incursions into areas hitherto untouched by scientific intervention. The medicalization of social space advanced steadily during the nineteenth century, challenging conventional beliefs and narratives, claiming jurisdiction over an expanding sphere of individual and collective life, positing itself as the superior prescription for a brave modern world.[16] The intensification of medicalization (sometimes understood as a form of "medical imperialism"[17]) was predicated on a more fundamental paradigmatic shift already underway by that time: developing a global approach to health and welfare that shifted the focus from individual to universal concerns. A stream of scientific studies about the natural world, initially confined to plants and animals, were gradually imported into the analysis of the human condition. This trend carried with it a potential for bringing into focus a wide range of human activities traditionally viewed as belonging to the private sphere and regulated by a different set of moral codes—for example the role of religion in influencing patterns of social conduct, sexual behavior, and reproduction, as well as in providing explanations for the descent of man.

At the same time, a sense of unbound opportunity, fuelled by the elevation of the prestige of science and a positivist belief in steady progress, provided new ammunition for utopian thinking about new forms of human agency, and teleological visions of society.[18] Making full use of modern forms of enquiry and technological progress, biomedical practitioners could enhance the scope of treatment while at the same time strengthening their ambitions with regard to the range of human conditions over which they could now claim jurisdiction. As more and more phenomena of alleged pathology were rendered accessible through human cognition and empirical analysis, medicine fuelled an optimistic belief that solutions to human pathos may be obtained through exploiting the full resources of the scientific paradigm (*Allmachtswahn*). Science challenged the foundations of social reformism, the belief that man-made social conditions, rather than immutable biological factors, were responsible for an array of situations that were "problematic" in one sense or another. While socialism and liberalism recognized and propagated the impact of specific conditions of nurture for the reproduction of individual life, certain branches of biomedical thought attempted to expropriate this space of knowledge by exposing the predominant role of nature and heredity in understanding human inequality.[19] Against the belief that the environmental and the social milieu could be of primary significance in eradicating hereditary forms of alleged pathology, a far more inflexible reading of heredity, based on ostensibly immutable biological conditioning, also gathered momentum in the late nineteenth and early twentieth centuries.[20]

The drive towards scientific empowerment, however, had a distinct defensive rationale, derived from a degree of uncertainty vis-à-vis the progress of modernity. Concerns about a growing disequilibrium between population and resources, unease about certain effects of modernity (such as urbanization), and an almost pathological fear of European "decadence," cast a long shadow upon the social optimism of human progress and collective betterment.[21] Notions of "danger" were radicalized as a result of the utopianism of new scientific discourses: new definitions of social pathology emerged as impediments to the realization of new utopian visions and as perversions of positive norms. The modern ethos of productivity recast idleness in strikingly more negative terms; emphasis on pro-natalism, as a counterbalance to the declining fertility rates across Europe, incriminated individual behavior that appeared reticent towards procreation. Similarly, competition for resources prompted a re-examination of the welfare axiom of aiding the "weak" as inherently dangerous for the health of the collective. Already at the turn of the century, the idea that the humanistic principle of aiding the physically and mentally "weak" could become a source of social "degradation" attracted enthusiastic disciples among the Social Darwinian scientific constituency.[22]

These debates ensued within a rapidly changing socio-political environment.[23] The drive towards imperialism in the last decades of the nineteenth century had revived a wider interest in the differences between white and non-white in both social and anthropological terms. Increasing contact with colonial populations fuelled an "us/them" mentality, harking back to earlier waves of imperialism. The idea of "white superiority" had always been inherent in the history of European colonialism; but it had been primarily articulated in moral, cultural and civilizational terms, and not as a racial-biological concept. Growing stress on genetic difference gave rise to fears about miscegenation, which in turn promoted discourses of exclusion and segregation.[24] In this respect, new colonial experiences proved instrumental in engendering discourses about biological (racial) superiority, rooted in heredity, and of similar eugenic "threats" to the biological health of the "white peoples."

Moreover, the escalation of nationalism as the crucial determinant of group membership and identity not only strengthened a modern trend towards the valorization of the collective body, but also redefined it as a *national* entity.[25] Towards the end of the nineteenth century, this assumption carried fundamental implications, since it also redefined the scope of the state's responsibility in terms of collective welfare—namely, the idea of collective social good became inextricably tied to national interest. Integration was still possible, but its prerequisites had become more inflexible, blending elements of social, cultural and political conformity while simultaneously rendering individual non-normative behavior potentially dangerous to the collective.

This strand of ethnically exclusivist nationalism operated on the basis of a strict inclusion-exclusion paradigm that gradually infiltrated the domain of scientific enquiry. Beliefs in national superiority, conventionally rooted in and derived from historical and cultural achievements, also fell under the expanding reach of biological enquiry. The relationship between a science fulfilling nationalist goals and a nationalism deploying science as a legitimizing factor was mutually dependent. The porous relation between race and nation in the late nineteenth and early twentieth centuries produced scientific discourses that nurtured a series of political claims including international domination, territorial expansion, internal hierarchies and so on.[26] What is of interest, however, is the way in which nationalist filters conditioned areas of enquiry, paradigms and prescriptions of scientific research. For example, when European anthropologists, biologists and geneticists started speaking about the alleged superiority of the Nordic or Aryan "race" in relation to other European sub-groupings (Alpine, Mediterranean, Slavic), they were applying (implicitly or explicitly) essentially nationalist concepts of distinction (e.g. Chamberlain's and, later, Günther's identification of Aryans with the Germans). Equally, criticisms of specific models of scientific enquiry were often conceptualized in distinctly national terms (for example, the critique of "Jewish science" and "Nordic racialism").[27]

As the discourse of race appeared to supply pseudo-scientific legitimacy to claims of national superiority, its political expediency increased dramatically, and so did its deployment by the official state. Even more important, the biological focus on heredity contradicted the foundations of the "integrationist" efforts of nation-states vis-à-vis their various minorities. If the minority groups' alleged difference was rooted in racial heredity and emanated from "inferior" genetic pools, then exclusion was potentially a one-way path. Medicalizing the notion of "race" became a crucial source of bio-power (that is, power exercised over a defined collective) with the overriding goal of preserving the continuity and optimal health of the nation. Presenting race as an allegedly scientific and immutable determinant of group membership, and thus an overriding axiom of inclusion or exclusion entailed a further permission ("licence") to segregate, exclude and potentially eliminate on a mass basis. But the medicalization of race remained part of a wider process of historical prejudice, rooted in longstanding conceptions of difference—but crucially updated by modern definitions of "conformity" and "deviance." It was precisely on this basis that the health of the national body became a matter of both external and internal defence: medicalized racial-national prejudice provided the external dimension of exclusion, while medicalized social prejudice pitted the "racial health" of the community against its own internal forms of "deviance."

In its most ideal-typical form, bio-power represented an open-ended call for regulating human life at both the individual and the collective level. It was driven by an allegedly pure scientific ethos and unhindered by moral caveats. Its realization, however, whether partial or total, depended on the empowerment of political institutions, above all the state as the supreme guardian of individual and collective life. Herein lay an opportunity and a problem.

On the one hand, the same period witnessed a rapid expansion of the pastoral power of the modern state, which emerged as the crucial institutional force of legitimization for grand social projects. This responsibility was perfectly congruent with the rationale of bio-politics, in the sense that modern states exercised their authority over nationals on the rational basis of ensuring and facilitating the continuity of life. It equipped state authorities with an expanding scope of intervention, from demographic and population policies to public health projects, environmental interventions, and social obligations. At the same time, however, the state as sole guarantor of collective life could not be deprived of the right to exclude and eliminate an external or internal "threat." This right involved various forms of disciplinary containment but did not preclude a more direct form of elimination, hence the potential right to remove life, to kill on the basis of defending and strengthening the health of the group.[28]

Yet on the other hand, such an escalation of state power came at a time of increased preoccupation with liberal discourses on individual rights. Thus, while most radical scientific claims had already been articulated by the first two decades of the twentieth century, their translation into concrete action clashed with an apparent reverse emphasis on individual self-determination and the freedom of the private sphere. Where the revolutionary biomedical counter-paradigm pointed to the alleged significance of nature and heredity in determining social trends and explaining various forms of social pathology, the political prescriptions of the liberal state remained aligned to discourses of nurture and progressive social intervention. Where biomedicine alluded to inherent inequalities, liberal democracy upheld the notion of equality. Any form of totality, whether political or scientific, was antithetical to the process of liberalization because it presupposed a reversal of the trend towards individual empowerment, self-determination and consensus. The liberal state controlled legislative arrangements, the allocation of resources and strategies of social engineering. Its vote of confidence for the decidedly anti-concessionary biomedical vision, a crucial "licence" for the wholesale practical implementation of the radical scientific counter-paradigm, appeared increasingly unlikely. As Bertrand Russell (1872–1970) noted in the 1920s, "what stands in the way (of introducing eugenic measures) is democracy."[29]

Interwar Germany and National Socialism

A political experiment in Germany began in the aftermath of the First World War, one committed to advancing liberalization, shielding individual freedoms, and providing progressive solutions to a spate of socioeconomic phenomena. A categorical distinction, however, between liberal and totalitarian political systems in the context of social engineering may be dangerously misleading. The modern state, even in its liberal guise, was becoming increasingly interventionist and powerful at the beginning of the twentieth century. Making full use of the scientific apparatus of modernity, it promised social improvement and attempted to deliver it by interfering more and more in areas of social life. Even if the most extreme "negative" variant of bio-politics had been more or less effectively constrained by a political commitment to individual rights and welfarism in the 1920s, this does not indicate that the liberal state itself was immune to illusions of human and social perfectibility. In fact, most social welfare experiments undertaken by authorities in the Weimar Republic remained anchored to the belief that social problems could be eradicated by comprehensive state action.[30] Yet the stress remained on integrative solutions, not exclusion or elimination. The reality of the "gardening state," and the illusion of state omnipotence, encompassed both scenarios: a progressive one, rooted in individual rights and in a wider integrationist drive; and a more radical one, predicated on the necessity of coercive measures in the name of collective "health."[31]

The fact that pressure for the introduction of negative measures (such as compulsory sterilization and euthanasia) failed to deliver any tangible legislative initiatives or wide political support in the 1920s reveals the strength of public discourses in the Weimar Republic. However, the idea that the liberal state alone is a sufficient safeguard against extreme forms of bio-politics is highly problematic. Indeed, how are we to interpret the introduction of sterilization in "progressive" political cultures such as Switzerland, Scandinavia and the US in the first decades of the twentieth century?[32] Equally, how should we interpret political support for compulsory sterilization in the 1920s, or the generous funding of the Kaiser Wilhelm Institute for Racial Anthropology from the Weimar federal budget, with the support of the SPD (*Socialdemokratische Partei Deutschlands*)?[33] If extreme negative eugenic trends had been effectively neutralized by 1929, then the sub-

sequent economic crisis made the expansion of social welfare spending increasingly onerous and less immune to attack even from within
the Weimar establishment. Given the problems intrinsic to the allocation of scant resources, views on the "human value" behind a normative
retrenchment of social welfare persisted.[34] It was no accident that the
Prussian Staatsrat asked the government in January 1932 to reconsider
its previously generous spending on those perceived as having a lower
"racial value" (*Minderwertigen*).

Undoubtedly, what happened after Hitler's appointment as Reich
chancellor in 1933 was a radical transformation of the Weimar ethos,
away from individual rights and any humanistic "gardening" principle
geared towards integration, in favour of a strict, exclusionary definition
of the "national body." National Socialism was instrumental in remov-
ing institutional and conventional moral impediments that had prevented the adoption of specific measures in the preceding decade—including, most notably, eugenic abortion or sterilization. The socially and
politically totalitarian ambitions of the new regime matched the scope
of vision and the ruthlessness of its prescriptions, as advocated by the
most radical wing of negative eugenics in pre-1933 Germany. In its
quest for total control over society, the Nazi regime recognized that
medical totalitarianism was largely congruent with its own vision of
solving "racial problems" through exclusion, and of fostering the unity
of the national community through radical collective measures. But,
above all, it also realized the potential for a crucial extension of state
power through the implementation of a radical wholesale program of
bio-politics advocated by the most extreme wing of eugenicists.[35] In
this sense, biomedical totalitarianism was fashioned as an ideological
concomitant to the socio-political totalitarianism of the National Socialist state. Technocracy would overwhelm traditional moral caveats, but
the legitimacy of those changes would derive from, and remain under,
the auspices of state authority.

Bio-politics encouraged scientific and political experts to think in
terms of defined groups, over which they could wield their authority.
Concerns about social and national "health" dovetailed with the perceived unifying basis of "race." In this way, eugenics and racial
nationalism became two sides of the same currency of an ideal "race-
nation." Race provided the common political and scientific rationality
through which physical or social "deviance," on the one hand, and
physical or cultural "otherness" on the other, could be conceptualized

as primary concerns of bio-power, functioning in a constant state of war against its opponents.[36] The shared idiom of "race-nation" was at the very heart of the alliance between biomedicine and National Socialism in interwar Germany. This vision of biomedical totalitarianism required both the very sense of omnipotence derived from modernity and the political sanction of National Socialist Regime in order to become systematic state policy.

Where did the apparent "uniqueness" of the National Socialist biomedical totalitarian vision originate? Was it the result of a scientific *Sonderweg*, an adjunct to the wider perception of a "special path" in the course of modern German history?[37] Did it result from a unique cultural endorsement of physical, aggressive "eliminationism" directed at specific "alien" groups? Was it the product of a particular National Socialist agency that set German fascism apart from other contemporary manifestations across Europe?[38] The unease which post-war historiography has experienced in attempting to deal with either the modernity of the National Socialist vision of racial nationalism, or the excesses of biomedical totalitarianism, has been evident in attempts to present the German case as an exceptional distortion of both modernity and science.[39] The paradigmatic importance, however, of the German case does not lie in the alleged uniqueness of any of the components that underpinned the policies of the post-1933 period. Race and aggressive (ethno-exclusive) nationalism had been in wide circulation across Europe for decades before the rise of fascism. Discourses about the medicalization of social space proceeded alongside modernity, in both authoritarian and progressive directions.

The shift from individual to collective responsibility, from private to public choices, from partial to wholesale solutions for perceived pathologies were all inherent in the nature of the modernizing project in the late nineteenth and early twentieth centuries. Fears of degeneration, as well as a heightened sense of "danger" from both "alien" elements and social pathologies, constituted a widely shared precondition for the emergence of biomedical totalitarianism across interwar Europe. Definitions of what constituted "(non-)normative" social behavior abounded in all European states, at a time when the debates about integration or exclusion were far from concluded one way or another. An escalation of the nation-state model rested on modern radical redefinitions of both its components: exclusive nationalism and enhanced state power in all spheres of social life. Taken separately, all these were

ominous trends but they could be mitigated by some perceived safeguards: individual freedoms, the survival of traditional allegiances and
moral codes that militated against any form of totalitarianism, scientific negations of extreme notions of race and heredity, and self-restraint
in the absence of any international precedence. Ironically, most of
these guarantees appeared in Germany during the 1920s, perhaps
much more so than in other liberal countries.

The biomedical totalitarian vision presupposed the scientific and
corporate agency of experts in relevant fields. This is the association
that links the modernity of the Weimar Republic with the different and
abhorrent modernity of systematic segregation, of mass population
transfer, and, indeed, of the concentration camp itself. What appeared
infeasible, impractical, and unethical less than three decades previously was now regarded as intelligible, realizable, and, most of all, desirable. The Weimar Republic should thus be regarded as a crucial, if
unwitting, laboratory of this fundamental shift in perception. Once the
significance of this empowering influence is appreciated, the radicalization of both medicalization and nation-statism in the 1930s becomes
a side-effect rather than an anomaly. The Nazi recasting of ethno-exclusive nationalism in "racial-anthropological" terms, and its subsequent
prescription of the elimination of "the other," borrowed heavily from
the imagery of modern *Allmachtswahn*, with its intoxicating sense of
unbound opportunity and its focus on external and internal danger,
which had to be eradicated. The articulation of social health in racial
hygienic terms by biomedical experts nurtured the "gardening" principle, drew new lines of social exclusion, and permitted aggressive technologies of elimination. Under the veneer of medicalization and legalization, the bio-political state could regulate life and death in the name
of, and on behalf of, a bio-national mythic community.

The Failure of Biomedical Totalitarianism in Interwar Europe

What happened in National Socialist Germany in the 1930s supplied
an aura of permissibility to thinking in terms of a biomedical totalitarianism administered by a "total" political will exercised by the state.
Just as German radical experts on racial hygiene had invoked the American precedent in the 1920s as a scientific and ethical legitimizing

model for similar measures in Europe, National Socialist Germany itself became paradigmatic subsequently in rejecting the notion of racial hygiene.[40] Emboldened by the German case, eugenicists now dared to suggest a more holistic model of population and racial-hygienic policies for their own countries. They felt justified to quote selectively, adopt liberally, adapt fractionally, cherry-pick out of context, and propagate the advance of the modern project as both a form of counter-paradigmatic redeployment and as a regression to atavistic notions of racial purity. Political priorities as well as cultural and ideological caveats, however, produced a multitude of disincentives for the emergence of a biomedical totalitarian vision as a viable alternative course of action.

The case of interwar Italy is more indicative of both the tensions involved in the modernity project and the interaction between fascist/totalitarian visions, and traditional beliefs. Although Fascist population policies included negative measures such as financial penalties for bachelors and restrictions on immigration, there was a fundamental pro-natalist consensus that bound together Mussolini's "il numero come forza" (force in numbers)[41] with the Catholic opposition to anti-concessionary measures on birth control.[42] One cannot exaggerate the role of the Catholic Church in producing a cultural and ethnic milieu that rejected both neo-Malthusian principles of population control as well as more radical "negative" eugenic measures like sterilization and euthanasia.[43] With the 1929 political alliance between the Fascist State and the Vatican, the field of population policy became even less available to totalitarian experimentation, in spite of pressures to introduce radical measures from a small constituency of admirers of the Nazi racial model within the Fascist National Party (*Partito Nazionale Fascista*), such as Roberto Farinacci (1892–1945), former secretary of the party; Giovanni Preziosi (1881–1945), an arch anti-Semite and editor of the Italian translation of the *Protocols of the Elders of Zion*; and Telesio Interlandi (1894–1965), editor of the *Difesa della Razza* (The Defence of Race).[44] But the perceived "Nordic" origins of this brand of eugenic thinking generated significant hostility within Fascist Italy and eventually produced a counter-narrative, propagating the idea of a distinctive "Mediterranean *razza*."[45]

Yet what was even more interesting in the Italian case was the way in which cultural caveats had already restricted the potential of the scientific field to produce viable totalitarian visions. By the time that

Mussolini came to power and embarked upon a project of constructing a genuine *stato totalitario* in the 1920s, there had already been a significant corpus of native scientific and biological arguments that pointed to the direction of biomedical totalitarianism. In the second half of the nineteenth century, the distinguished professor Cesare Lombroso (1835–1909) from Pavia University had attempted to shift the focus of criminology away from sociological factors of crime and towards evolutionary and genetic studies, particularly with the publication of *L'Uomo Delinquente* (The Delinquent Man) in 1876. By 1911–1912, the scene had already been set for the first Congress of Eugenics in Italy. This provided the opportunity for a number of scholars with wide-ranging interests, backgrounds, and scientific prescriptions to articulate their views on the "nature-nurture" subject.[46]

It became clear very quickly that the main debate in Italy would revolve around the question of birth control. Two different strategies were articulated. On the one hand, neo-Malthusian scientists, concerned with the growing disparity between population and natural resources, advocated interventionist measures aimed at both arresting the rate of population increase and improving the quality of the "genetic stock" of the national community. On the other hand, a strange alliance of Social Darwinism and radical nationalism produced the equation "number = strength," thus prioritizing pro-natalism at any cost. While the former group (Angelo Zuccarelli, Ettore Levi, Ferdinando De Napoli, Umberto Gabbi) advocated anti-concessionary measures in order to forestall negative biological and social developments, the latter category of scientists drew their legitimacy from highly disparate scientific, political, moral and economic arguments.[47] There was a group of Catholic commentators that contributed to the debate, making their moral aversion to involuntary negative eugenic measures the basis for a comprehensive paradigmatic rejection of scientific totalitarianism. Giuseppe Moscati (1880–1927), Mario Mazzeo (1889–1973), Giuseppe De Giovanni (1876–1967), and particularly Agostino Gemelli (1878–1959), all agreed on the absolute desirability of keeping scientific and political intervention out of matters of life and death, including procreation and euthanasia.[48] This support, however, was qualified by the morally "orthodox" behaviour of the individual. Gemelli, Moscati and Mazzeo, for example, qualified their support for pro-natalism with the idea of chastity outside marriage. Thus, while uninhibited pro-natalism was eulogized, voluntary self-control (chastity and abstention) was also praised outside marriage.

It is tempting to draw a line between those scientists ideologically or institutionally aligned to Catholicism or generally inspired by strict Catholic values, and those operating on the basis of more secular, empirical scientific norms. This distinction, however, is problematic in the case of Italy, given the embedded cultural and social influence of religion on society. The example of the so-called Latin eugenicists is emblematic of the permeability of these two spheres, and of the impossibility of a "value-free" scientific discourse. A distinguished member of the group, Enrico Morselli (1852–1929), found no apparent contradiction between his advocacy of a biological hierarchy of races, racial segregation, and his attack on miscegenation as destructive for the "racial stock" of the Italians on the one hand, and the moral castigation of euthanasia and sterilization as "inhuman" and "morally unacceptable" on the other.[49] Similarly, Leonardo Bianchi (1848–1927) criticized "negative" eugenic measures on purely ethical grounds, in spite of their potentially "positive" effect in the direction of improving the Italian nation. Even dedicated neo-Malthusians felt that they had to qualify their radical anti-natalist prescriptions with moral conformity. Achille Loria (1857–1943), for example, advocated extensive birth control in order to improve the "quality of the genetic stock," but only through voluntary means.[50] As a result, a law scheme underlining the necessity of birth control in Italy was discussed in 1922 and once again at the 1924 Congress of Social Eugenics, but came to nothing due to the fact that the majority of the participants insisted on the "involuntary" character of any such initiative.[51] In fact, even De Napoli, who had been instrumental in the preparation of the scheme in the first place, warned against "anti-concessionary" measures, branding them a dangerous erosion of individual liberty and invasion of the private sphere. [52] This argument, while derived from secular and liberal norms, pointed to the same notion of scientific and political non-intervention in matters of life and death upheld by Catholics.

For his part, Benito Mussolini expressed his support of the "numero = forza" equation in his May 1927 *Discorso dell'Ascensione* (Ascension Day Discourse), in which he criticized neo-Malthusianism and legitimized the drive towards pro-natalism.[53] To be sure, Il Duce was aware of the neo-Malthusian warning about a growing disequilibrium between demographics and natural resources: this was linked politically to the "battle for wheat" and the reclamation of land, in order to increase productivity and thus support the growing population.[54] Negative eugenic measures were ruled out, not so much for moral reasons as for their incompatibility with the priority of demographic expansion. Yet the distinction between the private and public spheres was becoming increasingly irrelevant and inconsistent with the regime's ideological orientation. Umberto Gabbi, editor of the *Archivio Fascista di Medicina Politica* (The Fascist Archive of Political Medicine), encapsulated the victory of an organic nationalist discourse by emphasizing the idea that the nation and the state should take priority over the individual in promoting national interest.[55]

This was a crucial point for the future of any biomedical totalitarian potential in interwar Italy. The debate as to who was better placed to improve the fortunes of the nation and, equally importantly, how, was in full motion in the second half of the 1920s. With the increasing etatist orientation of the Fascist regime, articulated in the official programmatic publication *Dottrina del fascismo* (The Doctrine of Fascism), the institutional role of the state in legitimizing political prescriptions and providing patronage in return for compliance increased.[56] In the context of this unitary state doctrine, aimed at bestowing legitimacy upon the regime as the only institution capable of representing national interest in its totality, the 1929 Concordat with the Catholic Church became a powerful disincentive to any introduction of anti-natalist measures, or indeed to enhancing the interventionist powers of secular groups in the traditional domain of marriage, procreation and death. Given the regime's non-negotiable stance on pro-natalism, eugenics was deprived of a crucial asset in the struggle for hegemony: it was negative, not positive or voluntary measures that opened up opportunities for a more wholesale intervention. Even those scientists who had resisted the pro-natalist, agnostic orthodoxy of the 1920s realized that time had run out. Professional pride had accentuated the polemics between Italian and "Anglo-Saxon" or "Nordic" eugenics.[57] With Mussolini himself openly critical and even scornful of German racial thinking, the cause of negative eugenics in Italy was effectively subverted in two different ways: first, by the scientific community itself; and second, by the absence of political will on the part of the Fascist regime. Even after 1936, when a combination of a newfound racial consciousness—derived from the regime's colonial pursuit in Ethiopia—and a recourse to anti-Semitism produced a racialist platform for the exercise of more extensive bio-political control over the population, the project lacked the clarity and open-ended ambition of the totalitarian combination of eugenics and racial nationalism witnessed in National Socialist Germany.[58]

Conclusions

With hindsight, it becomes clear that the Weimar Republic failed to curb radical, counter-paradigmatic trends, or to ensconce sufficient safeguards able to function as a bulwark against them. By allowing a supremely modern space of seemingly unbound opportunity and freedom, it perversely nurtured dangerous and illusory beliefs in the fringes of social space. The notion that the Weimar republic functioned as a laboratory of extreme scenarios—kept at bay due to the overall political orientation of republican elites up to 1929—is crucial for understanding the swiftness of the transition to the totalitarian models of the post-1933 period. In other words, the Weimar period witnessed the charting of a plethora of modern but decidedly contradictory possibilities for the future. Some were congruent with the liberal experiments of the 1920s, while others moved in the opposite direction. Sexual liberation, the novel sense of social tolerance, political pluralism, and progressive social engineering all carried their own nemesis: their swift implementation during the 1920s afforded their radical negations ammunition for a counter-attack, even on the level of rhetorical exchanges, perfectly in accordance with the spirit of democratic freedom. The tenuous consensus vis-à-vis social integration that characterized the Weimar period was uncontested neither by radical nationalist undercurrents, nor by a progressive authoritarianism of state power or bio-politics. If negative eugenics represented an escalation on the scientific level, the rise of National Socialism articulated an escalation of nationalism and statism that worked in a similar direction. The shift from pluralism to singularity, from the individual to the national body, from multidimensional social intervention to principles of prescription, all became possible on the basis of an intersection between different strands of totality, each feeding off and nurturing one another.

The absence of scientific and political support for extreme negative eugenic measures outside Germany was generally matched by wider social opposition to the ethical code underpinning them. The realization of the totalitarian potential of Nazi bio-politics sidestepped public opposition—in the case of measures such as the T-4 ("euthanasia") by shrouding the project in utter secrecy—which meant that the complicity of the scientific establishment was more or less guaranteed.[59] Nevertheless, clear-cut distinctions are problematic. While the totality of the National Socialist eliminationist vision was neither entertained nor accepted elsewhere, particular lethal prescriptions found enthusiastic disciples in Central and Southeast Europe. Racial measures proposed and introduced in one European country were often suggested or emulated in other countries. It is precisely at this point that one may speak of the National Socialist case as the purveyor of a moral and political "licence" to eliminationism across Europe. What was missing elsewhere, however, was the holistic, limitless scope of intervention—exemplified in Germany under National Socialism. It was there and then that a unique combination of factors came together. It was there that the disorienting effects of "high" modernity (the violent reaction to the disintegration of the cosmic order that it bred, the sense of a loss of meaning and direction in which individual and collective existence could be anchored) and an explosive surplus of discordant energies that it had hosted pointed in the direction of a revolutionary re-ordering of history. A process of advanced, but uneven and precarious, liberalization afforded radical utopia. Neither the extremity of the solutions nor the weakness of their various counterbalances were particular to inter-war Germany; what was unique was their lethal coincidence under the authority of a regime transfixed by the prospect of the realization of the bio-political totalitarian chimera.[60]

Endnotes:

[1] Michael Burleigh and Wolfgang Wippermann, *The Racial State. Germany 1933–1945* (Cambridge: Cambridge University Press, 1991); and Raul Hilberg, *The Destruction of the European Jews*, 3 vols. (New York: Holmes and Meier, 1985).

[2] Götz Aly, Peter Chroust, and Christian Pross, eds., *Cleansing the Fatherland: Nazi Medicine and Racial Hygiene* (Baltimore: Johns Hopkins University Press, 1994), x.

[3] The term "eliminationism" is borrowed from Daniel Goldhagen. See Daniel J. Goldhagen, *Hitler's Willing Executioners: Ordinary Germans and the Holocaust* (London: Vintage, 1997). The broader applicability of the term, comprising different forms of physical extermination of ethnic, religious, and cultural groups (*democide*), crimes perpetrated by the state against its citizens (*democide, politicide*), as well as indirect forms of persecution with a view to eliminating "deviant" behaviour (segregation, expulsion, forced assimilation), was not intended by Goldhagen, who used it in the context of anti-Semitism in modern Germany. However, such a broad scope is more suited to biomedical totalitarianism, in its attempt to medicalize an array of racial-hygienic and racial-anthropological "deviances," and to prescribe their elimination on a holistic basis. See also Karl Schleunes, *The Twisted Road to Auschwitz. Nazi Policy Toward German Jews, 1933–1939* (Urbana, Il: University of Illinois Press, 1970). For an overview of the intentionalist-structuralist debate on the Holocaust, see Ian Kershaw, *The Nazi Dictatorship: Problems and Perspectives of Interpretation* (London: Arnold, 2000), esp. chapters 5 and 9.

[4] On the plurality of theories and visions in this respect, see Paul J. Weindling, *Health, Race and German Politics between National Unification and Nazism, 1870–1945* (Cambridge: Cambridge University Press, 1989).

[5] See Thomas S. Kuhn, *The Structure of Scientific Revolutions* (Chicago: University of Chicago Press, 1996), especially chapters 7 and 9.

[6] Karl Binding and Alfred Hoche, *Die Freigabe der Vernichtung lebensunwerten Lebens* (Leipzig: Felix Meiner Verlag, 1920).

[7] Peter Weingart, "German Eugenics between Science and Politics," *Osiris* 5 (1989), 260–282.

[8] Robert J Lifton, "Medicalized Killing in Auschwitz," *Psychiatry* 45, 11 (1982), 283–297; see also Robert J. Lifton, *The Nazi Doctors: Medical Killing and the Psychology of Genocide* (New York: Basic Books, 1986).

[9] Milan L Hauner, "A German Racial Revolution," *Journal of Contemporary History* 19, 4 (1984), 669–687.

[10] Zygmunt Bauman, *Modernity and the Holocaust* (Oxford: Blackwell, 1993).

[11] See, for example, Yehuda Bauer, *Rethinking the Holocaust* (New Haven: Yale University Press, 2001); and Yehuda Bauer, "Conclusion: The Significance of the Final Solution," in David Cesarani, ed., *The Final Solution: Origins and Implementation* (London: Routledge, 1996), 300–309.

[12] See, in this volume, Roger Griffin, "Tunnel Vision and Mysterious Trees: Modernist Projects of National and Racial Regeneration, 1880–1939," 417–458.

[13] Werner Sohn and Herbert Mehrtens, eds., *Normalität und Abweichung. Studien zur Theorie und Geschichte der Normalisierungsgesellschaft* (Opladen: Westdeutscher Verlag, 1999).

[14] Michel Foucault, *History of Sexuality* (New York: Vintage Books, 1985); Peter Conrad and Joseph W. Schneider, *Deviance and Medicalization: From Badness to Sickness* (Philadelphia: Temple University Press, 1980), especially chapters 1 and 2.

[15] Malcolm Bull, "Secularization and Medicalization," *The British Journal of Sociology* 41, 2 (1990), 245–261. The idea that medicine replaced religion as the dominant ethical guardian of modern society is argued by Bryan S. Turner, *The Body and Society* (Oxford: Blackwell, 1984).

[16] Alan Peterson and Robin Bunton, *Foucault, Health and Medicine* (London: Routledge, 1997). On the concept of "medicalization" (and the critique thereof), see Robert Nye, "The Evolution of the Concept of Medicalization in the Late Twentieth Century," *Journal of the History of the Behavioral Sciences* 39, 2 (2003), 115–129.

[17] Peter Conrad and J. Schneider, "Looking at Levels of Medicalization: A Comment on Strong's Critique of the Thesis of Medical Imperialism," *Social Science and Medicine* 14, 1 (1980), 75–79; Peter Conrad and J. Schneider, "Medicalization and Social Control," *Annual Review of Sociology* 18, 2 (1992), 209–232.

[18] Edward Ross Dickinson, "Biopolitics, Fascism, Democracy: Some Reflections on Our Discourse about 'Modernity'," *Central European History* 37, 1 (2004), 1–48.

[19] See, for example, Karl Pearson, *Nature and Nurture. The Problem of the Future* (London: Dulau and Co., 1910).

[20] Alfred Kelly, *The Descent of Darwin. The Popularization of Darwinism in Germany, 1860–1914* (Chapel Hill: University of North Carolina Press, 1981), especially chapter 1; John P. Jackson, Jr. and Nadine M. Weidman, *Race, Racism, and Science. Social Impact and Interaction* (Santa Barbara, CA.: ABC-Clio, 2004).

[21] Daniel Pick, *The Faces of Degeneration: A European Disorder, 1848–1918* (Cambridge: Cambridge University Press, 1989); Nancy Stepan, "Biological Degeneration: Races and Proper Places." in J. Edward Chamberlain and Sander L. Gilman, eds., *Degeneration. The Dark Side of Progress* (New York: Columbia University Press, 1985), 97–120; and Neil Macmaster, *Racism in Europe, 1870–2000* (Basingstone: Palgrave, 2001), 33.

[22] Sheila Faith Weiss, *Race Hygiene and National Efficiency. The Eugenics of Wilhelm Schallmayer* (Berkeley: University of California Press, 1987), especially 45–97.

[23] David Roberts, *The Poverty of Great Politics. Understanding the Totalitarian Moment* (London: Routledge, 2005), especially chapters 1 and 3.

[24] Macmaster, *Racism in Europe*, 124–129.

25 Ute Planert, "Der dreifache Körper des Volkes: Sexualität, Biopolitik und die Wissenschaften vom Leben," *Geschichte und Gesellschaft* 26, 4 (2000), 539–576; and Hans-Uwe Otto and Heinz Sünker, "Volksgemeinschaft als Formierungsideologie des Nationalsozialismus: Zur Genesis und Geltung von 'Volkspflege'," in Hans-Uwe Otto and Heinz Sünker, eds., *Politische Formierung und soziale Erziehung im Nationalsozialismus* (Frankfurt-am-Main: Suhrkamp, 1992), 50–77.

26 Michael Banton, *Ethnic and Racial Consciousness* (London: Longman, 1997); and John Rex, *Race and Ethnicity* (Milton Keynes: Open University Press, 1986).

27 See Ritchie Witzig, "The Medicalization of Race: Scientific Legitimization of a Flawed Social Construct," *Annals of Internal Medicine* 125, 8 (1996), 675–679.

28 Michel Foucault, *Power. Essential Works of Michel Foucault, 1954–1984*, vol. 3, ed. by James D. Faubion (New York: New Press, 2000), 410.

29 Bertrand Russell, *Marriage and Morals* (London: George Allen & Unwin, 1929), 263.

30 Cornelie Usborne, *The Politics of the Body in Weimar Germany. Women's Reproductive Rights and Duties* (Basingstoke: Macmillan, 1992); and Young-sun Hong, *Welfare, Modernity, and the Weimar State* (Princeton: Princeton University Press, 1998).

31 James C. Scott, *Seeing Like a State: How Certain Schemes to Improve the Human Condition Have Failed* (New Haven: Yale University Press, 1998); Giorgio Agamben, *Homo Sacer: Sovereign Power and Bare Life* (Stanford: Stanford University Press, 1998), 119–180; and Astrid Hedin, "Stalinism as a Civilization: New Perspectives on Communist Regimes," *Political Studies Review* 2, 2 (2004), 166–184.

32 See Philippe Ehrenstrom, "Eugénisme et santé publique: La stérilization légale des malades mentaux dans le canton de Vaud (Suisse)", *History and Philosophy of the Life Sciences* 15, 2 (1993), 205–227; Nils Roll-Hansen, "Eugenics before World War II: The Case of Norway," *History and Philosophy of the Life Sciences* 2, 2 (1980), 269–298; Stefan Kühl, *The Nazi Connection. Eugenics, American Racism, and German National Socialism* (New York: Oxford University Press, 1994); and Gunnar Broberg and Nils Roll-Hansen, eds., *Eugenics and the Welfare State. Sterilization Policy in Denmark, Sweden, Norway, and Finland* (East Lansing, MI.: Michigan State University Press, 1996).

33 Paul J. Weindling, "Weimar Eugenics: The Kaiser Wilhelm Institute for Anthropology, Human Heredity and Eugenics in Social Context," *Annals of Science* 42, 3 (1985), 303–318.

34 See Aristotle Kallis, "Race, 'Value' and the Hierarchy of Human Life: Ideological and Structural Determinants of National Socialist Policy-Making," *Journal of Genocide Research* 6, 1 (2005), 5–30.

35 On the connection between fascism and the "escalation" of state power see Michael Mann, *Fascists* (Cambridge: Cambridge University Press, 2004).

[36] Michael Elden, "The War of Races and the Constitution of the State: Foucault's 'Il faut défendre la société' and the Politics of Calculation," *Boundary 2*, 29 (2002), 125–151.

[37] See Paul J. Weindling, "The Sonderweg of German Eugenics: Nationalism and Scientific Internationalism," *British Journal for the History of Science* 22, 74 (1989), 321–333; and David Blackbourn and Geoff Eley, eds., *The Peculiarities of German History. Bourgeois Society and Politics in Nineteenth-Century Germany* (Oxford: Oxford University Press, 1984).

[38] See, in particular, Zeev Sternhell, *The Birth of Fascist Ideology. From Cultural Rebellion to Political Revolution* (Princeton: Princeton University Press, 1989), 1–9. See also Burleigh and Wippermann, *The Racial State*, 305–307.

[39] For example, Jeffrey Herf, *Reactionary Modernism. Technology, Culture and Politics in Weimar and the Third Reich* (Cambridge: Cambridge University Press, 1984). For a further discussion see Paul J. Weindling, *Nazi Medicine and the Nuremberg Trials. From Medical War Crimes to Informed Consent* (Basingstoke: Palgrave Macmillan, 2004).

[40] Robert Proctor, *Racial Hygiene. Medicine under the Nazis* (Cambridge, Mass.: Harvard University Press, 1988), 97–99.

[41] Benito Mussolini, "Il numero come forza," *Gerarchia* (9 September 1928), 1. For a translation of a short excerpt see Benito Mussolini, "Strength in Numbers," in Roger Griffin, ed., *Fascism* (Oxford: Oxford University Press, 1995), 101–103.

[42] Carl Ipsen, *Dictating Demography. The Problem of Population in Fascist Italy* (Cambridge: Cambridge University Press, 1994).

[43] See, for example, Egilberto Martire, "Il fattore religioso," *Archivio fascista di medicina politica* 2 (1928), 315–321.

[44] Giampiero Mughini, *A via della Mercede c'era un razzista. Pittori e scrittori in camicia nera. Un giornalista maledetto e dimenticato. Lo strano "caso" di Telesio Interlandi* (Milan: Rizzoli, 1991); and Giorgio Israel and Pietro Nastasi, *Scienza e razza nell'Italia Fascista* (Bologna: Mulino, 1998), 221.

[45] Claudio Pogliano, "Scienza e stirpe: Eugenica in Italia (1912–1939)," *Passato e Presente: Rivista di Storia Contemporanea* 5, 1 (1984), 61–79.

[46] Alberto Aquarone, *L'organizzazione dello stato totalitario* (Turin: Einaudi, 1965); Renzo De Felice, *Mussolini il duce. II: Lo Stato totalitario 1936–1940* (Turin: Einaudi, 1981). For an excellent discussion of Italian "totalitarianism" see Emilio Gentile, *La via italiana al totalitarismo. Il partito e lo Stato nel regime fascista* (Rome: Carocci, 2001).

[47] See Roberto Maiocchi, *Scienza italiana e razzismo fascista* (Florence: La Nuova Italia Editrice, 1999), 16–20.

[48] See Agostino Gemelli. *Le dottrine eugeniche sul matrimonio e la morale cattolica* (Milan: Vita e pensiero, 1931).

[49] Enrico Morselli, *L'uccisione pietosa (l'eutanasia) in rapporto alla medicina, alla morale ed all'eugenica* (Turin: Bocca, 1923), 87–92.

[50] Achille Loria, *Una scienza nuova. l'eugenica* (Naples: Pensiero Sanitario, 1927), 1–20.

[51] Maiocchi, *Scienza italiana e razzismo fascista*, 20–21.

[52] Ferdinando De Napoli, *La visita prematrimoniale nel primo Congresso di Eugenetica* (Naples: Pensiero sanitario, 1924).

[53] Benito Mussolini, "Discorso dell'Ascensione—il regime fascista per la grandezza d'Italia pronunciato il 26 maggio 1927 alla Camera dei deputati," in *Opera Omnia di Benito Mussolini*, ed. by Eduardo Susmel and Duilio Susmel (Florence: La Fenice, 1927), 360–390.

[54] Steen Bo Frandsen, "'The War That We Prefer:' The Reclamation of the Pontine Marshes and Fascist Expansion," *Totalitarian Movements and Political Religions* 2, 1 (2001), 69–82.

[55] Umberto Gabbi, "Per la visita prematrimoniale," *Archivio fascista di medicina politica* 1, 1 (1927), 88–97.

[56] Benito Mussolini (with Giovanni Gentile), *La dottrina del fascismo* (Milan: Treves, 1932).

[57] See Aaron Gillette, *Racial Theories in Fascist Italy* (London: Routledge, 2002), 50–99.

[58] Luigi Goglia, "Note sul razzismo coloniale fascista," *Storia Contemporanea* 19, 6 (1988), 1223–1266; and Gillette, *Racial Theories in Fascist Italy*, 50–59.

[59] Michael Burleigh, "Racism as Social Policy: The Nazi Euthanasia Programme, 1939–1945," *Ethnic and Racial Studies* 14, 4 (1991), 453–473; and Donald Dietrich, "Catholic Resistance to Biological and Racist Eugenics in the Third Reich," in Francis R. Nicosia and Lawrence D. Stokes, eds., *Germans against Nazism: Nonconformity, Opposition and Resistance in the Third Reich* (New York: Berg, 1990), 137–155. See also Aristotelis Kallis, *Fascism and Genocide* (London: Routledge, 2007), especially chapters 6–8.

[60] Stein Larsen, "Was there Fascism outside Europe? Diffusion from Europe and Domestic Impulses," in Stein Larsen, ed., *Fascism Outside Europe* (New York: Columbia University Press, 2001), 705–817.

Tunnel Visions and Mysterious Trees: Modernist Projects of National and Racial Regeneration, 1880–1939

Roger Griffin

The No could not be so powerful if there were not simultaneously in our midst a Yes worth fighting for that is lethal to it; if beneath the veils separating us from life, below the nihilism of the new age, there did not stir all the time a force unknown to our morality or imagination, one which has been constantly thwarted by every sort of fear and obstacle. It is via this hazardous path that almost everything has reached us, fleeing a world inhospitable to life so as to find refuge in us as gardeners of the most mysterious tree, a tree which has yet to grow. In us alone the light still burns while earth and heaven collapse all around: the supreme creative, philosophical moment has arrived.

Ernst Bloch, *The Spirit of Utopia,* 1916[1]

With these words Ernst Bloch (1885–1977), later to become famous as one of the most original thinkers in the history of Marxism, articulated a paradoxical, defiant optimism just as the First World War and the fate of the entire Western world seemed to be reaching an apocalyptic climax. Anticipating the theme of his monumental *Principle of Hope* (1954–1959), written in exile from Nazism two decades later, *The Spirit of Utopia*, couched in an abstruse metaphysical register, presents the Marxist project to create a just social order not as the product of material factors, but as a palingenetic, Nietzschean leap beyond the void of contemporary nihilism.[2] However, its immediate relevance to contextualizing the variants of European eugenics and racial nationalism that are the subject of this book lies in Bloch's visionary sense of belonging to an elite whose mission is not just to usher in an era of certainty and light where there is now chaos and darkness, but to act as gardeners; that is arrogating the power to decide what shall be planted and encouraged to thrive, and what shall be cut back or rooted out on behalf of a society otherwise doomed to self-annihilation. This conclusive chapter probes into the deeper historical substrata underlying these metaphors in the context of early-twentieth-century European modernity in the hope of casting fresh light on the complex relationship between nationalism and the modernist revolt against decadence. In doing so, the chapter hopes to illuminate a hidden stratum of the ideological and cultural factors that shaped the aspirations of cultural, scientific and political elites in Europe, urging them to establish humanity's future on more robust and healthy foundations through forms of eugenically conceived nationalism.

The starting point for this analysis is the recurrence of two conspicuous features of cultural and political nationalism's confluence with ethnography, anthropology, demography, public health policies and eugenics. The first is the harnessing of Enlightenment rationalism, humanism, and science to the mythic imperative of offering humanity the prospect of purification and renewal. Closely bound up with this is the hybrid discourse in which each variant of bio-politics is constructed, one that fuses science and myth, academic scholarship and populism, and the cult of knowledge and progress with "atavistic" assumptions about the existence of an ethnic essence attached to an organic nation. This is a discourse which Eric Voegelin (1901–1985) described, in a seminal essay published in the aftermath of the Second World War, as "scientism." He considered the interweaving of the "advancement of science" with "the rationality of politics" in a way that concealed the dynamics of these projects, which he was convinced were both rooted in the secularized perversion of a mystic, "Gnostic" quest to transcend the physical universe.[3]

The thesis explored here is that the origins of scientistic nationalism lie not, as Voegelin believed, in the contamination of natural science and rationality by the atavistic metaphysics of Gnosticism. Instead, such a contamination is to be sought in the paradoxical synthesis promoted by modernity, particularly of rationalist and scientific notions of progress with archetypal longings for cyclic rebirth, which far transcend the parameters and perimeters of European culture. In order to contrast this argument with traditional assumptions that racial nationalism is the product of a reactionary nostalgia for "how things were," this chapter is broken into two parts. The first establishes the revolutionary dynamic at the heart of modernism, while the second focuses on the modernist dimension of racial and eugenic forms of nationalism that emerged with particular vigor within the technocratic circles of interwar Europe.

I. "The Yes worth fighting for": The Palingenetic Dynamics of Modernism

This chapter does not treat politicized nationalism and racism as the exclusive fruits of modernity, let alone as phenomena peculiar to Western Europe. What it does suggest is that "high modernity,"[4] which resulted when the corrosive impact of modernization on traditional European society gained increased momentum in the late nineteenth century, played a formative role in bringing about the fusion of a purportedly scientific understanding of race with the programmatic mission to bring about national renewal, which is one of the defining features of interwar variants of ultra-nationalism.[5] It is this synthesis that endowed visions of social regeneration and a new order, shaped by racial nationalism, with a destructive potential of unprecedented ferocity if they were ever to be enacted by a modern industrialized state as the basis of a new social order.

The corollary of everything associated with progress in liberal, capitalist, scientific, and technological terms was modernity's growing power to erode religious certainties, undermine customs, disturb the homogeneity of cultural communities, sap the vitality of religious faith, and destroy both the sense of unbroken continuity with the past and the predictability of the future. This expressed itself phenomenologically as the diffusion in society of a "malaise,"[6] an ill-described sense of loss of spiritual bearings,[7] of exile from some primordial existential home.[8]

The pioneers of modern sociology found different ways to make this protracted cultural malaise accessible to enquiry by Enlightenment rationalism. Max Weber, for instance, concentrated on the process of rationalization itself, the concomitant of which was the progressive "disenchantment" of the world and the privatization of metaphysical experience, to a point where "the individual can pursue his quest for salvation only as an individual."[9] Convergent ways of conceptualizing the erosive impact of modernity are evident in Ferdinand Tönnies' thesis that the relatively cohesive organic "community" (*Gemeinschaft*) was being replaced by a loose-knit, atomized "society" (*Gesellschaft*);[10] Émile Durkheim's theory of the breakdown of "mechanical solidarity" and the spread of anomie;[11] Georg Simmel's investigation of the psychological impact of the rise of materialism and urban life; and Walter Benjamin's memorable image of modernity as a "storm of progress" whose force drives the "Angelus Novus" back from the steadily rising detritus that it produces.[12]

Far from rejecting such approaches, modern cultural theory has built on and enriched them further. Thus the literary critic Friedric Jameson writes of modernity as a "catastrophe" that "dashes traditional structures to pieces, sweeps away the sacred, undermines immemorial habits and inherited languages, and leaves the world as a set of raw materials to be reconstructed rationally."[13] Zygmunt Bauman approaches the same phenomenon from a different angle by stressing the way in which modernity radically fragments the relative cohesion of pre-modern societies, installing "dysfunctionality as its functionality," and opening up the alternative of experiencing the world as either order or chaos in untold conceptions or combinations of both concepts. The result is a pervasive sense of ambivalence and the irresolvable fuzziness of all the categories and boundaries that once enabled human beings to interpret experience and negotiate the world communally as a stable, meaningful entity.[14]

For the model of modernity that we are constructing here, the temporal dimension of its "dysfunctionality" is of central concern. The pioneer of this line of investigation (though curiously omitted from several more recent works on the subject) is Mircea Eliade (1907–1986), who—after his own disillusionment with the prospects of the spiritual rebirth of society offered by the Romanian Iron Guard—wrote a number of seminal texts documenting the universality of the experience of sacred time and space in pre-modern societies. He contrasted this with Western European modernity that exposes its inhabitants to the "terror of history," the sense of being engulfed by a totally desecrated, undifferentiated, meaningless "chronos," or profane linear time.[15] Anthony Giddens is unwittingly treading in Eliade's footsteps when he writes of the "disembedding" impact of high modernity that "empties out" time and space.[16] In the same vein, Elissa Marder has used the works of Baudelaire and Flaubert as documents of the "temporal disorders" of modernity, whose bitter fruit is a decentred, deracinated space, and a disjointed time without transcendence. It is the prevalence of this "dead time" which prevents individuals from translating immediate sensation (what Walter Benjamin called *"Erlebnis"*) into *"Erfahrung,"* or meaningful experience, as part of a coherent narrative.[17] Marder's exposition of the deeper connotations of Baudelaire's recurrent image of the sky within the cycle of poems "Ideal and

Spleen" draws attention to the way modernity has destroyed the firmament "which shapes the world and gives it dimensions; thus it provides
a home for those who live under its shelter."[18] The result has been the
collective fall into a disenchanted "chronos" that forms a central motif
of the 1857 *Fleurs du Mal* (The Flowers of Evil), one that encapsulates
the evil which he strives to convert into beauty through the alchemy of
poetry.

Such exquisitely aestheticized pessimism epitomizes the lush
metaphorical jungle that has grown up in modern poetry, fed by underground streams of metaphysical energy liberated by the loss of a terra
firma (*coelum firmum*?) on which to ground the modern experience of
the sacred. The result for many is a protracted, insatiable *ennui* punctu-
ated by epiphanies, unsustainable moments of lightness of being and
glimpses of higher states of consciousness, which only make the fallen
state of everyday life more intolerable.

The Luxuriance of Cultural Decadence

In the decades following the publication of *Les Fleurs du mal,* the
impression of accelerating cultural disenchantment and disintegration
became so widely diffused in Western Europe that Baudelaire came to
be treated as one of the first articulators of the anomic experience of
modernity. In the twentieth century, Franz Kafka (1883–1924) has
come to occupy a similarly iconic status in the collective imagination
precisely because he explored in naturalistic detail the oneiric experience of being in the "wrong" place and asking the "wrong" questions
when confronted by events that tantalizingly elude rational comprehension and control. In one of his published diary notes, he vividly
evokes the sense of disorientation and the "emptying out" of time and
space under high modernity with a sustained metaphor of passengers
surviving a crash in a railway tunnel. We are unable to make out which
is the exit and which is the entrance, but "all around us we can see,
because of the confusion of our senses or the extreme sensitivity of our
senses, unearthly creatures, and a kaleidoscopic play of images which,
according to the mood and the degree of injury of the individual, is
either entrancing or exhausting."[19] In his short prose piece "Conversa-
tion with a supplicant," Kafka describes a man praying with ostenta-
tious piety in a church, who finally confesses to the narrator: "There

has never been a time in which I have been convinced from within myself that I am alive. You see, I have only such a defective awareness of things around me that I always feel they were once real and are now falling away. I have a constant longing, my dear sir, to catch a glimpse of things as they may have been before they show themselves to me. I feel that then they were calm and beautiful. It must be so, for I often hear people talking about them as though they were."[20]

In the present context, however, what is important about the passage from Kafka is that it uses terms whose etymology alludes to the dominant metaphor of the late nineteenth century for evoking the disorienting, anomic nature of modernity among an entire generation: decadence. The term derives from the Latin "to fall," and indicates another analogy, deeply rooted in the human metaphorical faculty, between falling and the loss of reality, vitality, or health, producing in English such words as "decay" (from *decadere* via French), collapse (from the Latin stem *labi*, to slip, slide, or fall), and decline (from the Latin for "sinking away"). Likewise in German, there is an audible link between *fallen* (to fall) and *verfallen* (to decay), and in Italian between *cadente* (falling), *scadente* (defective, shoddy), and *decadenza* (decadence). In the passage from Kafka cited above, "falling away" renders the German *versinkend* (literally "sinking away"), while the phrase "defective awareness" translates *hinfälliges Bewusstsein*, that is, the "sense of things falling away."

"Decadence" appears to be an archetype of human experience, and certainly not one confined to the late nineteenth century, or to any one aspect of reality. An obsession with Rome's decline or imminent destruction and its need for renewal (*renovatio*) haunted the Romans centuries before its actual collapse as the centre of imperial power.[21] The enduring fame of Gibbon's *Decline and Fall of the Roman Empire*, written at the height of the Enlightenment, may be attributed to the link it makes in the human historical imagination between moral and social decadence and the waning of political power. It is within the "cultural field" of modernity, however, that "falling" acquires particularly rich connotations in the artistic imagination.[22]

The hidden nexus between a generalized sense of cultural dissolution and the fall into secularized, profane time is exposed in forensic detail in the writings of E. M. Cioran (1911–1995). In particular, his 1948 *Précis de décomposition* (A Short History of Decay) can be seen as the twin of *La chute dans le temps* (The Fall into Time) of 1964.

These soliloquies on the human condition are unsurpassed in their sardonic dissection of the decadent state of modern civilization and the putrefaction of modern life. These texts stand in marked contrast to the optimism he expressed for the possibility of Romania's salvation from decadence when he was an enthusiastic supporter of the Romanian Iron Guard.[23] The connotations of primordial sin which "falling" acquires in the Christianized West (also haunting Baudelaire's poetry) were dramatized in Friedrich Dürrenmatt's Kafkaesque short story of 1952, *Der Tunnel* (The Tunnel), in which a train enters a railway tunnel and, instead of crashing, begins to career ever more vertically downwards into the bowels of the earth. A terrified passenger yells: "What are we to do?" The reply comes: "Nothing. God let us fall, and so we are plunging down to him," implying "Modern Man's" collective fall from grace, a fate that cannot be escaped through technological advance or by anaesthetizing existential time with the commuter's time, yet another poisoned fruit of rationalization.

It was in the latter part of the nineteenth century that the combined work of countless European artists produced a glittering display of kaleidoscopic images to evoke the modern fate of living out the death throes of an effete civilization and an accompanying mood of personal exhaustion, one alternating with desperate bouts of hedonism bent on extracting perverse sensuous pleasures as the world collapsed on all sides. The prospect that contemporary history was literally the fin de siècle, not just in a calendrical but in an eschatological sense, became a leitmotiv of the times to a point where "decadence" became a European buzzword, subsequently entering cultural history as a period concept.[24] The decadents themselves were far from having any sort of common program or organizational centre. In retrospect, however, all their work can be seen as highly personal attempts to find verbal, visual, musical, or plastic form for the unprecedented cultural malaise, induced by the sense that Western European civilization, while progressing economically and technologically in ever more spectacular fashion, was actually—unbeknown to the vast majority of its inhabitants—sinking ever more deeply into a process of spiritual decline and moral bankruptcy. As an art movement, decadence thus expressed the phenomenological reality of the disenchantment, anomie, and breakdown of community being conceptually modeled by the major sociologists of the day. In different ways, the "perversion" explored by Gustave Flaubert's *Salammbô* (1862), or Joris-Karl Huysman's *A Rebours* (1884) can be seen as elaborate, super-refined "objective correlatives" of a growing disjuncture between the European society's increasing technological, material, military, scientific, imperial, and capitalist power, and its cultural (moral, ideological, religious, spiritual, artistic) cohesion.

The Dark Side of the Moon

Once hypostatized, "Decadence" is generally associated with an artistic movement. However, it was in the sphere of the natural sciences that a growing concern with the pathology of modern progress was to have its most profound social consequences. In 1857, the same year that Baudelaire published *Les Fleurs du mal* (itself a composition on decomposition), the appearance of Bénédict Morel's *Traité des dégénerescences physiques, intellectuelles et morales de l'espèce humaine* (Treatise of the Moral, Intellectual and Physical Degeneration of the Human Race) portrayed the main threat to the progress of civilization as stemming not from external enemies and inferior races abroad, but from the unchecked proliferation of hereditary pathological conditions within France's own population. Two years before, the Comte de Gobineau's *Essai sur l'inégalité des races humaines* (Essay on the Inequality of Human Races) had issued a dire warning about the long-term impact racial mixing had on the Aryan qualities of Europeans, a process which would inevitably lead to a decline in racial vigor and the ultimate collapse of Western civilization.

In the second half of the nineteenth century, the alarm of both cultural and scientific elites at the pernicious effects of physical and moral decay can be seen as a direct corollary of the dominant ideology of the Europeanized world—namely, the unbroken linear ascent of humanity from the primitive to the peak of civilization, the shadow of the Enlightenment view of progress.[25] Paradoxically, the fear of degeneration, rather than being the fruit of a healthy climate of scientific objectivity, was itself the product of morbid fears and pathological fixations. Accordingly, the dream of using science to promote a "healthy race" itself contained unhealthy components, much as psychoanalysis was for the Viennese critic Karl Kraus (1874–1936) "an illness for which it claims to be the cure."[26]

In his *Faces of Degeneration: A European Disorder c. 1848–c. 1918*,

Daniel Pick has charted the development of growing concern in scientific circles with genetically determined pathological conditions in the decades preceding the First World War, and the threat they posed to the "healthy" population by the allegedly physiological degeneracy of entire social groups which predisposed them to lives of dissolution.[27] Inevitably, fears of the threat posed by degeneracy to European civilization gave rise to concerns about the health of the nation in particular. The concern of French artists and positivists with degeneration is inextricably connected to the obsession with national decline that followed France's defeat in the Franco-Prussian War (1870–1871), and the exacerbated mood of *revanchisme* arising from the loss of Alsace-Lorraine that fed currents of ultra-nationalism and racism.[28] The most iconic text for the interweaving of the themes of cultural, physiological, psychological and national health in this period was written in Paris, not by a Frenchman but by a former medical student of Jewish-Hungarian background, Max Nordau (1849–1923). His 1883 *Die konventionellen Lügen der Kulturmenscheit* (The Conventional Lies of Society), and especially the 1892 *Entartung* (Degeneration), encapsulated the late-nineteenth-century *Zeitgeist* so well that they both achieved international bestseller status with their broad-brushed panorama of the pathological symptoms affecting Europe's fin-de-siècle. One critic has summarized the central theme of *Entartung* in terms that dovetail perfectly with Zygmunt Bauman's description of the Age of Ambivalence: "forms have become blurred, boundaries upset, order forsaken, logic and values abandoned—all in favor of undisciplined chaos."[29]

In a populist register that merged medical, socioeconomic, political and cultural discourses with a fervor that at times smacked of a hysterical, neurosthenic disposition, Nordau depicted modern society as threatened by a rising tide of madness, suicide, crime, deviancy, hysteria and neurasthenia.[30] These were associated with other symptoms of a culture that had lost its way, such as the fad for occultism, subversive forms of utopian politics, and the outpouring of modish literature and painting in which a taste for the perverse, the erotic, the sick, the morbid, the prurient, the sadistic, the distorted and the debauched prevailed over classical norms of serenity, rationality and the sublime. In short, Nordau was convinced civilization was threatened by a "severe mental epidemic," one comparable to a "Black Death of degeneration and hysteria."[31]

Palingenetic Pessimism as a Response to Liminality

True to the ambiguous character of modernity, there was another side
to Nordau's vision of the world. *Entartung* ends not on a note of bleak
pessimism but triumphant optimism, to a point where one critic suggests that his book should have been called "regeneration."[32] The principles of Social Darwinism that informed his diagnosis dictated that, in
the coming century, those best adapted to modernity would survive, so
that those embodying life-asserting forces of health and civilization
would eventually prevail over physical, psychological and artistic
degeneracy. But there is a caveat: the "fittest" are warned in the last
pages that in order to prevail they may have to be prepared to beat to
death the anti-social "vermin" who threaten their world. It is perhaps
no coincidence that the term Nordau used for "vermin" was *Ungeziefer*,
precisely what the office-worker Gregor Samsa found he had turned
into after a night of restless dreams in Kafka's 1912 *Die Verwandlung*
(The Metamorphosis).

In 1897, Nordau was able to give full vent to the latent social utopianism that was the flipside of his pessimism when he gave a speech to
the Second Jewish Congress in Basel. In it, he suggested that Zionism
should be understood as a moral, national, and corporeal revolution in
which the awakening power of Jewish nationalism went hand in hand
with the emergence of a new breed—the "Muscular Jewry" (*Muskel-
judentum*)—the Jewish counterpart of the "muscular Christianity"
being preached at the time,[33] one which needed to measure up physically as well as spiritually to the heroic tasks involved in the creation
of a Jewish homeland in Palestine. This "new" Jew, the epitome of
masculine values, combined the mental alertness and courage of a warrior with the strength and stamina of an athlete, and was the diametric
opposite of the image of the sickly, physically repulsive "intellectual"
or "professional" Jew cultivated by the anti-Semitic propaganda of the
day, the product of a dysgenic urban life which deprived them of the
revitalizing contact with nature and the dignity of manual labor. Not
surprisingly, Nordau enthusiastically supported the foundation of the
newly founded Jewish Gymnastics Association (*Jüdische Turnerschaft*);
and, in 1900, he wrote an article for its newspaper which presented it
as a major step towards the physical regeneration of the Jewish race
that was the precondition of the fulfillment of the Zionist project.[34]

Most major thinkers and social commentators of the late nineteenth

century who discussed matters of racial health or the state of civilization were keen to promote renewal and revitalization in terms that
betrayed growing alarm at the proliferating symptoms of society's
moral and physiological degeneracy. Whereas, in the mid-nineteenth
century, Gobineau had made gloomy prognostications about the long-
term prospects of the white race, by the 1880s, Social Darwinist "racial
experts" such as Francis Galton, Thomas Huxley and Cesare Lombroso
were all convinced of the state's potential to introduce interventionist
measures based on the latest medical knowledge and scientific understanding that could, if not rid society of degeneracy, at least minimize
its debilitating effects on the healthy, and reverse national decline. In
fact, the striking contrast between Gobineau and Nordau, or between
Cioran in his fascist phase, when he believed in Romania's imminent
resurrection from decay, and the bottomless well of logorrheic despair
he fell into after the war, suggest the need to distinguish between two
types of pessimism: a fatalistic one in which decline and entropy are
seen as ineluctable, and a palingenetic variety which envisages the
possibility that civilization can pull out of its nose-dive towards destruction just in time for it to assume a cyclic shape imbued with the "telos"
of rebirth and renewed life.[35]

While research may have shown that optimism does not increase
the prospects for cure of lung-cancer patients, from the point of view
of evolutionary ethics, a case could be made for the survival value of
the human predisposition to adopt a "Positive Mental Attitude,"[36] to
create utopias through which to imagine ways out of an intolerable
present. Evidence for the existence of an archetypal human drive to
counteract threats to the community through the generation of utopias
is provided by what is known to anthropologists as "revitalization movements." These are described by the pioneering expert on the social
function of ritual, Victor Turner, as emerging at times of "marked cultural change and its accompanying personal psychological stress" as
an attempt "to revitalize a traditional institution, while endeavoring to
eliminate alien persons, customs, values, even material culture from
the experience of those undergoing painful change." When a tribe is
threatened by a more powerful one, "revitalistic movements tend to
take on a religious character" which involves the "ludic combination
of old and new cultural components," and the emergence of "new
prophets" whose dreams and visions experienced in shamanistic trance

become translated into new myths and rituals embodying the emergence of an "existential *communitas*."[37]

For Turner, a crucial factor in the dynamics of revitalization movements was that the objective crisis to the tribe's well-being and way of life had created a temporary liminal period in which an entire society moved from one state to another, and thus experienced the triadic process that is the feature of all *rites de passages*: separation from one state of social reality, entry into a marginal or threshold (*limen*) state, and finally reintegration within a new state. The imminent threat of cultural annihilation is warded off by improvising a new cultural basis for survival through a process of mythic recombination, synthesis, and "reaggregation," allowing the given tribe to experience collective rejuvenation (its rebirth or palingenesis). Ritual improvisation and mythic syncretism are thus integral to the human revitalization process, to the point where "many of the features found in liminal and liminoid situations come to dominate the new religion, drawing sustenance from many hitherto separate tribal conditions."[38] At the heart of such a process lies the mythopoeic drive to restore a primordial, pristine order of society.[39]

Liminality and Modernity

Victor Turner stresses the relevance to an understanding of pre-modern history of his analysis of tribal responses to crises, suggesting that Christianity and Islam both arose as revitalization movements, and that "millenarian movements arise in historical situations when society as a whole, or major groups in it, are in a transitional, liminal state."[40] Yet in an earlier book, he also fleetingly acknowledged that the progressive complexity of modern society means that "what was, in tribal society, principally a set of transitional qualities 'betwixt and between' defined states of culture and society, has become itself an institutionalized state."[41]

The explosive heuristic potential for understanding modernity contained in this throw-away remark is fully realized by the sociologist Arpad Szakolczai. The conclusion to his study of the competing constructs of modernization expounded by "reflexive historical sociologists" such as Weber, Durkheim and Voegelin is a sustained exploration of the implications of conceiving modernity as a condition of perpetual crisis. He argues that the proliferation of ideologies offering revolutionary alternatives to the present order is to be seen as "a response to the real-world large-scale liminal crisis of the 'ecumenic age'. (...) This utopian and eschatological mentality has become incorporated into the everyday reality of a world that has entered a phase of *permanent liminality*."[42]

Szakolczai was concerned primarily with the sociology of modernity. But applying the anthropology of tribal liminality to understanding modern culture in the broadest sense offers a fresh perspective on the various symptoms of modern society's dysfunctionality, identified by sociologists and cultural pessimists alike. It suggests that, at a certain point, Europeanized society entered a process of accelerating social "disaggregation" with no immediate prospect of spontaneous, organic reintegration ("reaggregation") through which a new homogeneity could be restored. In other words, an old *communitas* and cosmological order has been suspended, but no comparably homogeneous *communitas* or new order can emerge to replace it, so that the temporary, the provisional, the transitional have become permanent, and crisis has become normality. Modern human beings are thus unwitting participants in a *rite de passage* with no clearly delimited new state to reach naturally on the other side of the present chaos and flux, and no possibility of return to the preceding (largely mythicized) state of relative cultural integrity and homogeneity.

Combining Bloch's "principle of hope" with Szakolczai's "permanent liminality" makes it both intelligible and predictable that modernity stimulates an instinctive urge to flee away from decay and entropy and towards rebirth and renewal, thus predisposing its most disaffected inhabitants to palingenetic rather than entropic forms of pessimism. Whether the creativity remains hidden in the recesses of a person's inner world or is displayed in high-profile religious, social, or political movements, countless modern initiatives undertaken in the spheres of art, ideology, society and politics, from occultism to the scouting movement, from expressionism to psychoanalysis, can be seen as attempts to re-enchant and re-spiritualize the world, abolish anomie, rejuvenate society, foster a new type of *Gemeinschaft*, re-embed and re-root (or "re-enracinate") society, make time and space full once more, resolve ambivalence, create order, and find a new spiritual home. The hidden mainspring of cultural creativity in the thrall of modernity is the instinctive need to put time back on its hinges, to re-erect the vault of heaven so as to find shelter, even temporary shelter, beneath a new ontological firmament.

In the context of the genesis of bio-politics and racial nationalism, another feature of Kafka's work acquires particular significance, namely the way it defies neat classification into conventional taxonomic categories of literary genre: terms such as parable, allegory, short story, science fiction, fantasy, satire, aphorism, fairy tale; or categories such as symbolism, magic realism, surrealism, humor, tragedy, realism, irony, satire, fact, fiction all become approximations, while the boundaries between "notes," "fragments," "drafts" and the final text are frustratingly fuzzy. In Kafka's case, even the act of writing was carried out in a liminal space between the private and the public, between personal and historical time, between keeping a diary of his "dreamlike inner life" and the self-conscious creativity of a publishing author, so that literature and art meld with autobiography and psychosis in a way reminiscent of the hybrids of humans with animals that are a feature of his shorter prose pieces.

This illustrates an important point latent in the vision of modern human beings as "conglomerations" of disparate fragments of past and present realities. For the editors of the thirty-five articles comprising the seminal anthology *Modernism 1890–1930*, this observation goes to the very heart of modernism itself: "Modernism was in most countries an extraordinary compound of the futuristic and the nihilistic, the revolutionary and the conservative, the naturalistic and the symbolistic, the romantic and the classical. It was a celebration of the technological age and a condemnation of it; an excited acceptance of the belief that the old regimes of cultures were over, and a deep despairing in the face of that fear; a mixture of convictions that the new forms were escapes from historicism and the pressures of the time with convictions that they were precisely the living expressions of these things."[43]

This rampant syncretism is a pattern entirely consistent with Victor Turner's anthropological account of revitalization movements. These, as we saw earlier, emerge at times of "marked cultural change and its accompanying personal psychological stress" and "tend to take on a religious character," involving the "ludic recombination of old and new cultural components."[44] As a feature of revitalization movements, Turner also identified the emergence of "new prophets," whose dreams and visions experienced in shamanic trance become translated into new myths. Certainly, interwar Europe witnessed a proliferation of would-be shamans performing the artistic correlative of sacred dances and magic incantations to transform a reality grown unbearable. However, as long as the sense of progress and stability was sustained for the educated classes, and it remained for the bourgeoisie, at least, a belle époque, the surfeit of rival redemptive visions produced by the avant-garde remained ignored by the population at large, for whom the continuum of history was still unbroken. The revitalization movement that does not move, the shaman who works his magic in private are thus also symptomatic of modernity. As we shall see in the second part of this chapter, it was not in the realm of art and aesthetics that modernism was to have a cataclysmic effect on modern civilization, but in the alliance that its social and political manifestations formed with the natural sciences.

II. "Rooting away the noisome weeds": The Modernism of Racial Nationalism

> *Gardener:* Go, bind thou up yon dangling apricocks,
> Which, like unruly children, make their sire
> Stoop with oppression of their prodigal weight:
> Give some supportance to the bending twigs.
> Go thou, and like an executioner,
> Cut off the heads of too fast growing sprays,
> That look too lofty in our commonwealth:
> All must be even in our government
> You thus employed, I will go root away
> The noisome weeds, that without profit suck
> The soil's fertility from wholesome flowers.
>
> William Shakespeare, *The Tragedy of King Richard the Second*,
> Act III, Scene IV

Political Modernism

The thrust of modernism is to break out of the contemporary wasteland so as to live under a new firmament of time and space and, therefore, in a re-rooted, re-centred universe. It would be a contradiction in terms if this taboo and mould-breaking aspiration, both iconoclastic and icono*plastic*, were tethered to a clearly delineated sphere of aesthetics that, perhaps reassuringly for traditionalists, minded its own business. But by definition modernism cannot produce "art for art's sake" without wanting to aestheticize the universe. Just as the train accidents and earthquakes of high modernity at the turn of the nineteenth century hurled many an original artist or thinker into far-flung orbits of socio-political and metapolitical utopianism, so the most historically self-aware experts working at the cutting-edge of social, political, or technological change, tended to be infused with the regenerative, cyclic transformative ethos of modernism in ways that conflicted with the linear progressiveness inscribed into their Enlightenment heritage.

To be sure, modernism is a term encountered almost exclusively by students of literature and art history. It is to be considered a cultural phenomenon, but not if "culture" connotes no more than the safely domesticated appendage to "leisure and sport," to which it has been reduced in the schemes of contemporary liberal governments. Instead, modernism should be associated with the tectonic, totalizing connotations of culture familiar to anthropologists in the context of Aztec or ancient Greek culture, where the term denotes an entire socioeconomic, political, ritual and cosmological whole. It is a usage akin to Lewis Mumford's concept of the "megamachine," first expounded in *Technics and Civilization* (1934), when so much of Europe and the Americas had become a vast building site for the completion of gigantic projects of social metamorphosis. Under modernism, metaphysical ideals and projects of socio-political renewal leak into, and merge with, each other.

A glimpse of the psychological and ideological matrix behind modernism's bid to transform the present is provided by the chapter "The Shape (*Gestalt*) of the Inconstruable Question" in Ernst Bloch's *The Spirit of Utopia*, written at the height, or rather in the depths, of the First World War. His call for a Nietzschean, life-affirming "Yes" is presented as a visionary, visceral rejection of the encroaching nihilism of the age in which "the descending fog grows ever thicker, and with it a pervasive ambiguity about what marks the threshold to be crossed, confusing us about where to direct our energies once more and shrouding our goals in uncertainty." In an early expression of the "principle of hope"—whose exhaustive exploration as a human and historical phenomenon was to become the basis of his magnum opus (which provides the epigraph to this chapter)[45]—Bloch finds "the paradoxical courage to prophesy that light will burst forth from the fog."[46]

The passage proceeds to express Bloch's conviction—itself messianic and utopian given the state of European society and his total obscurity outside narrow Marxist circles at the time—that he was personally charged by unspecified higher forces to become a "gardener" with the mission of cultivating "the most mysterious tree." This organic metaphor evokes the emergence of an entirely new order that will subsume art, society, politics, and history itself. In this way, from a deeply idealist cast of mind, the modernist vision becomes a psychological *Gestalt* from which action inevitably flows.

The view of modernism emerging here is one that stresses the role played in it by palingenetic, "metaphysically constitutive" myth, and the way such myth is not just a feature of artistic creativity at the turn of the twentieth century, but one that informs ambitious projects of cultural and socio-political transformation. Both revolutionary and counter-revolutionary forms of political modernism may be seen as attempts to combat the liminality of modernity, and the accompanying sense of chaos and dissolution; its complexity and amorphousness inspire those who long for a completely comprehensible, ordered, controllable world.

Technocratic Modernism

But totalitarian politics is not the only conspicuous manifestation of the modernist drive to achieve a new temporality that resolves the ambivalence and decadence of the present. A substantial section of Peter Osborne's *The Politics of Time* (1995) meticulously reconstructs the process by which a major academic philosopher such as Martin Heidegger, working within the "cultural force field"[47] peculiar to modernism, could become, however temporarily, a convinced and proactive Nazi.[48] Once he moved from contemplating the phenomenology of modernity's ontological crisis to considering the historical process of spiritual decline that was allegedly producing it, he was fatally attracted into the orbit of a form of revolutionary politics that seemed to share his diagnosis and offer practical solutions to its "cosmic" implementation.

This philosophical modernism had its counterpart in other spheres rarely associated with culture in the narrow aesthetic sense of the word. We have already suggested that phenomena such as the occult revival or the cults that developed round figures such as Ibsen, Nietzsche,

D'Annunzio, Wagner, and Bergson, should all be seen as symptoms of "cultural modernism," one conceived as a countervailing force to the anomie bred by permanent liminality. However, this cultural modernism itself had even wider social reverberations. The confluence between the experience of modernity and developments in the fields of technology and modern physics is well-charted territory;[49] but, more recently, Ronald Schleifer has demonstrated the reciprocal relationship between aesthetic modernism and radical changes of thinking within early-twentieth-century mathematics, physics, analytical philosophy, and even economics.[50] The resulting perspective emphasizes how misconceived it is to assume that there is some sort of basic contradiction of mindset or temporality at work when educated elites associated with some of the iconic progressive activities of modernity, such as subatomic physics, civil and mechanical engineering, genetics, medicine or town planning, willingly lent their expertise to promoting forms of right-wing extremism generally associated with reaction. An example of such flawed logic underlies Jeffrey Herf's thesis that the espousal of Nazism by some of Weimar's leading figures in science and technology combined a progressive with a regressive current of thinking, which he summed up in the phrase "reactionary modernism." Yet within the virulently anti-traditional ethos of modernity, any form of radical reaction against the status quo is itself imparted with a futural momentum. As Peter Osborne points out, the terms "conservative revolutionary" or "reactionary modernism" do not point to conflicting but to complementary temporalities: reaction is itself a revolutionary force.[51]

It is entirely consistent with this line of argument that in the early twentieth century there should have arisen a powerful current of programmatic and technocratic modernism bent on reversing decadence and decay, with an elective affinity not just with a wide range of aesthetic modernisms, but with political modernisms of the liberal centre, the socialist Left or the nationalist Right. In fact, it is precisely those trained to see the world as transformable through scientific and technological expertise that are likely to be the most predisposed to "celebrate and identify with the triumphs of modern science, art, technology, economics, politics; with all the activities that enable mankind to do what the Bible said only God could do: to 'make all things anew'."[52] This is nowhere more evident than in the sphere of architecture and town planning. In his classic survey of artistic modernism, *The Shock of the New*, Robert Hughes observed "The generation of northern

European architects who came to professional age between 1910 and 1920 was profoundly charged by a sense of the millennium—the literal renewal of history at the end of a thousand-year cycle, the beginning of the *twentieth* century."[53] It was in this spirit that, in 1919, Walter Gropius (1883–1969) used the launching of the official manifesto of the Marxist Work Council for Art in Berlin to proclaim that: "There are no architects today, we are all of us merely preparing the way for him who will build gardens out of deserts and pile up wonders in the sky."[54] However, while Gropius was drawn to the revolutionary Left, Charles-Edouard Le Corbusier (1887–1965), another visionary of architectural transformation, was drawn to the fascist Right,[55] while Ludwig Mies van der Rohe (1886–1969), on the other hand, produced some impressive architectural designs for the Third Reich before going on to create a number of skyscrapers in the US that later became emblems of Western capitalism and the "Free World."[56]

In the aftermath of the First World War, when both politics and technocracy were infused by the palingenetic spirit of modernism, an endless supply of architects and planners eagerly made themselves available to the totalitarian regimes of the twentieth century in order to create the spaces and structures that would both physically and symbolically eliminate the decadence of the past and embody the new age.[57] However, the roots of this development lay in the late nineteenth century when "decadence" gave way to "modernism" as the dominant avant-garde mood, and when the cosmopolitan metropolis began to be seen as a fomenter of physical and moral degeneracy able to be overcome only through a type of resolute action consonant with a rapidly accelerating modernity.

It was a sign of the times, then, that the concept of the garden city as a source of regeneration was pioneered almost contemporaneously by the Englishman Ebenezer Howard (1850–1926) in his *To-morrow: A Peaceful Path to Real Reform* (1898), and by the German Theodor Fritsch (1852–1933), whose *Die Stadt der Zukunft* (The City of the Future) appeared two years earlier.[58] For both men, town planning was destined to assume the role in the social engineering of public or national health that Max Nordau hoped would be played by the combination of athleticism and nationalism in overcoming the degeneracy of the Jews as a people. The difference was that, while Howard's utopianism remained firmly within the tradition of Enlightenment humanism, Fritsch's was informed by an obsession with racial hygiene and the preservation of the *völkisch* essence of the German people, an obsession that made him into one of the most prolific publicists of anti-Semitism in history.

Racial Nationalism as a Modernist Hybrid

Once set within the conceptual framework of modernity that we previously established, the various permutations of racial nationalism examined in this volume can be seen in a fresh light. Far from being expressions of anti-modern reaction, an atavistic regression to barbarism, or the desperate mystifications and populist hysteria of a beleaguered capitalism, scientific racism can be viewed as manifestations of the modern revolt against decadence which fused two distinct currents of modernism: ultra-nationalism and eugenics. The last decade has seen a bewildering proliferation of theories about the dynamics of populist nationalism, known under such codenames as primordialism, perennialism, ethno-symbolism, pre-modernism, industrial modernism, Marxist modernism, Marxist postmodernism, and feminism.[59] What has been generally obscured in the flood of competing constructions of the topic, however, is that the emergence of ultra-nationalism in late-nineteenth-century Europe (also known as Romantic, organic, identity, ethnic, ethnocratic, or tribal nationalism) is an outstanding manifestation of the instinctive collective attempt to erect a bulwark against the savagely deracinating and disembedding impact of the "storm of progress"; or, rather, "the hurricane of modernity." It was as part of the revolt against encroaching liminality—and hence as a form of cultural modernism on a par with the occult revival or the Wagner cult—that the forces of ultra-nationalism turned highly mythicized versions of national traditions into the source of a mass-mobilizing political religion.

These modern revitalization movements sanctified their war on cosmopolitanism, materialism, Jews, atomization, hedonism, rising immorality, and various other signs of degeneration by furnishing themselves with spontaneously generated creeds and rituals whose thrust was towards inaugurating a triumphant phase of renewal and regeneration after an age of decay.[60] This occurred against the background of cultural movements that celebrated the idea of a precious national essence, one silently haemorrhaging into the "asphalt culture"[61] of the modern world. Even before the First World War, ideologues of *völ-*

kisch nationalism and political anti-Semitism were hard at work in a number of European countries, forging a discourse viscerally opposed to communist collectivism and liberal individualism (both "decadent" phenomena). By the same token, they sought to turn the cause of national unity—whether irredentist, revanchist, secessionist, anti-imperialist, or imperialist—into the crusade of a political religion constructed through a combination of shared history, language, geography, religion, or culture.[62] It was a religion legitimated by its own martyrs, heroes, and charismatic leaders, and rationalized by rewriting and distorting the memory of the past to provide narratives of national rebirth.

Therefore, what drove the late-nineteenth-century explosion of nationalist mythopoeia was not primarily the past-oriented reaction of capitalism to socialism, or the ambition of conservative ruling elites, seized by the fear of being engulfed by the rise of democracy, to beguile the masses into supporting the status quo. Rather it was the subliminal terror of an utterly disenchanted "chronos," of the endless cul-de-sac of liminality, of an irrevocably decentred, rootless, unbounded, demythologized, anonymous future. Ultra-nationalism is thus sapped by the very forces that nourish liberal nationalism: rationalism, individualism, secularism, science, cosmopolitanism, inter-cultural exchange. It thrives on latent fears of decadence and dissolution. As Cioran sarcastically observed: "A nation dies when it no longer has the strength to invent new gods, new myths, new absurdities; its idols blur and vanish; it seeks them elsewhere and feels alone before unknown monsters."[63]

In its quest for revitalizing idols, some currents of ultra-nationalism became entwined with an originally quite distinct strand of modernism—technocratic rather than cultural—that we have already encountered: eugenics. Though often assumed to be a branch of the natural sciences, as a discourse, eugenics has a mongrel ancestry. On the scientific side, its rise is influenced by the attempts of eighteenth-century anthropologists to find a taxonomy for the different phenotypes of human beings discovered as a result of European geographical exploration; the development of physical and cultural anthropology; Darwinian theories of natural selection; and Mendel's groundbreaking empirical work on the laws of heredity, which established several basic principles of genetic research some forty years before the formulation of modern genetic theory in the early 1900s. On the mythic side, a decisive role was played by imperialism, Eurocentrism, Social Darwinism, the rise of positivism, and the belief in the mission of white, male, scientific, technocratic, and bureaucratic elites to manage civilization in the best interests of the untold masses of the poor at home and the "primitive" abroad. Of course, a decisive mythic ingredient was the result of a growing nineteenth-century obsession with decay and degeneracy, which fed countless expressions of scientifically rationalized cultural pessimism torn, at a mythic level, between fatalism and palingenetic utopianism.

The term "eugenics" was coined by Francis Galton (1822–1911) in the last chapter of his autobiographical *Inquiries into Human Faculty and Social Development* (1883), "Race Improvement." However, in the two-part article "Hereditary Character and Talent," written in 1864, he had already formulated the fundamental principle behind what was to become identified with the new science: the laws of heredity, already applied to breeding livestock, could serve as the basis for measures designed to improve the blood-stock of human beings. The prospect was that the "men and women of the present day" would be "to those we might hope to bring into existence what the pariah dogs of the streets of an Eastern town are to our own highly-bred varieties."[64]

It was against the background of the growing concern with degeneracy both within and outside academia that Galton developed his sense of personal mission to turn the utopia of breeding a regenerated human race into a reality, and that his theories achieved such extraordinary international resonance in the educated circles of the day.[65] Galton was himself dimly aware that eugenics was a hybrid of science and myth, and that his passionate commitment to promoting it was as much a matter of social engineering as of pure science. In an address delivered to the Sociological Society in 1904, he stressed the need for "persistence in setting forth the national importance of eugenics," and the need for it to be "introduced into the national conscience, like a new religion." He went on: "It has, indeed, strong claims to become an orthodox religious tenet of the future, for eugenics co-operates with the workings of nature by securing that humanity shall be represented by the fittest races. What nature does blindly, slowly, and ruthlessly, man may do providently, quickly, and kindly."[66]

The resonance of such a program with the mood of the time led to the formation of the Eugenic Education Society in 1907 (renamed the Eugenics Society in 1926), and the organization, in 1912, of the first International Congress of Eugenics, hosted by University College London. Two years later, John H. Kellogg (1852–1943) founded the

Race Betterment Society in the US. The study of any of the pre-1914 pioneers of eugenics such as Alfred Ploetz (1860–1940), founder in 1905 of the Society for Racial Hygiene (which later incorporated the term Eugenics), Ernst Haeckel (1834–1919), founder in 1906 of the Monist League that formulated a spiritual version of racist and anti-Semitic nationalism, or Charles Davenport (1866–1944), first director of the Eugenics Record Office at Cold Spring Harbor established in 1910, would reveal an inextricable mix of hard science with palingenetic cultural pessimism.

Much has been written about the pseudo-scientific nature of eugenics and its links to Nazism. What is still less well appreciated is its deep debt to the cultural climate of modernism.[67] Just as a major mobilizing factor in the dynamics of ultra-nationalism was the need to re-root and re-embed an increasingly deracinated and disembedded world, so eugenics was driven at a mythic level by the urge to reverse another aspect of its "permanent crisis," its pervasive ambivalence, irresolvable complexity, and the powerful psychological sense of loss of order and control, of life-threatening anarchy, and of chaos that this can engender. Eugenics was the scientific equivalent of totalitarianism in politics. In the peculiar climate of late-nineteenth-century Europe, the fear of flux and dissolution could all too easily be objectified and rationalized in terms of national decline, a loss of racial purity, and the growth of moral decadence and physical degeneracy. Deep psychic metaphors were being experienced as empirical, scientifically documentable, and remediable processes of degeneration and regeneration.

In short, the genesis of the eugenically informed variants of nationalism examined in this book can be heuristically located in the hybridization of the mobilizing myth of ultra-nationalism with eugenics—the latter itself a hybrid of myth and science. What made this hybridization possible, and what destined it to have such a powerful impact on inter-war Europe, was that the various permutations of bio-political nationalism and eugenic racism arising during that period all contained a mythic core of palingenetic pessimism. As a result, they naturally partook of the dynamism generated by the modernist mission to reverse the processes of cultural and racial decay and create a new world out of the ruins of the present one. As long as modernism was an avant-garde, cultural or scientific movement, however, the eugenic politics it had spawned remained marginalized by mainstream liberal capitalism.

The Rise of the Utopian "Gardening State"

In *Modernity and Ambivalence*, Zygmunt Bauman argues that the wide-spread favor that eugenics found among political, scientific, medical, and technocratic elites (and even some prominent artists) in early-twentieth-century Europe was closely connected to the utopia of ridding society of degeneracy and restoring its health and internal cohesion. The resulting "garden state" legitimated drastic policies of social engineering, whose unconscious rationale, whatever their conscious rationalization in scientific terms, ultimately depended on taking literally the organic metaphors of weeding, eliminating parasites and disease, pruning, cultivating, breeding, healing, and carrying out surgical intervention that were applied to the task of perfecting society. Politics was no longer a matter of administrating and regulating society but of actually regenerating it and ensuring its healthy growth. Bauman sees the currency of such metaphors, and the policies of social and racial hygiene they rationalized, as symptomatic of a "war on ambivalence," in which the state arrogates to itself the power of "design, manipulation, management, engineering" in the attempt to impose order, boundaries, and taxonomic cohesion on a world where modernization has been condemned to remain irreducibly chaotic, anarchic, and ambiguous.

Obviously, there is nothing new in the metaphorical use of the garden. When its importance within Zen Buddhist and Hindu cosmology is compared with the associations it acquired in Middle Eastern mythology and in the Christian conception of paradise (from the Old Persian *pairidaeza* meaning "walled enclosure, pleasure park, garden")—or with its importance within Jungian symbolism as the image for a *temenos*, the sanctuary crucial to individuation,[68] the garden appears as a genuine archetype for the creation of a bounded world of sacred time and space, of a time-transcending metaphysical order in the midst of the ephemeral material world.[69] As the "gardener scene" from Shakespeare's *The Tragedy of King Richard the Second* illustrates, there is nothing new in the use of horticultural imagery to rationalize the right of the state to act as a latter-day Procrustean, callously tailoring its citizens to fit its own yardsticks of health and the common good. What gives the conception of society as a garden to be tended by the state a new twist in the twentieth century, however, is the way it becomes articulated in a discourse which melds utopianism with the discourse of science and technology, in a way that subsequently legitimated projects of renewal to be carried out with ruthless efficiency on behalf of political masters who celebrate their all-embracing, totalizing vision of the new order.

Theodor Fritsch provides an eloquent case study in the modern mindset that underpins the "gardener state." Whereas his British counterparts saw the city as a solution to the problems created by the modern metropolis consistent with the liberal Enlightenment view of progress, Fritsch specifies that he sees his scheme as a way of warding off the "death of the nation,"[70] and refers his readers to a special issue (no. 83) of his periodical *The Hammer*, devoted to the theme of "the exhaustion of the *Volk*." This reference takes the researcher through a wormhole that leads from the airy realm of utopian town planning to the claustrophobic, underground world of late-nineteenth-century political racism and eugenics. The fortnightly appearance of *The Hammer* (founded in 1902), together with the success of his *Handbook of the Jewish Question* (1887), which by 1943 had run to forty-nine editions, not to mention the numerous other anti-Semitic tracts that poured forth from his publishing house, the Hammer Press—notably *The Protocols of the Elders of Zion*—established Fritsch as the most influential anti-Semitic publicist of the age. In 1912, he founded the *Reichshammerbund* to promote a political racism that claimed to be based on sound scientific and biological principles, a movement that became closely involved with the creation and promotion of two rural settlements with the utopian names of *Eden* (1893) and *Heimland* (Homeland) (1908), as practical experiments in the creation of an *Erneuerungs-Gemeinde* (community of renewal).

For Fritsch, measures to contain Jewish contamination of German culture and to counteract the effects of uncontrolled urbanization were clearly two fronts in the same war to revitalize an increasingly degenerate and dysgenic nation. There was a precise correlation between his fantasy of restoring Germany's mythic racial homogeneity and his attempt to impose the "spirit of order," symbolized in the perfect balance between city and country intended to be achieved through his rigidly symmetrical, circular town plans.[71] Nor was he the only eugenic racist to dabble in utopian town planning. Francis Galton dreamt up two utopias, *Laputa* (1865), which used social engineering to promote hard work, and *KantSayWhere* (1910), a fully-fledged utopian state where reproduction and migration were strictly controlled. By the turn of the century, moreover, powerful elective affinities existed between cultural modernism, urban renewal, anti-urbanism, vegetarianism, *völkisch* nationalism, racial hygiene, and anti-Semitism.[72]

It was not barbarism, then, but the conditions of high modernity that fostered the synthesis of ultra-nationalism with eugenics, some of whose individual permutations in Central and Southeast Europe are the subject of this book. They resulted in forms of bio-politics and racial nationalism that seemed to their protagonists to be consistent with a higher, modern stage of humanistic civilization, not a retreat from it. No matter how rooted they were in developments in science that started in the reign of Enlightenment reason and positivist science, their momentum was intensified by the irrational cultural dynamics of fin-de-siècle Europe, as researched from different angles by Daniel Pick and Zygmunt Bauman: the will to combat degeneracy, to resolve the ambivalence of the age, to create clear taxonomic categories of health and sickness, and to establish a new eugenic socio-political order able to reverse the decline of European civilization. The aspiration was not to perpetuate gradualistic, linear progress, but to induce an apocalyptic, cyclic rebirth in the fabric of time itself.

The emergence of biological and eugenic visions of the ideal society at the turn of the twentieth century thus expresses not a phase of cultural involution, collective regression, or mass psychopathology, but a deep paradox intrinsic to high modernity: the very success of Western science in its search for ordering principles (*nomoi*) based not on revelations of a sacred natural order but on human reason and empiricism that eventually eroded ontological certainties to the point where it created a nomic crisis, and with it the subliminal need for new mythic truths by which to live if social life was to retain meaning. Whatever their overt socio-political function, existentially these new myths served to rebuild, using the modern materials of science and social engineering, the ontological shelter and firmament whose original foundations had been provided by the religious cosmologies that modernity had damaged beyond repair. Scientistic racism, that unholy alliance between eugenic science and ultra-nationalism, thus offered a drastic remedy to the cultural decadence which it had helped bring about and of which it was, itself, a symptom.

Interwar Europe as an Incubator of Scientistic Racism

It was the traumatized Europe that emerged from the First World War that created the conditions in which eugenically conceived schemes of racial hygiene and racial nationalism of an overtly illiberal thrust could go mainstream and establish themselves as the basis of the alternative temporalities pursued by political modernism. The outbreak of the war in 1914 had encouraged those already convinced of Europe's cultural decadence to accept that the day of reckoning had arrived, and that the war would be the supreme test of strength, not just military but moral, both for individual nations and Europe as a whole, providing the chance to show whether the prevailing spiritual sickness was terminal, or whether modern society could be flung into a new historical trajectory. The resulting atmosphere led many artists, intellectuals, and political activists of every persuasion—in Italy even some Marxist syndicalists succumbed—to dedicate their creativity to the national cause, gripped by the heady archetypal fantasies of collective renewal and redemption that the hostilities excited.[73] It is symptomatic of this syndrome that Fritsch's *Reichshammerbund* greeted the outbreak of the First World War as Germany's salvation, banishing softness and "steeling body and soul."

Such sentiments were far from being the reactions of an "ivory tower" intelligentsia out of touch with the masses. They were echoed throughout Europe at a populist level and expressed behaviorally as a "war fever" that spread among ordinary people in several combatant countries, a fanatical belief in the value of the "supreme sacrifice" to the nation that was intensified rather than diminished by the unprecedented scale of the slaughter—in Italy, for example, the war became a popular cause after the disastrous defeat at Caporetto in the autumn of 1917. At the heart of this apparent mass heartlessness lies a palingenetic fervor of pandemic proportions, closely bound up at a psychodynamic level with the currents of aesthetic, cultural, and political modernism considered earlier. This helps explain why, for four long years, in the heartland of a civilization that considered itself at the forefront of world history, millions of young men throughout Europe exposed themselves willingly to the certainty of unimaginable discomfort and suffering, and the high probability of severe injury or death, for the sake of their nation (now consubstantiated with "God"). Moreover, they did so under the implacable gaze of the military high command, the intelligentsia, and the general public, who overwhelmingly saw their slaughter in heroic terms of sacrifice and the need to purge national life of its decadence.[74]

The collective belief that a Western Europe which had lost its bearings and grown soft might now be redeemed meant that the most militant wings of pro-war opinion duplicated the dynamics of archetypal revitalization movements. Its belief in the palingenetic magic of sacrifice overrode the traditions of Christian, Renaissance, Enlightenment, liberal, and positivist humanism, conforming instead to a deep-seated "pagan" psychological matrix which predisposes many human beings, once they have completed the necessary *rites de passage*, to be prepared to surrender their individualism within an ideological community able to combat the forces of chaos through a liturgical process of renewal. It is a process that demands blood to be shed in ritualized violence, whether through elaborate religious ceremonies or the chilling choreography of logistics and troop movements of the First World War.[75]

In these catastrophic circumstances, what had before been a spiritual malaise and a utopian longing for renewal, largely confined to the sphere of the intelligentsia, had now become an objective, socio-political crisis, forcing much of the continent, now at a populist level, willy-nilly into a protracted *rite de passage* between an "old world" which had irrevocably gone, and a new one whose contours were far from clear, and which could just as easily prove to be a continuation of the catastrophe or a more healthy, sustainable stage of civilization. In post-1918 Europe, Modris Eksteins detects a generalized "craving for newness" among "socialists and conservatives, atheists and fundamentalists, hedonists and realists (…) rooted in what was regarded by radicals as the bankruptcy of history and by moderates as at least the derailment of history"[76]—a phrase that unwittingly evokes Kafka's metaphor of the train crash of modern humanity. History itself was experienced as having entered a state of high liminality, creating a situation within Western culture as a whole in which "the search for a myth appropriate to modernity became paramount. (…) The myth either had to redeem us from the formless universe of contingency or, programmatically, to provide the impetus for a new project of human endeavor."[77]

It was this generalized sense of creative destruction that Wyndham Lewis (1882–1957), the high priest of the hyper-modernist Vorticist movement before the war, expressed in the imagery of the day, three days after the armistice: "So we, then, are the creatures of a new state of human life, as different from nineteenth-century England as, say, the Renaissance was from the Middle Ages. (...) Are the next generations going to produce a rickety crop of Newcomers, or *is the new epoch to have a robust and hygienic start-off?* A phenomenon we meet, and are bound to meet for some time, is the existence of a sort of No-Man's Land atmosphere. The dead never rise up, and men will not return to the Past, whatever else they may do. But as yet there is Nothing, or rather the corpse of the past age, and the sprinkling of children of the new. There is no mature authority, outside of creative and active individual men, to support the new and delicate forces bursting out everywhere today."[78]

As Lewis wrote these words, Communist Russia was already becoming a vast laboratory for the forging of a new society, a project that, given the cult of science and technology fostered the application of genetics and bio-politics to the task of breeding a "new Soviet Man."[79] But in Central and Southeast Europe where liberalism was weak, communism remained effectively marginalized, and national strength, not the overthrow of capitalism, was the mobilizing myth hosting scientistic fusions of eugenic racism with ultra-nationalism. A significant role in creating the right ethos for this habitat to emerge in the 1920s was played by the rise of fascism, and its establishment in Italy as a regime. In its many variants, this overtly palingenetic form of ultra-nationalism sought to resolve the liminality of modernity through a temporal revolution,[80] though not necessarily one in which eugenic concepts of race played a significant role. However, a much more immediate factor encouraging the rise of racial nationalism in the 1930s was the establishment of the Third Reich, the first regime to adopt racial hygiene and eugenics as the official basis of state policy.

Gardening in Jackboots

Nazism, both as an ideology and as a historical force, provides a practically inexhaustible supply of empirical case studies in the nexus that could arise in interwar Europe, between the sense of anomie and degeneracy produced by the permanent liminality of modernity, and the creation of a "gardening state." True to the sprit of political and technocratic modernism at its most radical, the Third Reich dedicated itself to ruthlessly implementing scientistic concepts of the nation as an organic people to be purged of decadence, which meant replacing liberal democracy with an ethnocracy based on a wide range of nationalist and racist ideologies that jostled for position under the umbrella of Nazism, some of which mixed eugenic and overtly mythic currents of nationalism.

It is significant for the thesis explored in this chapter that the NSDAP was only able to break through as an electoral force in the late 1920s in response to the Weimar Republic's economic collapse and the political paralysis that rapidly followed in the wake of the Wall Street Crash (1929). For a decade, Nazism had been consigned to the political ghetto, and only now, when it could present itself as the panacea to the nation's ills, could it achieve the level of popular support needed to legitimize its draconian experiment in totalitarian social engineering. A precondition to the sudden success of Hitler as the centre of a charismatic personality cult and the leader of a powerful mass movement was the acute spiritual, nomic, or "sense-making" crisis[81] that the sociopolitical calamity had unleashed upon a society already fraught with social, racist, class, ideological, and psychological tensions, and on which the storm of modernity had descended with particular fury even before 1914. Considerable light is thrown on the psychodynamics of this crisis by Hermann Broch's monumental study of the spiritual consequences of modernity, *The Sleepwalkers* (1932). Broch portrays a contemporary generation spiritually disoriented and isolated, not just by the immediate events of contemporary history but by the fragmentation of the unified cosmology of the Middle Ages into the countless competing logics of everyday life. Little wonder that many citizens of the Weimar Republic, no matter how educated or civilized, deeply experienced the "doubly strong yearning for a Leader to take him tenderly and lightly by the hand, to set things in order and show him the way; (...) the Leader who will build the house anew that the dead may come to life again; (...) the Healer who by his actions will give meaning to the incomprehensible events of the Age, so that Time can begin again."[82]

By the time *The Sleepwalkers* was published, there was only one political party that promised to rebuild the national house anew and restart time after the catastrophic *rite de passage* of the First World War and the Weimar Republic: the NSDAP. Having been systematically marginalized in elections before the Wall Street Crash—it received only 2.6 per cent of the vote in the 1928 Reichstag election—the NSDAP thereafter took off as a mass movement fuelled by the spontaneous charismatic energies and palingenetic expectations unleashed by the collapse of the Weimar system; consequently, in 1932, Nazism received 37.4 per cent of the vote. Once again, all this is curiously reminiscent of the pattern of revitalization movements observed by Victor Turner: Nazism emerged at a time of "marked cultural change and its accompanying personal psychological stress" and sought "to revitalize a traditional institution [the German *Volk*], while endeavoring to eliminate alien persons, customs, values, even material culture from the experience of those undergoing painful change."[83] More importantly, the Nazi ideology of national rebirth offered a "ludic recombination of old and new cultural components" orchestrated by a new prophet (Hitler) whose dreams and visions, experienced with a fanaticism reminiscent of the shamanistic trance, became translated into new myths which were to be lived out in the creation of a new *communitas*, the reborn *Volksgemeinschaft*. The central theme of all Nazi propaganda was the need to overcome decadence, restore health, and establish a new order.

The fact that some of its leaders perceived their task as not just establishing a new administration but eradicating the metaphysical ambivalence and liminality of the modern age, emerges clearly from the biblical tones of the introduction to Gottfried Feder's NSDAP program, which opens: "Today chaos reigns on earth, confusion, struggle, hate, envy, conflict, oppression, exploitation, brutality, egotism. Brother no longer understands brother. People have lost their bearings!" It is, of course, the NSDAP, inspired by "the will to give shape to the amorphous, the will to put a stop to chaos, to put in order a world out of joint" that was destined to create a new era under the banner of the swastika, "symbol of a life which is once more awakening."[84]

Although Feder was no eugenicist, the palingenetic climate he helped create with such texts lured not just *völkisch* campaigners for cultural rebirth, rabid anti-Semites, and self-appointed race experts like Fritsch, but a galaxy of university-educated professionals ranging from theologians, historians, philologists, art historians, and social scientists to experts in physical genetics, the heredity of plants and livestock, anthropology, eugenics, and racial hygiene. The latter were less concerned with the contagious spiritual evils spread by cosmopolitanism, internationalism, egalitarianism, and materialism, whether by Jews, liberals, or communists, than with the genetic basis of the "ferment of decomposition." The new *Volksgemeinschaft* would not only be created through education and solidarity, but also by purging and breeding; not just through ethics, but genetics too. Just as national decadence was dysgenic, so national rebirth needed to be eugenic. A new race of muscular, spiritually revitalized Germans would overcome degeneration and reverse Western Europe's decline.

One specimen of this class of converts was Richard Walther Darré (1895–1953) who, in the course of the 1930s, synthesized the knowledge he had acquired from a degree in agriculture and his specialist knowledge, both practical and theoretical, of horse-breeding, with his peculiar brand of *völkisch* nationalism, which was not only anti-Semitic but looked to a regenerated peasant "nobility" as the basis for Germany's racial and moral regeneration. In the pamphlet "Marriage Laws and the Principles of Breeding," which he published in 1930, the year he joined the Nazi Party, Darré stressed that the garden, "if it is to remain the breeding ground for plants," needs "the forming will of the gardener" who "carefully tends what needs tending and ruthlessly eliminates the weeds which would deprive the better plants of nutrition, air, light and sun." After all, "a people can only reach spiritual and moral equilibrium if a well-conceived breeding plan stands at the very *centre* of its culture."[85] Five years later, by which time Darré had become leader of the National Peasant Party and minister of food and agriculture, the Third Reich promulgated race laws which excluded German Jews from Reich citizenship and prohibited them from marrying or having sexual relations with persons of "German or related blood."

It was symbolically significant that the laws were promulgated not in a communiqué from some anonymous ministry building but by Hitler himself, in a 1935 speech delivered at Nuremberg to hundreds of thousands (and seen on newsreels by millions more). This was a supreme moment not just of policy, but of Nazi aestheticized, theatricalized politics and political religion. The once amorphous masses, now sculpted by the will of the führer into an ornamental symbol of order and discipline (the German *Zucht* translates as both discipline and breeding), stood symmetrically in serried ranks in a stadium consciously designed by Albert Speer to convey a sense of sacred space on a gigantic scale. Symbolically, the entire nation was being re-rooted and re-embedded in a ritualized moment of worldly transcendence in which decadence, ambivalence, and liminality were overcome.[86] It was the job of racial experts, the SS, and all those concerned with the moral and physical health of the nation to then see to it that the human weeds were eliminated as efficiently as possible so that the better plants would thrive more abundantly. Tragically for humanity, Bloch's "mysterious tree" had turned out to be a genetically modified scion of the German Oak from which all parasites and fungus were to be eradicated using the latest techniques.[87] Such a possibility was inscribed in Hitler's plans for the nation's rebirth from the very outset. In the opening section of *Mein Kampf*, he had written "Only when an age is no longer haunted by its own sense of guilt can it acquire the inner calm and outer strength necessary to set about brutally and ruthlessly cutting out the wild shoots and rooting out the weeds."[88]

West of Eden?

The essays presented in this volume provide detailed case studies in the modern discourse of eugenics and racial nationalism that emerged in Central and Southeast Europe between 1900 and 1940. They also reveal the dark side of the "principle of hope" postulated by Ernst Bloch. They show how easily the palingenetic brand of cultural pessimism it induces in those feeling themselves to be the victims of modernity can beget illegitimate couplings of science and myth, in which compensatory fantasies of national regeneration and racial purity are translated into a scientific, modern discourse. It is the scientistic guise assumed by archetypal drives to create a sacralized "homeland" to be defended at all costs from a demonized "other" that facilitated the implementation and bureaucratization of technocratic, ethnocratic racism as an official policy of any modern state whose ruling elites and experts set about, in totalitarian mode, the calculated "betterment" of the race and nation. Both the "national community" and the "homeland" at the centre of the Nazi *Weltanschuaung*—the "blood" just as much as the "soil"—were inextricable compounds of the physical and the mythic.

Except in the case of Germany, the racial nationalisms considered in this volume never became the basis of state policy; however, the practical implications of enacting them were demonstrated in all their blood-chilling horror by the Third Reich. The mass production of torture and murder, in which science, medicine, and technology became so deeply implicated under the Nazis, underlines the disturbing ambivalence of the instinctual human desire to grow mysterious trees of utopianism wherever they encounter a wasteland. It also demonstrates the equally profound ambivalence of modernity when its disturbed spiritual and psychological climate nourishes a utopian project to embark on the total makeover of early-twentieth-century modernism, rather than bringing gradual, sustainable improvements to what will always be an "imperfect garden."[89] It was, after all, only when he had renounced his hope of perfect happiness with Cunégonde and recognized the dangers posed as much by total optimism as by total pessimism, that Voltaire's Candide rediscovered the profound humanistic truth in a verse from Genesis: human beings were placed in Eden *ut laboret eum*, in order to work in it and on it. It is in this modest, gradualistic, melioristic spirit that Voltaire urged us to "cultivate our garden," an activity that, when it comes to races and nationalities, should be treated more like tending an allotment than running an industrial farm. Perhaps it is no coincidence that in Aldous Huxley's visionary 1932 novel of a society based on a comprehensive yet benign social engineering with dystopian results for humanity, *Brave New World*, the allotment garden becomes a means by which John, the "Savage," can elude the clutches of totalitarianism and withdraw into Thoreauian solitude: "He counted his money. The little that remained would be enough, he hoped, to tide him over the winter. By next spring, his garden would be producing enough to make him independent of the outside world."[90]

Endnotes

[1] Ernst Bloch, *Geist der Utopie* (Frankfurt-am-Main: Suhrkampf, 1991), 216. (First edition, 1918; passage written in 1916). For the English translation see Anthony Nassar, *The Spirit of Utopia* (Stanford: Stanford University Press, 2000), 171.

[2] The first volume of Oswald Spengler's *Der Untergang des Abendlandes* appeared in Vienna in July 1918, by which time the manuscript of *Geist der Utopie* was already finished.

[3] Eric Voegelin, "The Origins of Scientism," *Social Research* 15 (1948), 462–494.

[4] A term used by a number of social historians such as Anthony Giddens. See Anthony Giddens, *Modernity and Self-Identity: Self and Society in the Late Modern Age* (Cambridge: Polity Press, 1991).

[5] See Roger Griffin, *The Nature of Fascism* (London: Routledge, 1993).

[6] See, for example, Sigmund Freud, *Der Unbehagen in der Kultur* (Vienna: Internationaler psychoanalytischer Verlag, 1930); and Sigmund Freud, *Civilization and Its Discontents* (London: Hogarth Press, 1930). The German title translates literally as "The Malaise in (our/modern) Culture".

[7] For example, Carl G. Jung, *Modern Man in Search of a Soul* (London: Kegan Paul & Trubner Trench, 1933).

[8] Themes reflected in the titles of investigations of modernity include Erich Heller. *The Disinherited Mind Essays in Modern German Literature and Thought* (Philadelphia: Dufour & Saifer, 1952); H. E. Holthusen,. *Der unbehauste Mensch* (Munich: Piper, 1952); and P. L. B. Berger and H. Kellner, *The Homeless Mind: Modernization and Consciousness* (Harmondsworth: Penguin, 1974).

[9] Max Weber. "The Social Psychology of the World's Religions," in H. H. Gerth and C. Wright Mills, eds., *From Max Weber: Essays in Sociology* (New York: Oxford University Press, 1946), 282.

[10] Ferdinand Tönnies, *Gemeinschaft und Gesellschaft* (New York: Free Press, [1877] 1961).

[11] Émile Durkheim, *The Division of Labor in Society* (New York: Free Press, [1877] c. 1933).

[12] Georg Simmel, *The Metropolis and Mental Life* (1905), reprinted in *The Sociology of Georg Simmel* (New York: Free Press, 1950).

[13] Frederic Jameson, *The Seeds of Time* (New York: Columbia University Press, 1994), 84.

[14] Zygmunt Bauman, *Modernity and Ambivalence* (Cambridge: Polity Press, 1991).

[15] Mircea Eliade, *The Myth of Eternal Return: Cosmos and History* (Princeton: Princeton University Press, 1949); and Mircea Eliade, *The Sacred and the Profane* (New York: Harcourt Brace Jovanovich, 1959).

[16] Anthony Giddens, *The Consequences of Modernity* (Cambridge: Polity Press, 1990); and Anthony Giddens, *Modernity and Self-Identity* (Cambridge: Polity Press, 1991).

[17] This distinction is also made by Hans-Georg Gadamer, *Truth and Method* (New York: Continuum, 1993).

[18] For a "modern" sense of how the sky has ceased to be a firmament see the pivotal scene in Paul Bowles' *The Sheltering Sky* (New York: New Directions, 1949), 79, which emphasizes that behind the sky there is "Nothing. (...) Just darkness. Absolute night."

[19] Franz Kafka, "Diary entry, 20 October 1917," (Oktavheft G II, 2), in Heinz Politzer, ed., *Das Kafka-Buch* (Frankfurt-am-Main: Fischer Verlag, 1965), 247 (my translation).

[20] Franz Kafka, "Gespräch mit dem Beter," in *Sämtliche Erzählungen*, ed. by Paul Raabe (Frankfurt-am-Main: Fischer Verlag, 1970), 189 (my translation).

[21] Eliade, *The Myth of Eternal Return*, 75–76, 133–137.

[22] See Pierre Bourdieu, *The Field of Cultural Production* (Cambridge: Polity Press, 1991).

[23] For an eyewitness account of this vital phase in Cioran's life, see Mihail Sebastian, *Journal 1935–1944: The Fascist Years* (Chicago: Ivan R. Dee, 2000).

[24] Two classic texts exploring this phase of European history are Mario Praz, *The Romantic Agony* (Oxford: Oxford University Press, 1933); and K. W. Swart, *The Sense of Decadence in Nineteenth-Century France* (The Hague: International Archives of the History of Ideas, 1964).

[25] See J. Edward Chamberlain and Sander L. Gilman, *Degeneration: The Dark Side of Progress* (New York: Columbia University Press, 1985).

[26] This remark, made by Kraus in 1913, is quoted in Thomas Szasz, *Anti-Freud: Karl Kraus's Criticism of Psychoanalysis and Psychiatry* (Syracuse: Syracuse University Press, 1990), 24.

[27] Daniel Pick, *Faces of Degeneration: A European Disorder c. 1848– c. 1918* (Cambridge: Cambridge University Press, 1989).

[28] See Robert Nye, *Crime, Madness and Politics in Modern France: The Medical Concept of National Decline* (Princeton: Princeton University Press, 1984).

[29] Todd Samuel Presner, "'Clear Heads, Solid Stomachs, and Hard Muscles': Max Nordau and the Aesthetics of Jewish Regeneration," *Modernism/modernity* 10, 2 (2003), 277.

[30] See the chapter "Grounds for Anxiety," in Peter Gay, *Schnitzler's Century* (New York: W. W. Norton), 129–154. For contemporary testimonies of the degree to which "nervousness" had become a general condition of the age, see George Beard, *American Nervousness: Its Causes and Consequences. A Supplement to Nervous Exhaustion (Neurasthenia)* (New York: Putman's Sons, 1881); and Bartolomäus von Carneri, *Der moderne Mensch. Versuch über Lebenserfahrung* (Leipzig: n.p., 1890).

[31] Max Nordau, *Degeneration* (New York: D. Appleton, 1895), 537.

[32] Holbrook Jackson, *The Eighteen Nineties: A Review of Art and Ideas at the Close of the Nineteenth Century* (London: Pelican, 1950). Jackson comments on the sudden invasion of the adjective "new" in the age of decadence to describe countless cultural events, journals, and styles.

33 See Donald Hall, ed., *Muscular Christianity. Embodying the Victorian Age* (Cambridge: Cambridge University Press, 1994).

34 See Presner, "'Clear Heads, Solid Stomachs, and Hard Muscles'," 270.

35 For a discussion of the importance to modernism of "cyclic time" see Louise Williams, *Modernism and the Ideology of History* (Cambridge: Cambridge University Press, 2002).

36 W. Clement Stone and Napoleon Hill, *Success through a Positive Mental Attitude* (Engelwood Cliffs, N.J.: Prentice Hall, 1960).

37 Victor Turner, ed., *Celebration: Studies in Festivity and Ritual* (Washington, Wash.: Smithsonian Institution Press, 1982), 211–212.

38 Turner, ed., *Celebration. Studies in Festivity and Ritual*, 212.

39 Turner, ed., *Celebration. Studies in Festivity and Ritual*, 214.

40 Turner, ed., *Celebration. Studies in Festivity and Ritual*, 215.

41 Victor Turner, *The Ritual Process. Structure and Anti-Structure* (Ithaca: Cornell University Press, 1969), 107.

42 My emphasis. See Arpad Szakolczai, *Reflexive Historical Sociology* (London: Routledge, 2000), 225. I would suggest "era" is preferable to "phases," since phases by definition have endings (unless "permanent liminality" makes phases permanent as well). A wonderful black comedy exploring the permanent liminality of modernity in the terms Szakolczai describes is Tom Stoppard's *Rosenkrantz and Guilderstern are Dead* (1968).

43 Malcolm Bradbury and James McFarlane, eds., *Modernism 1890–1930* (Harmondsworth: Penguin, 1976), 46.

44 Turner, ed., *Celebration. Studies in Festivity and Ritual*, 212.

45 Ernst Bloch, *Das Prinzip Hoffnung* (Frankfurt-am-Main: Suhrkamp, 1959); and Ernst Bloch, *The Principle of Hope* (trans. by Neville Plaice, Stephen Plaice and Paul Knight) (Cambridge, Mass.: MIT Press, 1986).

46 Bloch, *Geist der Utopie*, 216.

47 A term used by Theodor Adorno, *Aesthetic Theory* (Minneapolis: University of Minnesota Press, 1996).

48 Peter Osborne, *The Politics of Time* (London: Verso, 1995).

49 For example Marshall Berman, *All That Is Solid Melts into Air* (New York: Penguin, 1988).

50 Ronald Schleifer, *Modernism and Time* (Cambridge: Cambridge University Press, 2000).

51 Osborne, *The Politics of Time*, 164.

52 Berman, *All That Is Solid Melts into Air*, 33.

53 Robert Hughes, *The Shock of the New* (London: BBC, 1980), 177.

54 Hughes, *The Shock of the New*, 177.

55 See Mark Antliff, "*La Cité française.* Georges Valois, Le Corbusier, and Fascist Theories of Urbanism," in Mark Affron and M. Antliff, eds., *Fascist Visions. Art and Ideology in France and Italy* (Princeton: Princeton University Press, 1997), 134–170.

56 Elaine Hochman, *Architects of Fortune. Mies van der Rohe and the Third Reich* (New York: Wiedenfeld & Nicolson, 1989).

57 This was illustrated in the Art and Power Exhibition which toured major

European cities during the mid-1990s. See David Elliot, ed., *Art and Power, Exhibition Catalogue* (London: Hayward Gallery, 1996).

[58] See Dirk Schubert, "Theodor Fritsch and the German (*völkisch*) Version of the Garden City: the Garden City Invented Two Years before Ebenezer Howard," *Planning Perspectives* 19, 1 (January 2004), 3–35.

[59] See, for example, Philip Spencer and Howard Wollmann, *Nationalism. A Critical Introduction* (London: Sage, 2002).

[60] See George Mosse, *The Nationalization of the Masses. Political Symbolism and Mass Movements in Germany from the Napoleonic Wars through to the Third Reich* (New York: Howard Fertig, 1975), chapter 4.

[61] A phrase used to evoke urban decadence by the *völkisch* utopian Artamanen Movement founded in 1923, which aimed to promote the appearance of a race of "organic" human beings. See George Mosse, *The Crisis of German Ideology: Intellectual Origins of the Third Reich* (New York: Howard Fertig, 1964).

[62] It is significant that in his seminal *Imagined Communities* (London: Verso, 1991), Benedict Anderson argues that nationalism arose in response to the loss of ontological security brought about by the subversion of "sacred languages" and sacred time brought about partly as a result of the spread of the printed word. See also Marius Turda, *The Idea of National Superiority in Central Europe, 1880–1918* (New York: Edwin Mellen Press, 2005).

[63] E. M. Cioran, *A Short History of Decay* (Oxford: Basil Blackwell, 1975), 112.

[64] Published in two parts in *MacMillan's Magazine* 11 (1864), 157–166.

[65] See Pick, *Faces of Degeneracy*, 176–221.

[66] Francis Galton. "Eugenics: Its Definition, Scope and Aims," *American Journal of Sociology* 1 (1904), 456–450; 478–479.

[67] See Modris Eksteins, *Rites of Spring. The Great War and the Birth of the Modern Age* (Boston, Mass.: Houghton Mifflin Company, 2000). The thesis, proposed in this chapter, of modernism as a generalized, multidisciplinary, cultural revolution driven by a sense of crisis, has a profound resonance with many passages in Eksteins' book. The introduction to Ruth Ben-Ghiat's *Fascist Modernities* (Berkeley: University of California Press, 2001), 1–15, also stresses the central role played in Mussolini's Italy by a fusion of modernism, technicism, and the myth of the new man and of the renewal of the Roman heritage, while chapter 5 (148–157) focuses on Fascism's home-grown demographic programme of national renewal which, even before 1938, had a markedly eugenic component independent of any Nazi influence.

[68] Margaret Eileen Meredith, *The Secret Garden. Temenos for Individuation* (Toronto: Inner City Books, 2005).

[69] Henry Wesselman, *Journey to the Sacred Garden.* (Carlsbad, Calif.: Hay House, 2003). See also references to "garden" in George Elder, *An Encyclopaedia of Archetypal Symbolism* (Boston, Mass.: ARAS, 1996).

[70] Theodor Fritsch, *Die Stadt der Zukunft* (Leipzig: Hammer, 1912), 27.

[71] See Schubert, "Theodor Fritsch and the German (*völkisch*) Version of the Garden City," 11.

72 Schubert, "Theodor Fritsch and the German (*völkisch*) Version of the Garden City," 5–24.

73 Roland N. Stromberg, *Redemption through War. The Intellectuals and 1914* (Lawrence: Regents Press of Kansas, 1982).

74 This emerges clearly from works such as Robert Wohl, *The Generation of 1914* (Cambridge, Mass.: Harvard University Press, 1979); George Mosse. *Fallen Soldiers: Reshaping the Memory of the World Wars* (Oxford: Oxford University Press, 1991); and Eksteins' *Rites of Spring* (see particularly page 211, which links the role of the soldier as an "agent of destruction and regeneration" to his liminal position "on the edge of no man's land").

75 Richard Koenigsberg has highlighted this syndrome in a series of important essays, notably "Dying for One's Country: The Logic of War and Genocide;" "Virility and Slaughter: Battle Strategy of the First World War;" "As the Soldier Dies, So Does the Nation Come Alive: The Sacrificial Meaning of Warfare;" "Aztec Warfare, Western Warfare: The Soldier as Sacrificial Victim." These can be downloaded at http://home.earthlink. net/~libraryofsocial science/aztec.htm. The same syndrome is illuminated from a psychoanalytical angle in the context of Nazism by Klaus Theweleit, *Male Fantasies* (Cambridge: Polity Press, 1988), and is explored anthropologically in the case of the United States in Carolyn Marvin and David Ingle, *Blood Sacrifice and the Nation. Totem Rituals and the American Flag* (Cambridge: Cambridge University Press, 1999).

76 Eksteins, *Rites of Spring*, 257. The "democratization" of the avant-garde experience of crisis previously confined to the avant-garde is also sustained by Sherry Vincent, *Ezra Pound, Wyndham Lewis, and Radical Modernism* (New York: Oxford University Press, 1993), 278.

77 Harvey, *The Condition of Postmodernity*, 30–31. Harvey's book presents many scintillating insights into the nature of modernity and modernism that are congruent with the thesis explored in this chapter but which cannot be developed here.

78 My emphasis. See Wyndham Lewis, "The Children of the New Epoch," *The Tyro* 1 (1921); reprinted in C. Harris and P. Wood, *Art in Theory 1900–1990* (Oxford: Blackwell, 1992), 244–245.

79 On the major role played by scientism in Soviet Russia see Loren Graham, "Science and Values: The Eugenics Movement in Germany and Russia in the 1920s," *American Historical Review* 82, 5 (1977), 1136–1164; Valerii Soifer, *Lysenko and the Tragedy of Soviet Science* (New Jersey: Rutgers University Press, 1994); and Tzvetan Todorov, "Totalitarianism between Religion and Science," *Totalitarian Movements and Political Religions* 2, 1 (2001), 28–42.

80 Roger Griffin, "'I Am no Longer Human. I am a Titan. A God!' The Fascist Quest to Regenerate Time," in *Beginning Time Anew. Essays on the Fascist Revolution*, ed. by Matt Feldman (Wellingborough: Dogma Publications, 2006).

81 George L. Platt, "Thoughts on a Theory of Collective Action: Language, Affect, and Ideology in Revolution," in Mel Albin, ed., *New Directions in Psychohistory* (Lexington, Mass.: Lexington Books, 1980).

[82] Hermann Broch, *The Sleepwalkers* (New York: Grosset and Dunlap, 1964), 548.

[83] Turner, *Celebration. Studies in Festivity and Ritual*, 212.

[84] Gottfried Feder, *Das Programm der NSDAP* (Munich: Eher, 1933), 25–26, 35, 64.

[85] Walter Darré, "Marriage Laws and the Principles of Breeding," quoted in Bauman, *Modernity and Ambivalence*, 27.

[86] On the significance of political liturgy as a source of communal stability see Angela Hobart, *Healing Performances of Bali. Between Darkness and Light. Community Well-Being and the Religious Festival* (Oxford: Berghahn, 2003).

[87] Kaiser Wilhelm II reputedly referred to the Jews as "a poison fungus on the German oak". See Harrison, *Gaze of the Gorgon*, 72.

[88] Adolf Hitler, *Mein Kampf* (Munich: NSDAP Press, 1938), 30.

[89] Tzvetan Todorov, *Imperfect Garden. The Legacy of Humanism* (Princeton: Princeton University Press, 2002).

[90] Aldous Huxley, *Brave New World* (London: Chatto & Windus, 1977), 202.

Index